Geologic History of Florida

UNIVERSITY PRESS OF FLORIDA

Florida A&M University, Tallahassee
Florida Atlantic University, Boca Raton
Florida Gulf Coast University, Ft. Myers
Florida International University, Miami
Florida State University, Tallahassee
New College of Florida, Sarasota
University of Central Florida, Orlando
University of Florida, Gainesville
University of North Florida, Jacksonville
University of South Florida, Tampa
University of West Florida, Pensacola

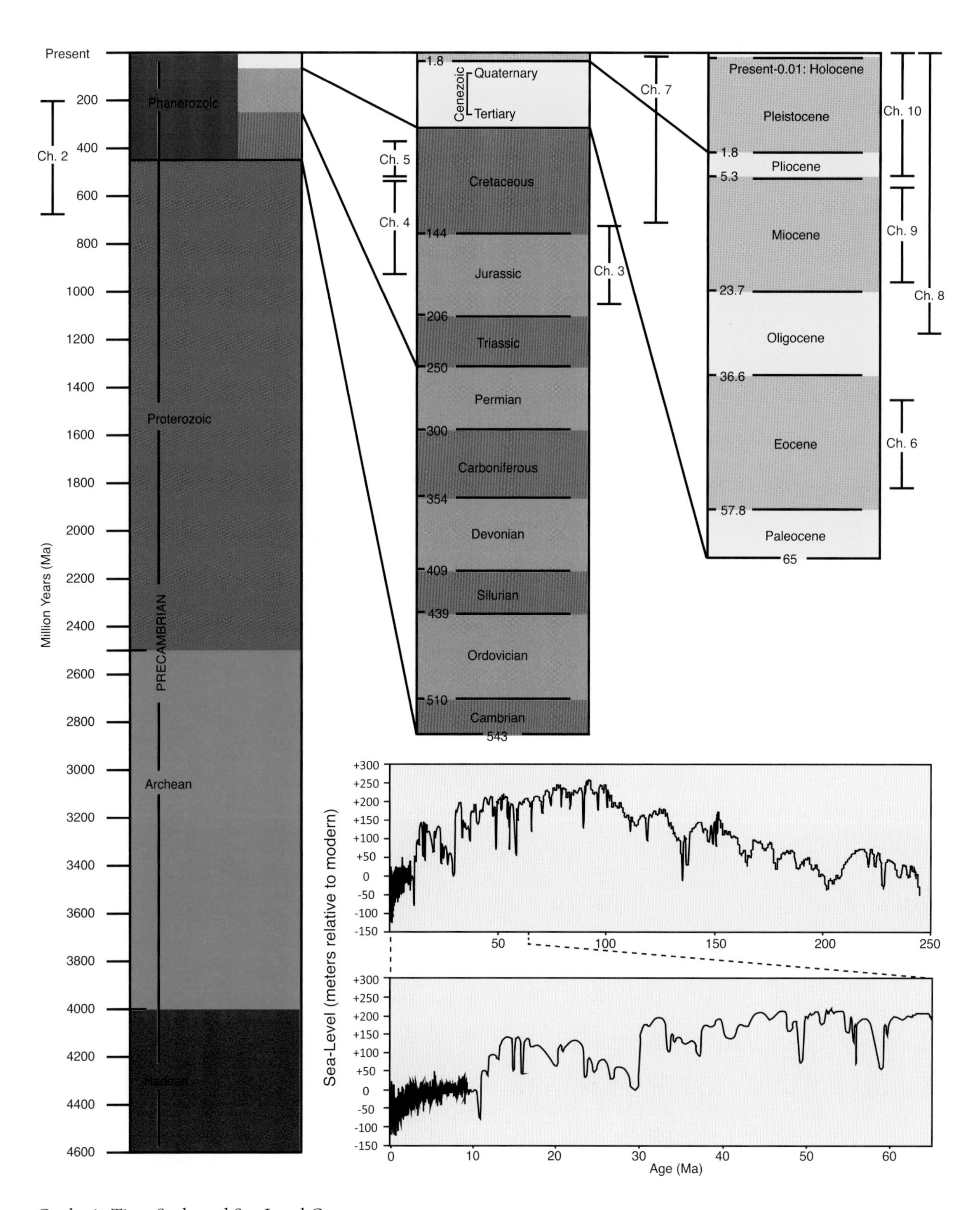

Geologic Time Scale and Sea Level Curve

Left and top: Geologic time scale with appropriate time slices addressed by various book chapters. *Bottom right*: Global sea level curve.

Geologic History of Florida

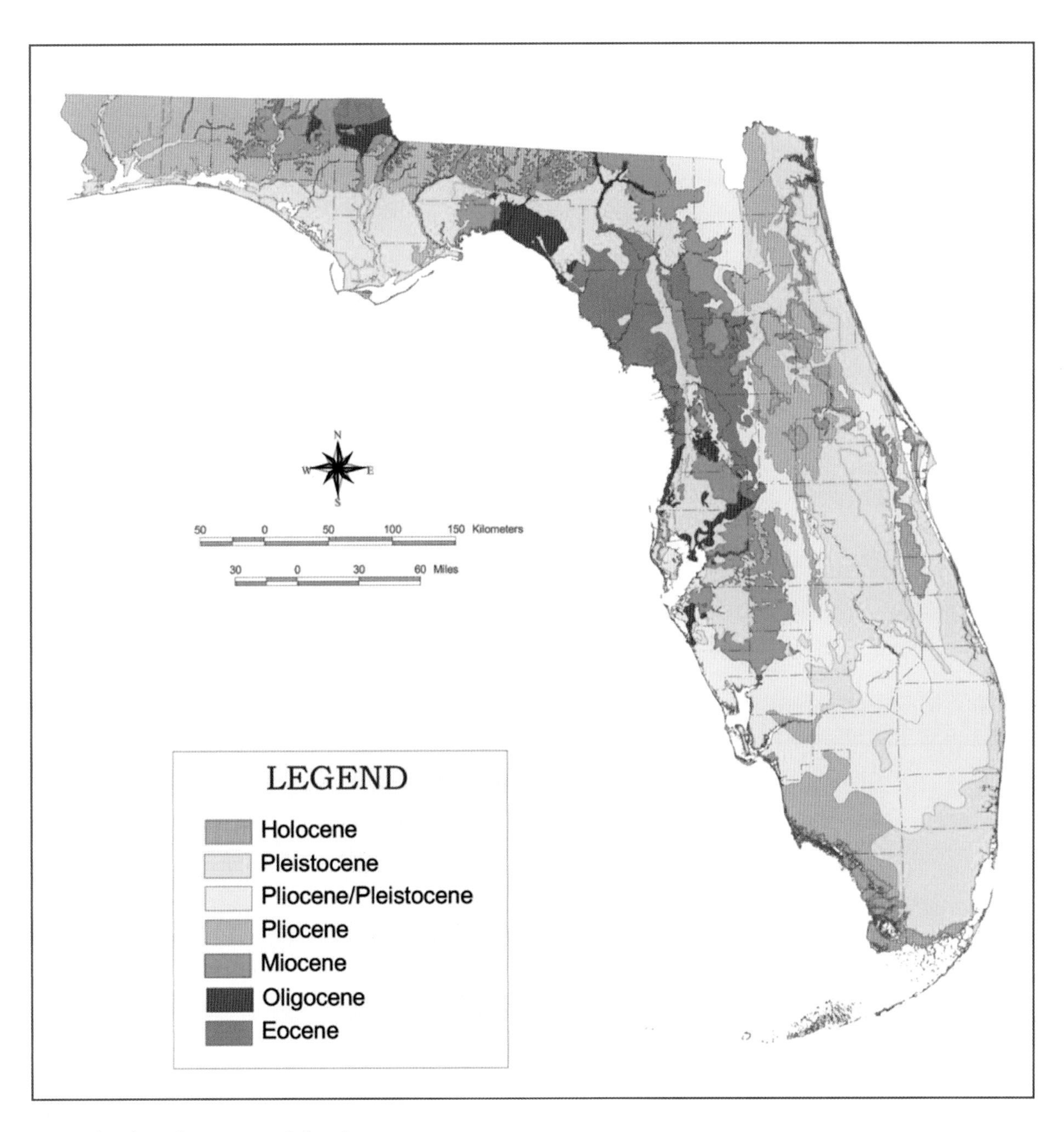

Generalized Geologic Map of Florida

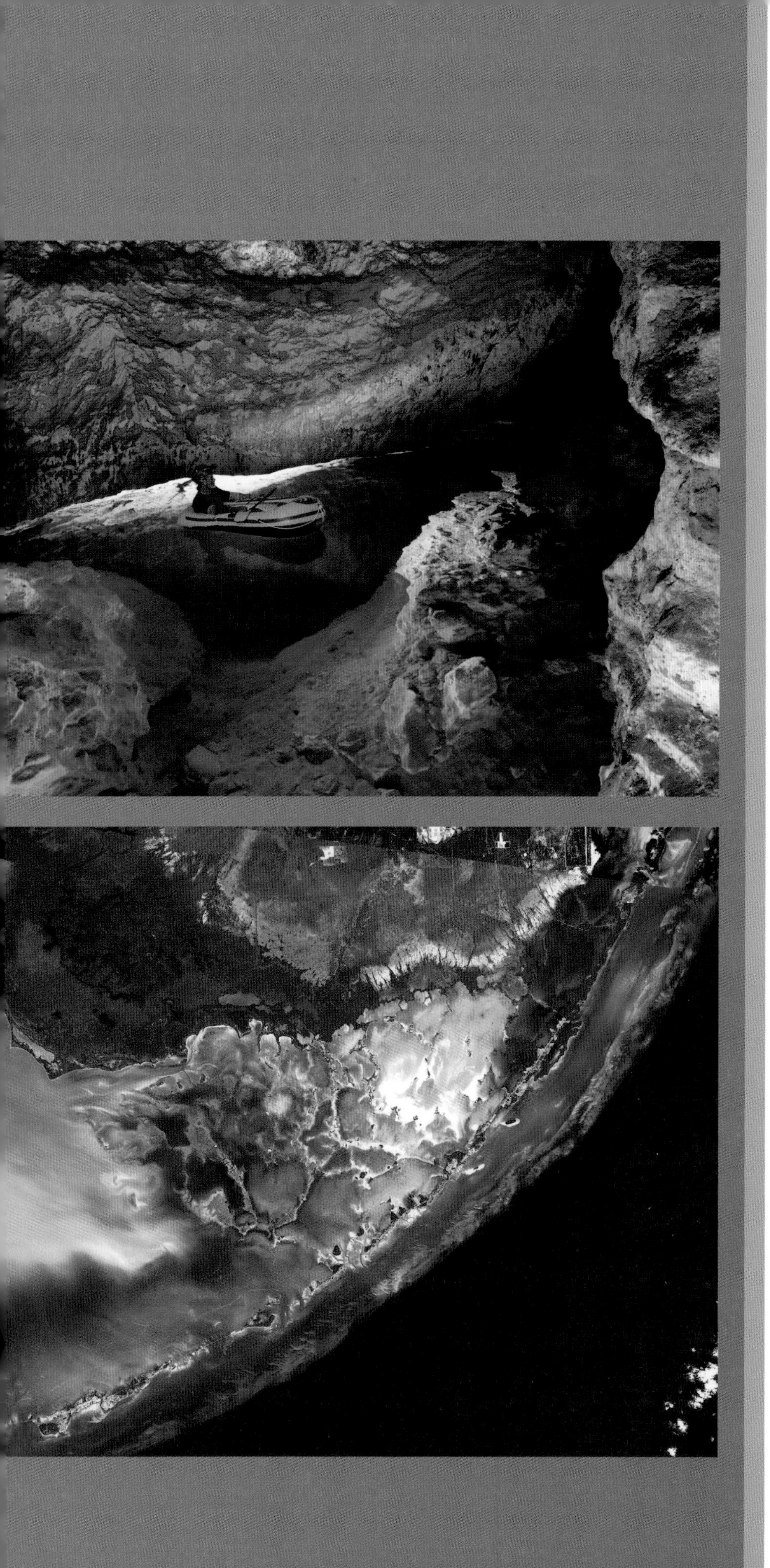

University Press of Florida
Gainesville
Tallahassee
Tampa
Boca Raton
Pensacola
Orlando
Miami
Jacksonville
Ft. Myers
Sarasota

Geologic History of Florida

Major Events That Formed the Sunshine State

Albert C. Hine

Illustrated by Carlie Williams

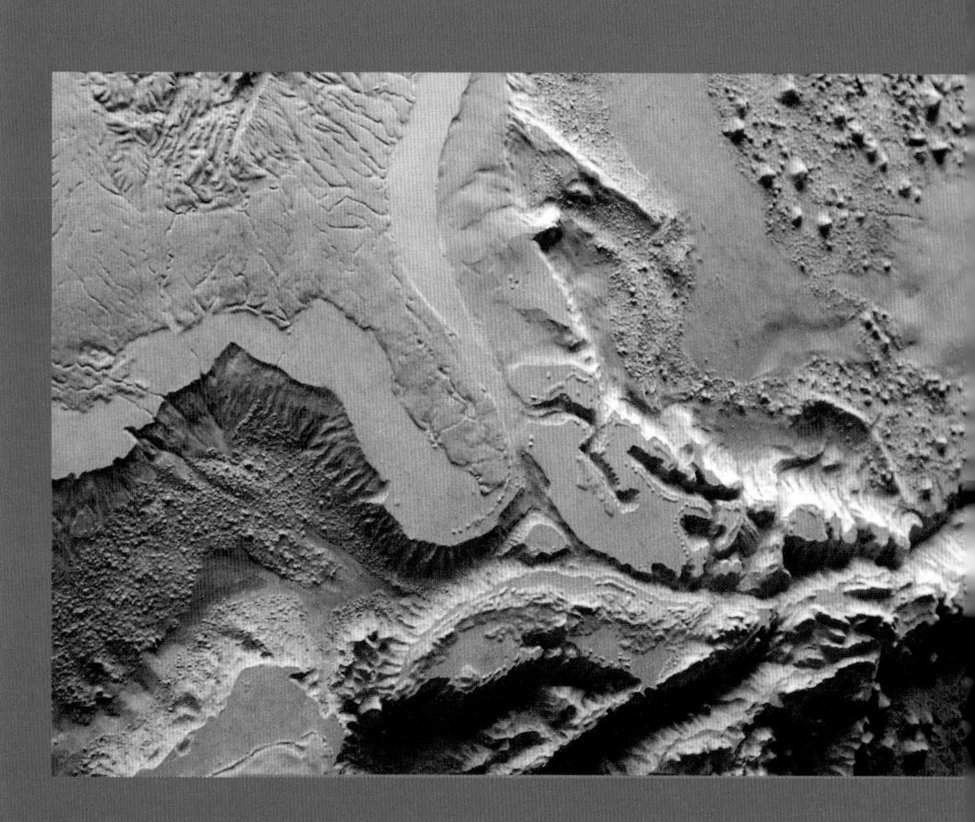

Contents

List of Illustrations / xi

Acknowledgments / xix

Introduction / 1

1 Florida Defined / 10

2 Florida Lost: Wandering the Globe and Finding Home (~700 Ma to ~200 Ma) / 23

3 The Big Split: Formation of Three Oceans and the Establishment of the Florida Basement (~225 Ma to ~140 Ma) / 35

4 The Carbonate Factory Cranks Up: Florida Being Born from the Sea (~160 Ma to Present) / 52

5 An Environmental Crisis: Drowning of the West Florida Margin and Development of the West Florida Escarpment (~100 Ma to ~80 Ma) / 70

6 Clash of Geologic Terrains: Colliding with Cuba (~56 Ma to ~40 Ma) / 93

7 Dissolution Tectonics: Sinkhole Development (~140 Ma to Present) / 107

8 Sands from the North: The Quartz Sand Invasion (~30 Ma to Present) / 132

9 Erosion in the Ocean, Marine Fertility, and Huge Sharks: The Florida Phosphate Story (~22 Ma to ~5 Ma) / 157

10 The Finish Line in Sight: Approaching Modern Florida and the Emergence of South Florida (~2.5 Ma to ~10 ka) / 185

Epilogue / 211

Index / 219

Illustrations

Frontispieces Geologic Time Scale and Sea Level Curve ii
Generalized Geologic Map of Florida iii

Table 1. English to Metric Conversion 8

Figures

I.1. Elevation map of Florida 2
I.2*A*. Light-dependent reef 3
I.2*B*. Aerial photo of Palm Beach 3
I.2*C*. Everglades alligator 3
I.2*D*. One of Florida's spectacular caves 3
1.1. Physiographic map of the United States 11
1.2. The Florida-Bahama Platform complex with individual Florida and Bahama Platforms 12
1.3*A*. Primary bathymetric features of the submerged portion of the Florida Platform 13
1.3*B*. Generalized cross section of entire Florida-Bahama Platform 13
1.4. Sea level curves of past events in the Pleistocene 14
1.5*A*. Earth's present tilt at 23.5° 15
1.5*B*. Precession has a 23 kyr cycle 15
1.5*C*. Eccentricity has 100 kyr and 413 kyr cycles 15
1.6. Four images of Florida illustrating area of inundation with sea level rises of 1, 2, 4, and 6 m respectively 16
1.7. Physiographic map of significant local relief in eastern Gulf of Mexico, Florida, and the Bahamas 17
1.8. Map showing full relief of main island of Hawaii, making it the tallest single feature on Earth 17
1.9. Map showing plate tectonic setting of Florida 19
2.1. View of spiral galaxy from the Hubble telescope 24
2.2*A*. Paleogeographic reconstruction illustrating distribution of continents at ~ 650 Ma 25
2.2*B*. Paleogeographic reconstruction ~300 Ma during the extended collision between Laurentia and Gondwana 26
2.2*C*. Paleogeographic reconstruction of supercontinent Pangea at ~200 Ma 26

2.3. Rifting of Rodinia supercontinent ~700 Ma illustrating the series of promontories and embayments 27
2.4. A Wilson cycle 28
2.5. Final assembly of the Pangean supercontinent showing additional components 30
3.1*A*. Pangea before breakup in Late Triassic 36
3.1*B*. Pangea breakup in Middle Jurassic 36
3.1*C*. Pangea breakup in Late Jurassic 37
3.1*D*. Pangea breakup complete by Late Jurassic 37
3.2. Paleogeographic map at ~160 Ma of proto Gulf of Mexico and proto Caribbean Sea 39
3.3. Map of eastern North America illustrating distribution of early rift basins 40
3.4. Paleogeographic map ~140 Ma illustrating formation of new North Atlantic Ocean, proto Caribbean Sea, and the fully developed Gulf of Mexico 41
3.5. Depth to basement map illustrating the shape of the Peninsular Arch and the location of smaller structural features 44
3.6. Paleogeographic map illustrating similarity and connectivity of rock beneath peninsular Florida and NW Africa 45
3.7. Transect across outer margin of west Florida illustrating faulted basement structural boundary 46
4.1. Idealized section extending W–E starting from NE Florida out across the Blake Plateau and off the Blake-Bahama Escarpment to the east 53
4.2. Paleogeographic map showing Florida Platform separated from North America by Georgia Seaway Channel 54
4.3. Map of the early opening of the Tethys Ocean forming the initial North Atlantic Ocean at ~160 Ma (Late Jurassic) 55
4.4. Paleogeographic global map of Tethys Ocean in the mid-Cretaceous 55
4.5*A*. Nonskeletal ooid sand grains in normal light 57
4.5*B*. Photomicrograph of ooid sand grains with cement 57
4.5*C*. Outcrop of rudist bivalves forming a solid rock mass or reef 57
4.5*D*. Photo of suspended carbonate muds being transported off the Great Bahama Bank into the deep Straits of Florida 57
4.6*A*. Space image of the Bahamas 58
4.6*B*. Two-dimensional transect across Great Bahama Bank illustrating distribution of carbonate sedimentary facies 58
4.7. Three-dimensional view across a reef-dominated carbonate platform margin showing lateral distribution of carbonate sedimentary facies 59

4.8*A*. Halite (NaCl) being precipitated in Salt Lake in the Coorong region of south Australia 61
4.8*B*. Gypsum outcrop in southern Spain 61
4.8*C*. Impermeable gypsum crystals "locked together" 61
4.9. Subsidence curves for a cooling passive margin 62
4.10. The carbonate platform complex along the east coast and Gulf of Mexico coast during the combined Jurassic and Cretaceous 64
4.11. The extent of the Late Jurassic/Early Cretaceous carbonate gigaplatform 64
4.12. Cross section over the eastern North American continental margin showing buried components of the Jurassic gigabank 65
5.1*A*. The seaward portion of the ramp forming the west margin of the Florida Platform 71
5.1*B*. Detailed seismic prolife illustrating the West Florida Escarpment 72
5.2. The correlation between the high sea level and the expanded ocean crust production in the mid-Cretaceous 73
5.3. Paleogeographic map of much of the northern and western hemispheres at the mid-Cretaceous 74
5.4. At the higher latitudes in the mid-Cretaceous, the ocean's surface was significantly warmer 75
5.5. Seismic profile indicating as much as 6 km of erosion occurred along the West Florida Escarpment 79
5.6*A*. Bathymetric map of middle section of the West Florida Escarpment illustrating scalloped erosional indentions 80
5.6*B*. Multibeam image of upper portion of the West Florida Escarpment and west Florida margin 80
5.7. Seismic line and line-drawing interpretation of buried detachment surface along west Florida margin 81
5.8. Bathymetric map of SW Florida margin illustrating large submarine canyon complex along lower section of West Florida Escarpment 82
5.9. Cross section showing brine seepage that led to canyon development along the base of the SW portion of the West Florida Escarpment 83
5.10*A*. Artistic rendition of meteorite striking Earth at the Chicxulub Crater site on the Yucatan at 65.5 Ma 84
5.10*B*. The Chicxulub Crater buried beneath the Yucatan Peninsula 84
5.11. Paleogeographic map of the Yucatan-Florida-Bahama Platform complex and Greater Antilles including Cuba soon after impact 85
5.12. Outcrop of brecciated, broken rocks of the Cacarajícara Formation that probably resulted from submarine landslides destabilized by the impact of the meteorite 86
6.1. Map of Greater Antilles and surrounding geography 94
6.2*A*. Detailed map of western Cuba 94

6.2*B*. Mogotes in the Vinales area of western Cuba with red soils 95
6.2*C*. Mogotes in Vinales area showing some mature karst details 95
6.3*A*. Paleogeographic reconstruction of Caribbean Sea area ~84 Ma 98
6.3*B*. Paleogeographic reconstruction of Caribbean Sea area ~72 Ma 98
6.3*C*. Paleogeographic reconstruction of Caribbean Sea area ~59 Ma 98
6.3*D*. Paleogeographic reconstruction of Caribbean Sea area ~36 Ma 98
6.4. Volcanic rocks in western Cuba folded as a result of the collision between the volcanic island arc and the carbonate platforms 100
6.5. N–S cross section of western Cuba and southern Straits of Florida showing folded thrust sheets and location of potential hydrocarbon accumulations 100
6.6. Paleogeographic map showing amount of the Florida-Bahama Platform that "overlapped" Cuba 101
6.7*A*. Bathymetric map of the Straits of Florida 102
6.7*B*. Seismic cross section in Straits of Florida revealing buried carbonate platform 102
7.1. The Guadalupe Mountains in west Texas 108
7.2*A*. Divers explore Diepolder Cave, located on Sand Hill Boy Scout Reservation near Brooksville, Florida 109
7.2*B*. Map of Devil's Eye and Ear cave system 110
7.3*A*. One of the complex interior chambers of the Carlsbad Caverns formed within the Permian carbonate platform of west Texas and New Mexico 111
7.3*B*. Inside a Carlsbad Cavern chamber 111
7.4*A*. Winter Park sinkhole 112
7.4*B*. Oblique view of the Winter Park sinkhole 112
7.5*A*. Map showing lateral extent of the Floridan aquifer and its changing thickness 113
7.5*B*. Cross section down peninsular Florida showing aquifer system, including the Boulder Zone in southern Florida 114
7.6. Map of top of Boulder Zone in SE Florida 115
7.7*A*. Surficial karst in Eocene age limestone along the Big Bend marsh coast near Ozello, Florida 116
7.7*B*. Close-up of surficial karst illustrating small, cavernous, fluted nature of limestone bedrock surface 116
7.8*A*. Cartoon block diagram of the surface aquifer system 117
7.8*B*. The making of a sinkhole 117
7.8*C*. Sinkholes in Florida since 1954 117
7.9. Cartoon illustrating subsurface mixing zone of brackish and freshwater and surficial karst features 119
7.10*A*. The Big Bend coastline near Ozello, Florida 120
7.10*B*. Small sinkhole in the exposed Eocene limestone 120

7.10*C*. Rocky nubs or flat pinnacles of Eocene limestone that once supported vegetation 121
7.11. Probe-rod profile across section of marsh illustrating rugged nature of buried karst 121
7.12. Oyster reefs in Big Bend area 122
7.13. Fracture patterns in rocks along the east-central portion of Florida 123
7.14. The Kohout convection process 124
7.15. High-resolution seismic reflection profile across mouth of Tampa Bay showing folds, warps, sinkholes, and sags 125
7.16. Interpretations of seismic lines from Charlotte Harbor estuary illustrating subsurface deformation features 126
7.17. Map depicting depth to the top of the deformed limestone beneath Tampa Bay 127
8.1*A*. The foredune ridge and beach at Caladesi Island 133
8.1*B*. Sanibel Island 133
8.2*A*. Classification system of Florida coastline 134
8.2*B*. The Big Bend marsh coastline 134
8.2*C*. Barrier islands along Florida's northern coastline 134
8.2*D*. Florida's mangrove dominates the Ten Thousand Island coastline 135
8.3. Microscope view of mostly carbonate sand from a beach consisting of broken shells 136
8.4. Two views of sieved quartz sand 136
8.5*A*. Sediment thickness map of inner shelf off part of Pinellas County 137
8.5*B*. Two cross sections traversing shelf off Pinellas County 138
8.6*A*. Mt. Everest 139
8.6*B*. Mt. Mitchell 139
8.7*A*. Trench dug into slope of Blue Ridge Mountains and showing heavy weathering 141
8.7*B*. Heavily weathered outcrop of folded strata in the Blue Ridge Mountains 141
8.8*A*. Vertical infrared image of the South Carolina and Georgia coast 142
8.8*B*. Image of Caladesi Island along Florida's west-central coast 142
8.9. The siliciclastic transport pathway from the southern Appalachian Mountains to the Straits of Florida 144
8.10*A*. Surficial geologic map of central peninsular Florida showing paleo sea level features such as shorelines and scarps 147
8.10*B*. Vertical image of eastern Florida showing linear features that are paleo-shorelines deposited during periods of high sea level 147
8.11. Map of quartz-rich sediment transport pathway via prograding river deltas south of the Lake Wales Ridge 148

10.11*B*. Aerial photo of sedimentary bedforms (sand waves or underwater dunes) indicative of strong water flow onto the Bahama Banks 198
10.11*C*. Outcrop Miami Limestone showing cross-bedding primary sedimentary structures in the rock resulting from migration of sand waves 199
10.12*A*. Multibeam image of drowned barrier island shoreline at Pulley Ridge along the SW margin of the Florida Platform in ~65–70 m water 200
10.12*B*. Bottom photo at Pulley Ridge revealing corals, encrusting red algae and green fleshy algae growing on top of the ancient barrier island 201
10.12*C*. Bottom photo of Pulley Ridge revealing similar features as shown in fig. 10.12*B* 201
10.13. Image of paleo-shorelines shown in the Cape Canaveral area 202
10.14*A*. A former cliffed shoreline in the Miami Limestone called the Silver Bluff 203
10.14*B*. Modern cliffed, rocky shoreline of carbonate island in the Bahamas 203
10.14*C*. Underwater photo of drowned cliffed shoreline 203
10.15*A*. Computer-generated, merged topographic and bathymetric map of the Tampa Bay area showing former islands that existed during previous sea level highstand 204
10.15*B*. Computer-generated, merged topographic and bathymetric map of the entire Tampa Bay drainage basin illustrating sinkholes and former sea level highstand shorelines 205
E.1. Oil and gas fields in the Gulf of Mexico, the Caribbean, and northern South American regions 215

Acknowledgments

I had to rely on the expertise of many to write this book. Understanding Florida's geologic history requires broad expertise in physical and historical geology, plate tectonics, sedimentary geology and sedimentary processes, coastal geology, structural geology, stratigraphy, climate, sea level, oceanography, paleoceanography, mineralogy, biogeochemistry, geomorphology, groundwater geology, regional geology, and many more of the Earth science systems and subdisciplines. No one geo-scientist has all of this expertise. So I have been very fortunate to receive assistance from many colleagues, friends, and students. This work barely scratches the surface scientifically. I realize that others have different views of some topics, probably leading to some controversy.

To keep me scientifically correct and honest, I have depended on advice from Drs. Jon Arthur (director and state geologist, Florida Geological Survey), Dan Belknap (University of Maine), Gregg Brooks (Eckerd College), Kevin Cunningham (USGS), Grenville Draper (Florida Atlantic University), Ben Flower (University of South Florida, USF), Pamela Hallock-Muller (USF), Bob Halley (USGS, retired), Peter Harries (USF), David Hollander (USF), Alex Isern (National Science Foundation), Manuel Iturralde-Vinent (Museo Nacional de Historia Natural, Cuba), Jack Kindinger (USGS), Barbara Lidz (USGS), Stan Locker (USF), David Mallinson (East Carolina University), Don McNeill (University of Miami Rosentiel School of Marine and Atmospheric Science), Ellen Prager (Earth2Ocean), Jeff Ryan (USF), Stan Riggs (East Carolina University, retired), Lisa Robbins (USGS), Tom Scott (Florida Geological Survey, retired), Karen Stewart (Florida Institute for Phosphate Research), and Dean Whitman (Florida Atlantic University).

I am very grateful to the following for helping me obtain figures: Don McNeill (RSMAS–University of Miami), Stan Locker (USF), Noel James (Queens University, Canada), Gene Shinn (USF), Barbara Lidz (USGS), Lisa Robbins (USGS), Dean Whitman (Florida Atlantic University), Gren Draper (Florida Atlantic University), Tom Missimer (King Abdulla University of Science and Technology, Saudi Arabia), Bob Weisberg (USF), Lianyuan Zheng (USF); Norman Kuring (NASA), Chuanmin Hu (USF), Charles Kerans (University of Texas–Austin), Lee Florea (Ball State University), Allen Yarborough (Florida Southwest Water Management District), Paul Mann (University of Texas Institute of Geophysics), Larry Mayer (Center for Coastal Ocean Mapping at the University of New Hampshire), James Mauch (MD), and Sean Roberts (Florida Museum of Natural History). I thank Dr. Ron Blakey of Colorado Plateau

Geosystems for his wonderful imagery shown in figures 5.3 and 10.3, and I thank the National Geographic Society for the use of Wes Skiles's freshwater spring image shown in figure 7.2*A*.

I thank the following graduate students and staff members who can provide the best advice of anyone: Kevin Bradley, Shane Dunn, Dawna Ishler, Patrick Schwing, Beau Suthard, Ann Tihansky (USGS), Carlie Williams, Monica Wilson, and Carrie Wall.

I reserve special thanks to Linda Kelbaugh, who helped me enormously with editing and formatting earlier drafts.

I am most appreciative of the technical advice provided to me by Karen Stewart of the Florida Institute for Phosphate Research, Barbara Lidz (especially for editorial advice), and Betsy Boynton of the U.S. Geological Survey.

I am most grateful to the following for reading through the entire text to assess continuity, consistency, and overall effect: Drs. Ray Arsenault (USF-St. Petersburg), Ernie Estevez (Mote Marine Lab), Bill Hogarth (Florida Institute of Oceanography), Gary Mormino (USF-SP), A. C. Neumann (professor emeritus, University of North Carolina at Chapel Hill), Paul Enos (University of Kansas), Bob Potter (USF), and Gene Shinn (USF). I also very much appreciate the comments from an anonymous formal reviewer. Paul Enos's review was amazingly thorough and helpful! Thanks, Paul!

The following institutions in Florida have also been extremely helpful: Florida Museum, Florida Geological Survey, Florida Institute for Phosphate Research, Florida Institute of Oceanography, U.S. Geological Survey, Windley Key State Fossil Park, Dry Tortugas National Park, the Mulberry Phosphate Museum (Jessie Ward, director), and the University of South Florida, College of Marine Science.

I thank the staff at the University Press of Florida, especially Meredith Morris-Babb (director), Michele Fiyak-Burkley (managing editor), and Elaine Otto (copy editor).

I am most indebted to one of the hardest working graduate students I have ever known, Carlie Williams, a doctoral student in the USF College of Marine Science at this writing, who worked with me over many months to get the illustrations into publishable shape. I could not have written this book without her help. I thank the USF College of Marine Science for providing her support to assist me in this project.

Geologic History of Florida

Introduction

To those teachers who spark imagination and ignite fires of the mind.
To those students, regardless of age, who stoke those fires and have never lost their love of learning.

Having grown up in New England and majored in the earth sciences as an undergraduate, I was constantly amazed to learn that geology could explain the scenery whether it be the mountains resulting from some tectonic plate collision in deep time or more recent features such as Cape Cod and Long Island, freshly made by the last mile-thick continental ice sheet that once covered the ground where we stood. The configuration and elevation of the Earth's surface, the plants that are distributed on this topography, and indeed, the nature of human habitation are controlled by the underlying geology.

Additionally, the common appearance of huge outcrops exposed in ever-increasing highway excavations, particularly in the more rugged parts of New England, allowed us to peer, at least a short distance, into the Earth's crust to see rocks that had been contorted and folded by unimaginable and unseen forces of the past. These rocks were once many kilometers beneath the Earth's surface, forever changed by intense heat and pressure, yet now they are exposed at the surface. How could that be? The Earth's geologic history was literally in our face, and the mysteries of the past that the science allowed us to unravel were too amazing to have been invented in even the most imaginative of minds. So I became hooked on the study of the Earth—the whole Earth, not just the continental land masses, but the interaction of members of the "sphere" family (lithosphere, atmosphere, hydrosphere, cryosphere [ice], and the biosphere).

We can add the "final frontier"—the extraterrestrial sphere whereby temporal variations of the Sun's radiant energy reaching the Earth due to fluctuations in our star's interior furnace and cycles of the Earth's orbit and rotation about its axis have had profound effects. Lastly, every once in a great while (even on geologic time scales), screaming out of the cosmos come ~10 km wide rocky meteors or icy comets speeding at ~10–50 km/sec whose impact and subsequent release of enormous energy become immediately transformative, causing global extinctions of life. The integration of all these processes is now called earth systems science—the next generation of scientists will be known as earth system scientists, rather than merely oceanographers, geologists, and atmosphere

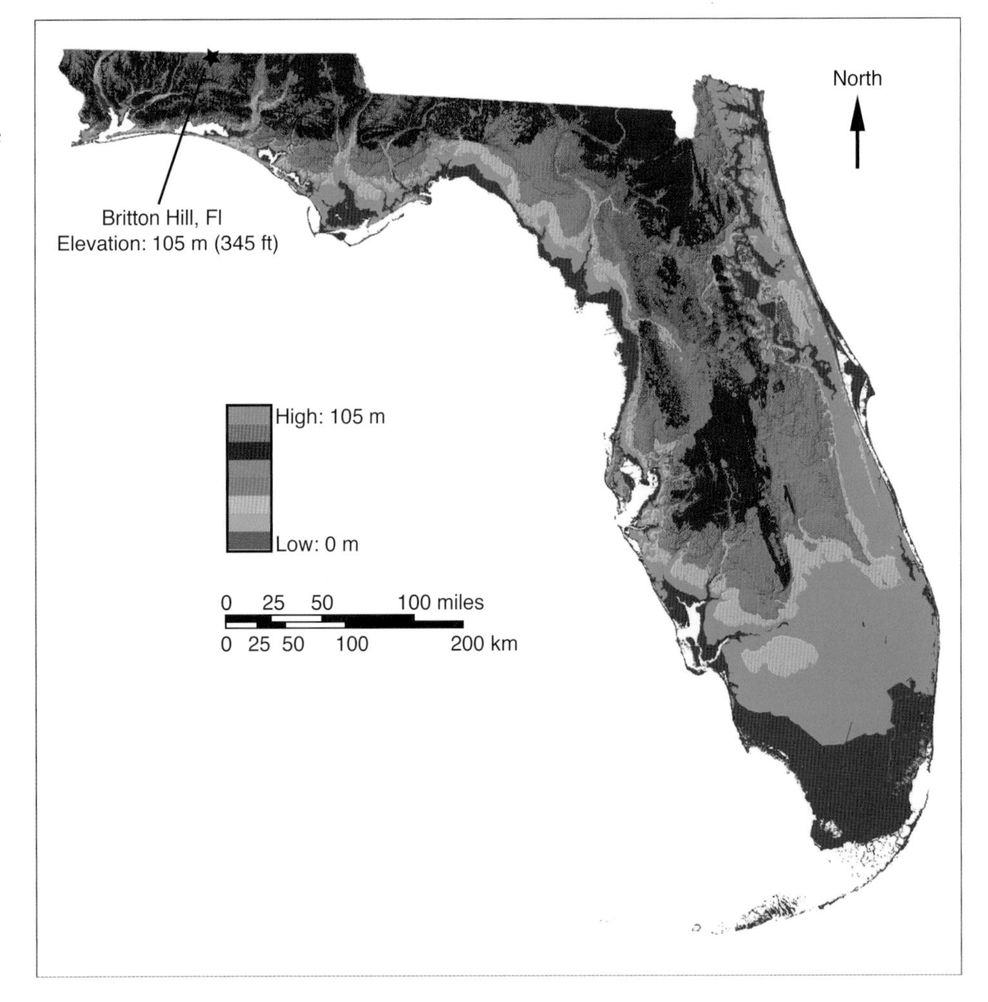

Figure I.1. Elevation map of Florida. The Panhandle has the highest elevations in Florida. Britton Hill, the highest natural point in Florida, has an elevation of 105 m (345 ft). (Source: Florida Geological Survey.)

scientists. Connectivity is the word *du jour*, that is, all parts of our planets are intimately linked and affect each other.

After doctoral and postdoctoral work in the Carolinas, my wife and I moved to St. Petersburg, Florida, where I had accepted a faculty position at the University of South Florida in what was the Department of Marine Science (now a College of Marine Science). By then, I had become a student of the geology of the ocean—a geological oceanographer—thus fulfilling a boyhood wish to study the marine realm.

So Florida with its huge coastal ocean and coastline was a great place to start one's career in geological oceanography. However, the land geology of Florida was not in your face, so to speak. The highest natural point in Florida is only ~105 m (345 ft; fig. I.1) and that is in the panhandle (Britton Hill—northern Walton County). Britton Hill is the lowest highest point of any of the fifty states. Actually, the highest point in Florida is a hotel in Miami topping out at 239 m! Overall, rocky outcroppings are relatively rare in Florida. By comparison, Mt. Mitchell in the North Carolina Appalachian Mountains is 2,025 m (6,684 ft)—the highest point east of the Mississippi River. And the former Mt. McKinley, now called Denali in Alaska, is a towering 6,194 m (20,440 ft) and is the highest point in the United States. Britton Hill would hardly be noticed if it stood next to it.

The attraction to Florida, at least for me and for many others, is the ocean. Florida, as we will see, was born from the ocean, and its geologic history lies beneath our feet, beneath the coastal ocean, deep underground, well out of sight, and therefore out of mind to most of us. But the subsurface geology ultimately explains Florida's highly visible modern topography and scenery. People come to Florida to see not the geologic events of the past but the geology of the present—the coastline and its beaches, the freshwater springs, the coral reefs, the Everglades, the swamps, and the wildlife associated with these environments (fig. I.2).

Figure I.2*A*. Light-dependent coral reef located in the Florida Keys. The Keys are as famous as the Everglades for their natural underwater scenery. (Source: NOAA.)

Figure I.2*B*. Aerial photograph of Palm Beach. Florida is famous for its beautiful quartz sand beaches. (Source: Michael Kagdis, Proper Media Group.)

Figure I.2*C*. The Florida Everglades is home to many indigenous animals and plants. (Source: National Park Service.)

Figure I.2*D*. One of Florida's spectacular caves. (Photo by Sean Roberts, Florida Museum of Natural History; used by permission.)

So the geologic history of Florida is a hidden secret—known only to a privileged few whose work, hobby, or both is to study Florida's geologic past.

Over the years, I thought that it was important for our marine science graduate students, particularly those interested in geological oceanography, to know something about Florida's geologic past. So I developed a full-length course, contributed a chapter to *The Geology of Florida,* and published a number of scientific papers on the details of selected portions of Florida's geologic history.

This book is not meant to be a textbook for my course (but it could be used as such). I have a broader audience in mind. As my own children were growing up, I would make a point of volunteering to come into their classrooms, much to their embarrassment, and talk about oceanography, earth science, and relate it to Florida's past. I would like to think that the kids learned something. However, I was always struck by the teachers telling me that they had no idea Florida had such an interesting geologic history. As a result, I have written this book with those teachers in mind as well as all of those out there who have never lost their desire to learn something new.

William Butler Yeats said, "Education is not the filling of a pail, but the lighting of a fire." I hope this book starts some fires. Unfortunately, much of our learning today is about filling heads (pails) with information required by standardized tests and not starting enough fires in the mind.

I realize that, to some, there is more here than you want to know. Others will feel that the book is not rigorous enough to use as a course textbook, leaving them unsatisfied by some explanations. There is simply not space to flesh out nuances or to present multiple interpretations. Rather, I have tried to provide the broad perspective and have approached this effort based upon the major events and the processes (chemical, biological, and physical) that have taken place, not just marching lock-step through geologic time explaining the land-based succession of rock/sedimentary (lithologic) formations. Ultimately, our understanding of the Earth is all about process. The products are important because they tell us what happened. However, to become good stewards of the Earth, we need to know how things happen, so we can live for sustainability and predict our impact on earth systems and prepare for what earth systems might have in store for us. Some of the interpretations presented are still under debate. I have taken the liberty of expressing my own view of the geologic past based on fragments of evidence, and I have tried to point out where such debate is still under way. I know that some of my geology colleagues will take issue with some of the points I have presented—such is scientific inquiry. Finally, like any field of inquiry, science is filled with terminology, and certain terms are impossible to avoid. I have tried to keep the arcane jargon to a minimum. However, some terms are required. To make reading this text a bit more understandable, I have *italicized* some words or key phrases that appear with an explanation at the end of each chapter.

Why Is the Geology Important?

People come to Florida to enjoy the modern environment. However, this environment is undergoing severe stress presented by the nearly 19 million people who now inhabit the state. Water quality and quantity problems, waste disposal, accidental spills of toxic substances, hurricane threats, other weather extremes (floods, droughts), beach erosion, phosphate and limestone rock mining, coral reef degradation, offshore oil drilling (remember the BP disaster of 2010?), live hard-bottom excavation, channel dredging, coastal wetlands and marine vegetation impacts, and harmful algal blooms are all subjects that contain an important geologic environmental component. A balanced, science-based assessment of these topics under one cover is a formidable task and is best left as a follow-on effort. However, an understanding of Florida's geologic past is a necessary first step to fully address all of these issues.

In the end, Florida is intimately and ultimately tied to global geologic events that synergistically link the atmosphere, ocean, crust, and all life contained therein. Florida's geologic history is not just a local phenomenon but one that is interconnected to Earth's interior and Earth's surface events. Plate motion, ocean circulation, climate, biotic evolution, and sea level changes are global phenomena whose integrated signal may appear differently in different places around the globe at different times. One of the great lessons of earth science is that any specific location is always linked to a much larger framework. It is our task as readers and interpreters of history to recognize and understand that linkage.

So take what you can from this effort. If you crave more, I have included some key references that may provide the required detail. More important, have fun learning about this amazing sequence of geologic events that have conspired to bring Florida to its present day. As I tell my students: "If it isn't fun, it isn't worth doing."

Thinking Geologically

Thinking about Time

Students new to geologic time and the Earth's great antiquity are amazed by the enormity of scale (see Geologic Time Scale on page ii). I have also used the 2009 Geological Society of America Geologic Time Scale (http://www.geosociety.org/science/timescale/timescl.pdf). Geologists toss around tens, if not hundreds, of millions of years as if we were talking about something that happened last week or last month. In a geologic sense, we are. Since the Earth is about 4.55 billion years old (Ga), what's a few million years here and there?

To put deep time into an even larger perspective, astrophysicists, through the technology provided to them from the Hubble telescope, have determined that the age of the universe is 13.8 Ga. So Earth is merely a young adult or even a celestial adolescent.

A word about abbreviations: herein 1 **ka** means at one thousand years or one thousand years ago; 1 **Ma** means at one million years or one million years ago; 1 **Ga** means at a billion years (1,000,000,000) or one billion years ago. Additionally, **kyr**, **Myr**, and **Gyr** express *geological duration*, that is, an event lasted 10 Myr or the event's duration was 10 million years long (B. Lidz, personal communication, and Christie-Blick, 2011).

To a "visual" person, converting time to distance may provide a perspective on deep time. For example, let's assume that 1 millimeter (.001 meter; ~thickness of dime) equals 1 year. Then, 1 million years (Myr) (13,333 consecutive human lifetimes at 75 yrs each; 40,000 consecutive human generations at 25 yrs each) equals 1 million millimeters (mm) or 1 kilometer (km). One kilometer is about .62 statute miles.

Now, how about a billion years (Ga)? Again, this is 1 billion millimeters, which is 1,000 km (~620 miles). So the age of the Earth at 4.55 Ga, if plotted on a graph, would be 4,550 km (~2,852 miles) away from the zero point (today; 0 years) if 1 yr = 1 mm. In this book, we travel back to about 700 Ma (700 km; ~438 miles) to consider Florida's geologic history with most of the action occurring during the last 200 Myr (200 km; ~124 miles; about 4 percent of the total age of the Earth). In this sense, Florida is but a toddler.

During some periods of geologic history, little may have happened during the passage of tens of millions of years. At other times, a short-lived but spectacular event such as the large meteor striking the Earth ~65.5 Ma (defining the famous K/Pg boundary; K means Cretaceous; Pg means Paleogene) can permanently alter Earth history even though the immediate event lasted no more than a few tens of seconds. This boundary used to be called the K/T boundary with T representing the Tertiary.

Obviously, some events take longer than others. The amount of time consumed during each event is not particularly relevant, but the consequences may be significant. As a result, the chapters below do not each represent an equal amount of geologic time, but each is important to the ultimate development of Florida. Florida's journey is about a specific sequence of events, not just a forced march through a seemingly endless period of time. But we do have to start at some point in "deep" time and finish in the geologic present.

Finally, we are able to "tell time" amazingly accurately in the geologic past. A recent article in a major geologic journal described an event 53 kyr in duration extending from 54.964 Ma to 54.911 Ma. That's astounding! Fifty-four million years ago geologists can discriminate 1,000 yr time slices—that is being able to discriminate the difference between events occurring at 54.911 Ma and those occurring at 54.912 Ma—an accuracy of <.0001 of the actual age. Fifty years ago, if geologists determined that events were within a few million years of each other, they were virtually simultaneous some 54 Ma back in time.

Our development of the modern Geologic Time Scale has brought both an accuracy and a precision to understanding the timing of Earth events to an unprecedented level. We can determine if two events are truly simultaneous or if one precedes or lags by very short periods—thus determining if one event caused the other or was the result of the other. So scientists have become very clever in figuring this out. Thinking temporally and having the Geologic Time Scale firmly implanted in one's thought process as second nature are essential to an earth systems scientist—not easily accomplished even after years of studying the Earth's past.

Thinking about Space

Another "trick" that geologists have to master is to think in three dimensions (3D) and over great distances in the subsurface. Since most of Florida's geologic past lies beneath our feet, we have to imagine the subsurface interconnectivity of strata, erosional surfaces, faults, folds, and other geologic phenomena. Additionally, we have to imagine what the topography of the Earth's surface must have been like at various points in time. Those shapes and surfaces lie in the subsurface, and we need to use whatever data we can obtain (seismic, gravity, magnetics, rocks from cores) and a creative imagination to reconstruct the shape and form of Earth at various points in the past. Additionally, we cannot forget that the surrounding geography, sometimes extending thousands of km from Florida, affected or controlled what went on in Florida.

Thinking about Linked Earth Systems

As mentioned earlier, the final element in thinking geologically is to understand the processes that shaped the Earth's surface—it just does not change by itself. In the past twenty-five years, students of the Earth (applied biology, chemistry, and physics) now think in terms of earth systems rather than just geology, marine science, oceanography, or atmospheric science. Earth systems consider that the two large fluid bodies (ocean and atmosphere) on the Earth and its crust are linked intimately with seemingly infinite feedback loops, ties, and connections. Additionally, students now have to understand processes at the Earth's core, variations of the Earth's orbit around the Sun, as well as other extraterrestrial processes to complete the whole linked Earth system. I try to bring as much as I can to this effort. It is not just about rocks—it is about how things work. It is about process—rates and intensities, amplitudes and frequencies.

Thinking Metric

Having grown up with English units as part of my DNA, it is still hard to visualize some measurements expressed in metric units, particularly temperature. Even after thirty-five years as a scientific researcher, I still do not know if I should wear a sweater if the outside temperature is 20°C. But it is a "metric world," and scientists in other countries use metric. So we all have to get used to it, the sooner the better. Use the conversion table provided—I do quite often.

English to Metric conversion table

English Unit	Symbol	Conversion Factor	Metric Unit	Symbol
Length				
inch	in	25.4	millimeters	mm
foot	ft	0.305	meters	m
yard	yd	0.914	meters	m
mile	mi	1.61	kilometers	km
Area				
square inch	in^2	645.2	square millimeters	mm^2
square foot	ft^2	0.093	square meters	m^2
square yard	yd^2	0.836	square meters	m^2
square mile	mi^2	2.59	square kilometers	km^2
acre	ac	0.405	hectares	ha
Volume				
fluid ounce	fl oz	29.57	milliliters	mL
gallon	gal	3.785	liters	L
cubic foot	ft^3	0.028	cubic meters	m^3
cubic yard	yd^3	0.765	cubic meters	m^3
Mass				
ounce	oz	28.35	grams	g
pound	lb	0.454	kilograms	kg
short ton (2000lb)	T	0.907	Megagrams (or metric ton)	Mg (or t)
Temperature				
Fahrenheit	°F	(F-32) × 5⁄9	Celsius	°C

Essential References to Know

Bjornerud, M. *Reading the Rocks: The Autobiography of the Earth*. Cambridge, MA: Westview Press, 2005.

Bryan, J. R., T. M. Scott, and G. H. Means. *Roadside Geology of Florida*. Missoula, MT: Mountain Press, 2008.

Buster, N. A., and C. W. Holmes. *Gulf of Mexico: Origin, Waters, and Biota. Vol. 3, Geology*. College Station: Texas A&M University Press, 2011.

Christie-Blick, N. "Geological Time Conventions and Symbols." *Geological Society of America Today* 22, no. 2 (2011): 28–29. doi: 10.1130/G132GW.1.

Cutler, A. *The Seashell on the Mountaintop: A Story of Science, Sainthood, and the Humble Genius Who Discovered a New History of the Earth*. 2nd ed. New York: Dutton, 2003.

Dobbs, D. *Reef Madness: Charles Darwin, Alexander Agassiz, and the Meaning of Coral*. New York: Pantheon Books, 2005.

Kastens, K. A., C. A. Manduca, C. Cervato, R. Frodeman, C. Goodwin, L. S. Liben, D. W. Mogk, T. C. Spangler, N. A. Stillings, and S. Titus. "How Geologists Think and Learn." *EOS Trans AGU 90*, no. 31 (2009): 265–66.

Press, F., R. Siever, J. Grotzinger, and T. H. Jordan. *Understanding Earth*. 4th ed. New York: W. H. Freeman, 2004.

Repcheck, J. *The Man Who Found Time: James Hutton and the Discovery of Earth's Antiquity*. Cambridge, MA: Perseus, 2003.

Ruddiman, W. F. *Earth's Climate: Past and Future*. 2nd ed. New York: W. H. Freeman, 2008.

Schopf, J. W. *Cradle of Life: The Discovery of Earth's Earliest Fossils*. Princeton, NJ: Princeton University Press, 1999.

1 Florida Defined

> The solid parts of the present land appear in general, to have been composed of the productions of the sea, and of other materials similar to those now found upon the shores.
>
> James Hutton, *Concerning the System of the Earth, Its Duration and Stability to Society*, 1785

What Is Florida?

This may seem like ridiculous question, but providing an answer prepares us to learn about the geologic history of Florida and assures that we are all on equal footing from the beginning.

Many kids growing up in the 1950s had as one of their first geography lessons a wooden puzzle whereby the pieces consisted of the 48 contiguous states. This was a good exercise in eye-hand coordination and spatial relationship learning. It was also a good way to discover how our country is put together. Additionally, the state capital was identified on each piece, providing another component to the geography lesson.

As a graduate teaching assistant, I was astonished to discover that many undergraduates did not know where certain states were located, nor could they properly locate the Mississippi River, the Great Lakes, or the Rocky Mountains. So, as an experiment, I asked them to draw in the states freehand on a sheet of paper (fig. 1.1). Actually, this is challenging to do without much erasing. But most of these young adults could hardly draw in and label the states east of the Mississippi River—much less the location of the Mississippi itself. The Midwest was a vast empty space, and only California, out west, was routinely correctly located and labeled. Perhaps they should have been introduced to that old wooden puzzle for some remedial training.

One can be challenged by the mostly "square" states like Wyoming, Colorado, Kansas, New Mexico, and perhaps Arizona. (They all look the same to an easterner!) But many states are highly distinctive in shape, and one knew

Figure 1.1. Physiographic map of the United States. I hope that the reader can find the Great Lakes, the Mississippi River, and the Rocky Mountains. Can you draw in the borders of the United States and the borders of each state? Note how easy it is to locate Florida due to its distinctive shape and location. It is the only state in the lower 48 that is mostly surrounded by the ocean. (Source: National Atlas of the United States, http://www.nationalatlas.gov/mapmaker?AppCmd=CU.S.TOM&LayerList=ShadedRelief&visCats=CAT-geo,CAT-geo.)

instinctively where they went. So you put those pieces down first and hoped that the other states would somehow fall into place properly to complete the problem.

No piece was more distinctive than Florida, however, due to its pistol-like outline with the panhandle forming the gun barrel and peninsular Florida forming the "grip." Besides, it stuck out into the ocean like no other state. So it was readily identifiable by name and location on the map. Even those college undergraduates, long ago, could readily identify Florida on that little test.

So the question "What is Florida?" seems self-evident. All one has to do is look at any map or any globe, and there it is. What's the big deal? It is a state located along the southeastern United States sticking out into the ocean. Enough said! But to the geologist this is a big deal. What we commonly think of Florida with its pistol-shaped outline is really only about 50 percent of the picture.

The Florida Platform and Sea Level

The state of Florida is defined by its shoreline and political boundaries to the north separating it from Georgia and Alabama. But Florida rests on top of what geologists call the *Florida Platform* (fig. 1.2), which is defined by geologic boundaries that lie deep beneath the present-day surface (more about these boundaries later). The surface of the Florida Platform includes the submerged *continental shelves* surrounding the emerged state of Florida, the *continental slope*, and a huge submarine cliff or *escarpment* that defines much of the western margin. The state really only covers ~50 percent of the Florida Platform (fig. 1.3).

The modern coastline is shaped by sea level. We all know about the ice ages of the past. If ice formed on land to create continental ice sheets, water must have been removed from the ocean and therefore sea level must have dropped. When the ice sheets melted, sea level rose. These cycles have been repeated hundreds

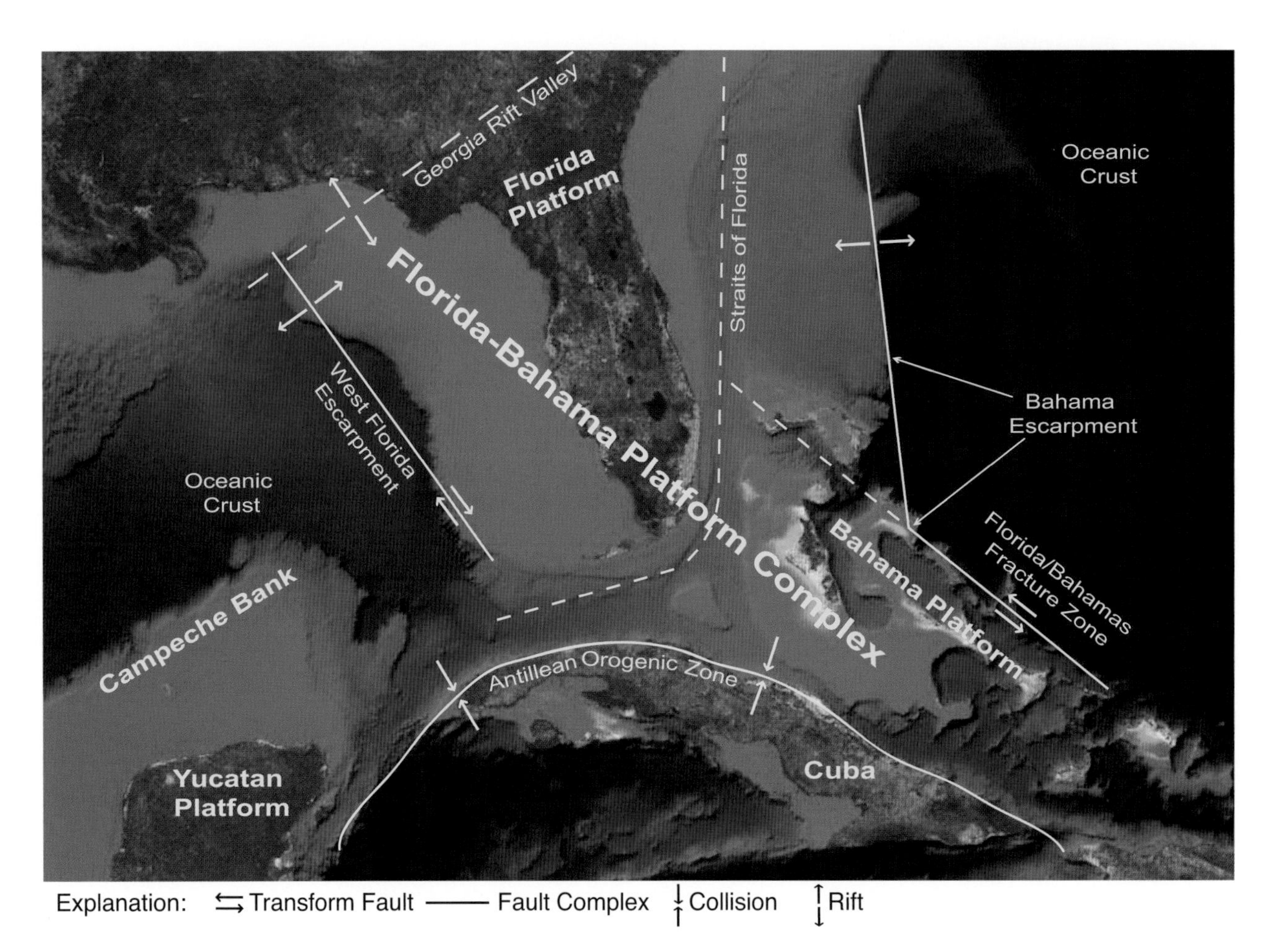

Figure 1.2. The Florida-Bahama Platform complex with individual Florida and Bahama Platforms. The key structural boundaries have been provided to demonstrate the distinct geologic entity of the Florida Platform of which approximately 50 percent is underwater. These structural boundaries lie deep in the subsurface but clearly have a significant bathymetric expression. (Source: Background map NASA, 2009.)

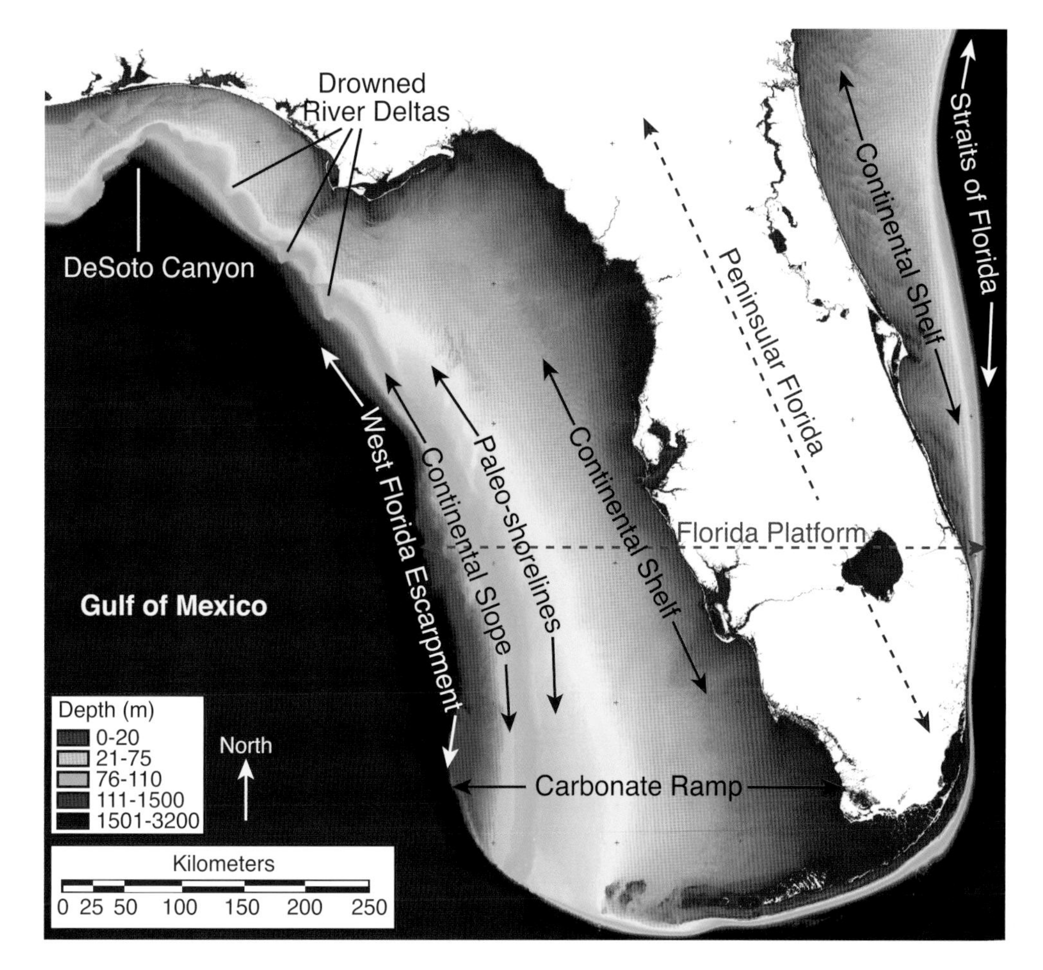

Left: Figure 1.3*A*. Primary bathymetric features of the submerged portion of the Florida Platform. (Modified from USGS Open File Report 2007-1397; courtesy of Dr. L. Robbins.)

Below: Figure 1.3*B*. Generalized cross section over the Florida-Bahama Platform illustrating stratigraphic succession down to basement rocks. Note the small size of the emerged land composing the state of Florida as compared with the entire platform complex. (Modified from Hine et al. 2003; used by permission from Elsevier.)

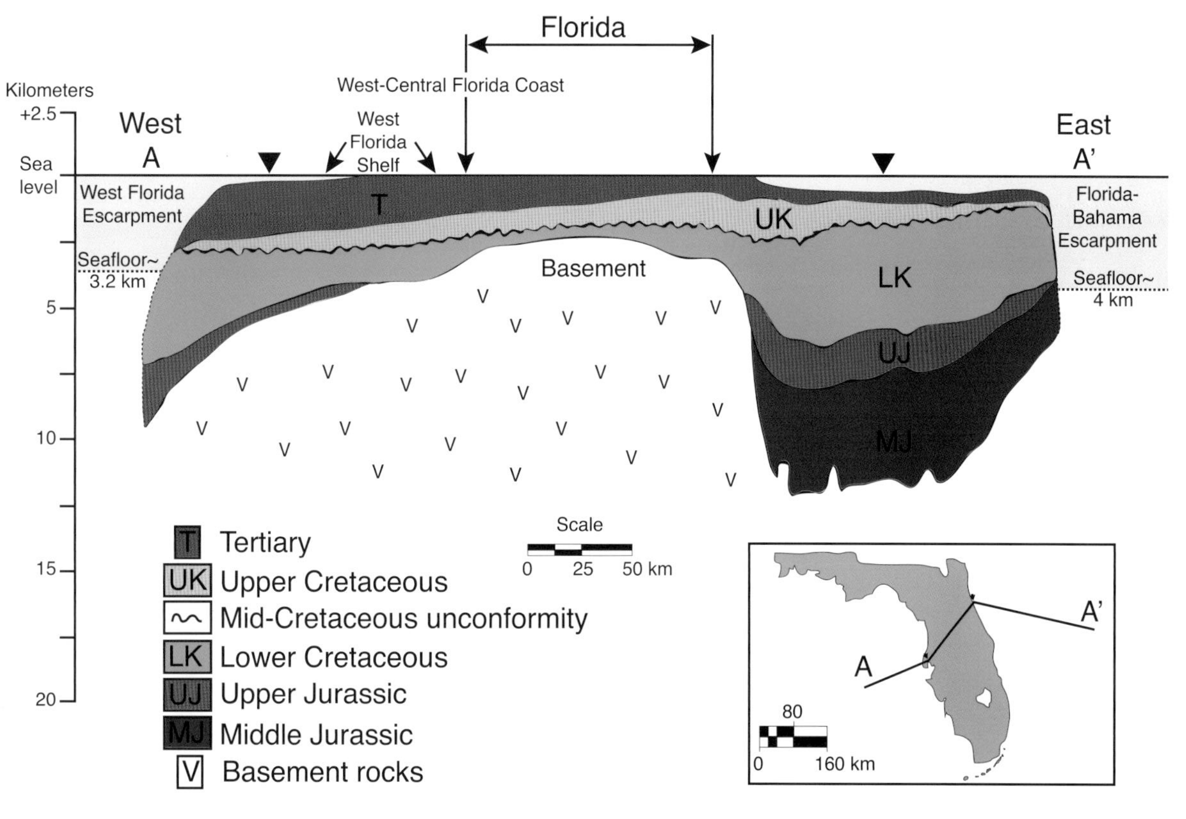

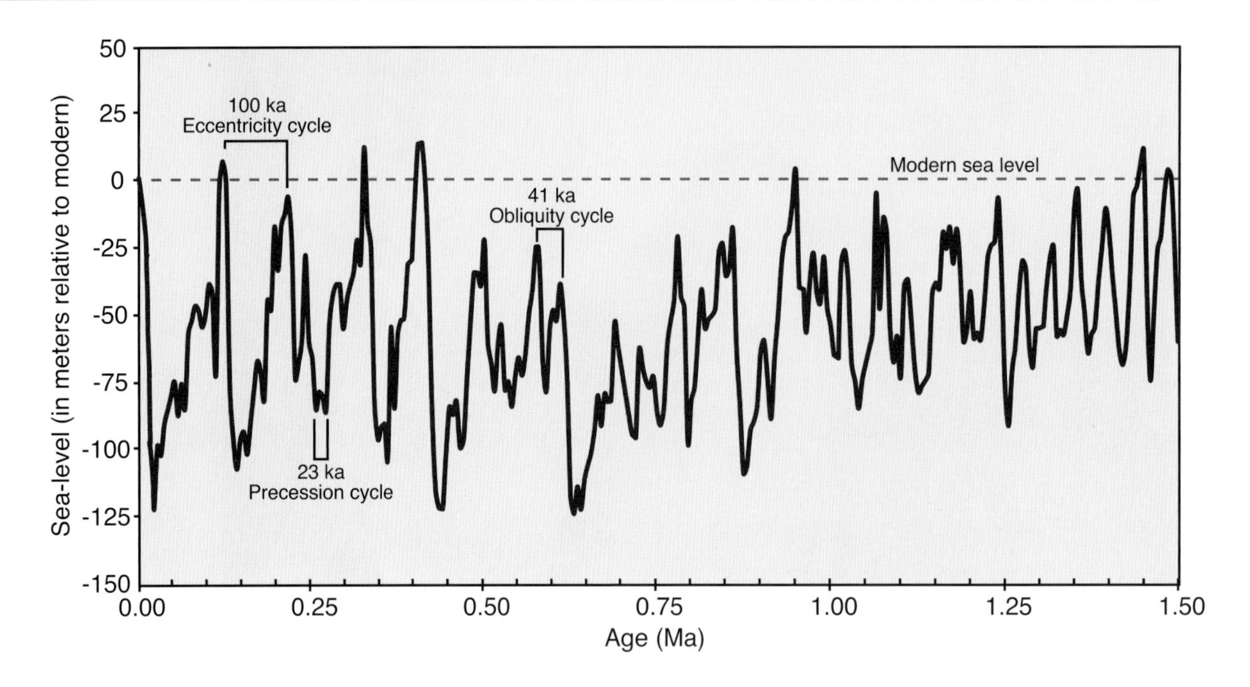

Figure 1.4. Sea level curves for the past ~1.5 Myr (most of Pleistocene) showing the responses to the glacial (ice ages) and interglacial global episodes. Note different periods of Milankovitch cycles, at 23 kyr, 41 kyr, and 100 kyr.

of times in the geologic past, having different amplitudes and frequencies and are known as Milankovitch cycles (fig. 1.4).

Milankovitch cycles play a key role in Earth's climate over time as they affect both the total amount of incoming solar insolation and, to an even larger degree, the distribution of the Sun's energy on Earth. There are three major orbital cycles that affect Earth's climate: precession, obliquity, and eccentricity (see fig. 1.5).

The Milankovitch cycles defined by variations in the Earth's rotation and orbit around the Sun create climate cycles of 23 kyr (resulting from interaction of two precession cycles), 41 kyr (obliquity), and 100 kyr (eccentricity), which in turn control sea level. The last ice age, which peaked about ~26.5–19 ka, leading to today's interglacial climate, is part of the last 100 kyr Milankovitch cycle.

Sea level fluctuations produced emergence and submergence of the land, thus constantly changing the shape of the map (fig. 1.6). The state of Florida, as defined by the location and shape of the present shoreline (exposed land) has completely disappeared in the geologic past due to *sea level highstands*—high enough to flood all of peninsular Florida. Additionally, the state of Florida has been twice its size in the past when its Gulf of Mexico shoreline was some 200 km off to the west on the west Florida shelf now lying in about 125 m of water. We know this because scientists have mapped the remnants of these *drowned shorelines* that still lie out there. So today's shape of the state of Florida, so easily recognizable and distinctive to kids, has been vastly different in the past. Today's shape is merely a snapshot in geologic time and is constantly changing.

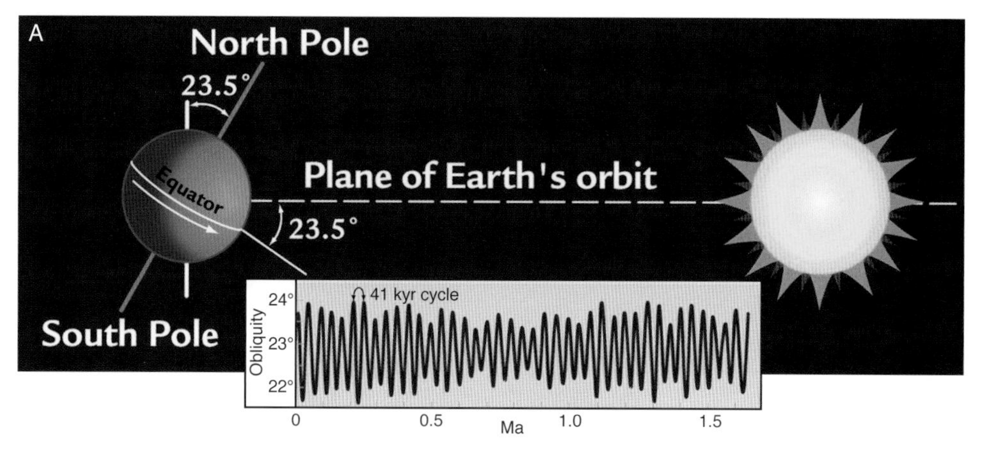

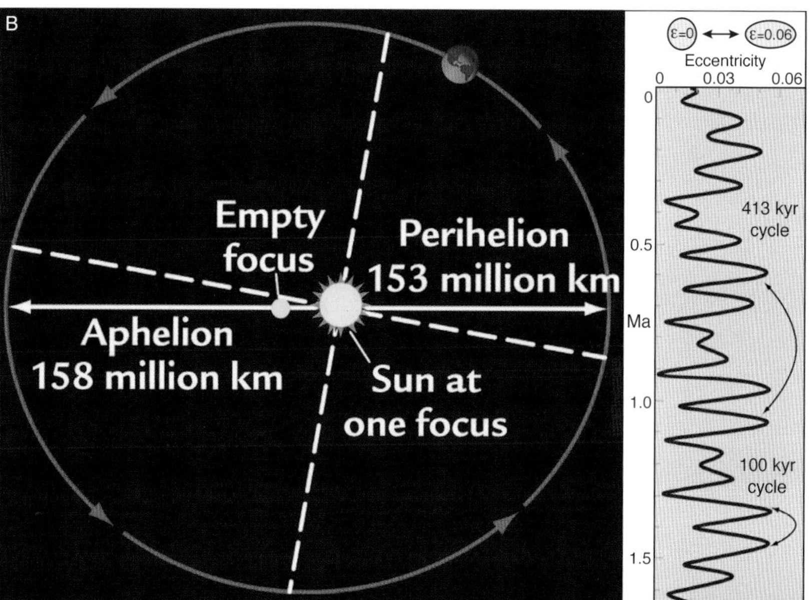

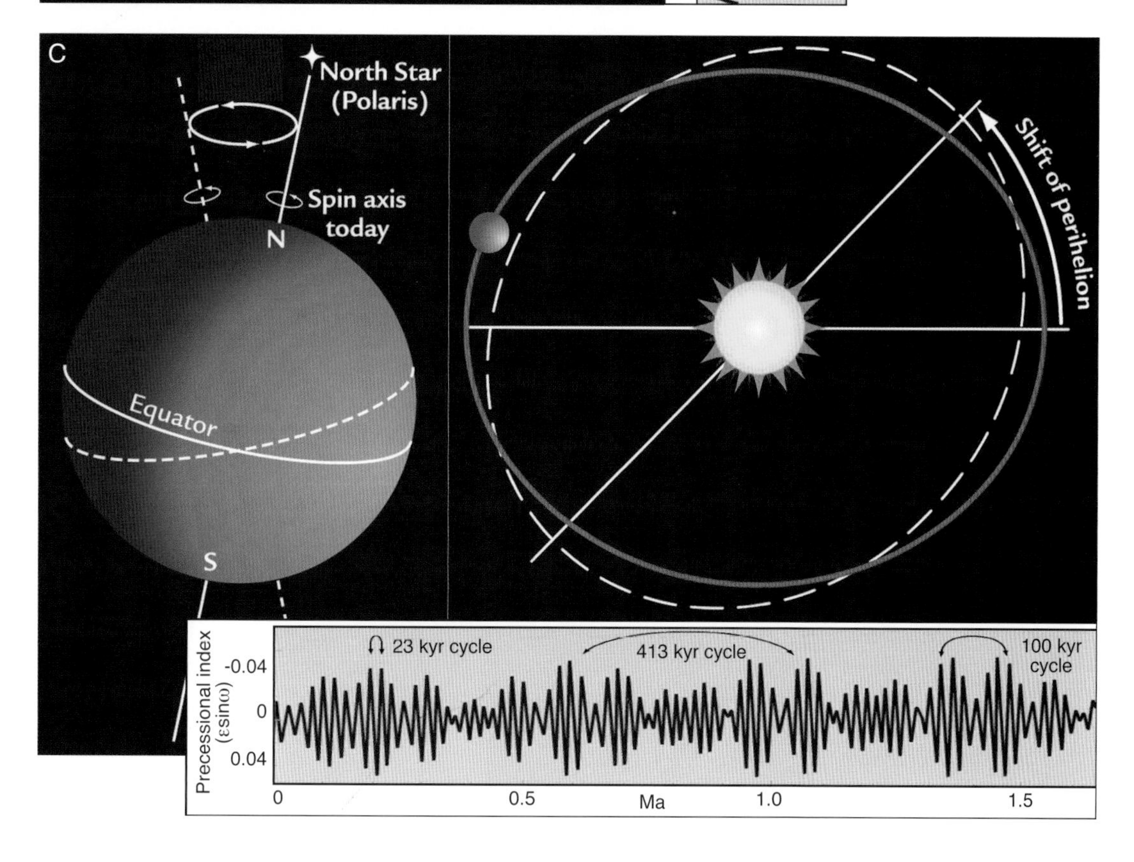

Top: Figure 1.5*A*. Obliquity has a 41 kyr cycle which results from the change in the angle of Earth's axis relative to the orbital plane from 22.5° to 24.5°. This figure shows the present tilt at 23.5°. (Source: Ruddiman 2008; used by permission from W. H. Freeman and Co.)

Middle: Figure 1.5*B*. Axial precession, which results from changes in the direction of the Earth's rotational axis, and elliptical precession, which is the slow movement of the Earth's orbit, combine to produce a 23 kyr cycle known as the precession of the equinoxes. Because elliptical precession is also affected by eccentricity, the amplitude of this 23 kyr cycle is modulated at 100 kyr and 413 kyr periods. (Source: Ruddiman 2008; used by permission from W. H. Freeman and Co.)

Bottom: Figure 1.5*C*. Eccentricity has 100 kyr and 413 kyr cycles and is the change in the ellipticity of Earth's orbit. (Source: Ruddiman 2008; used by permission from W. H. Freeman and Co.)

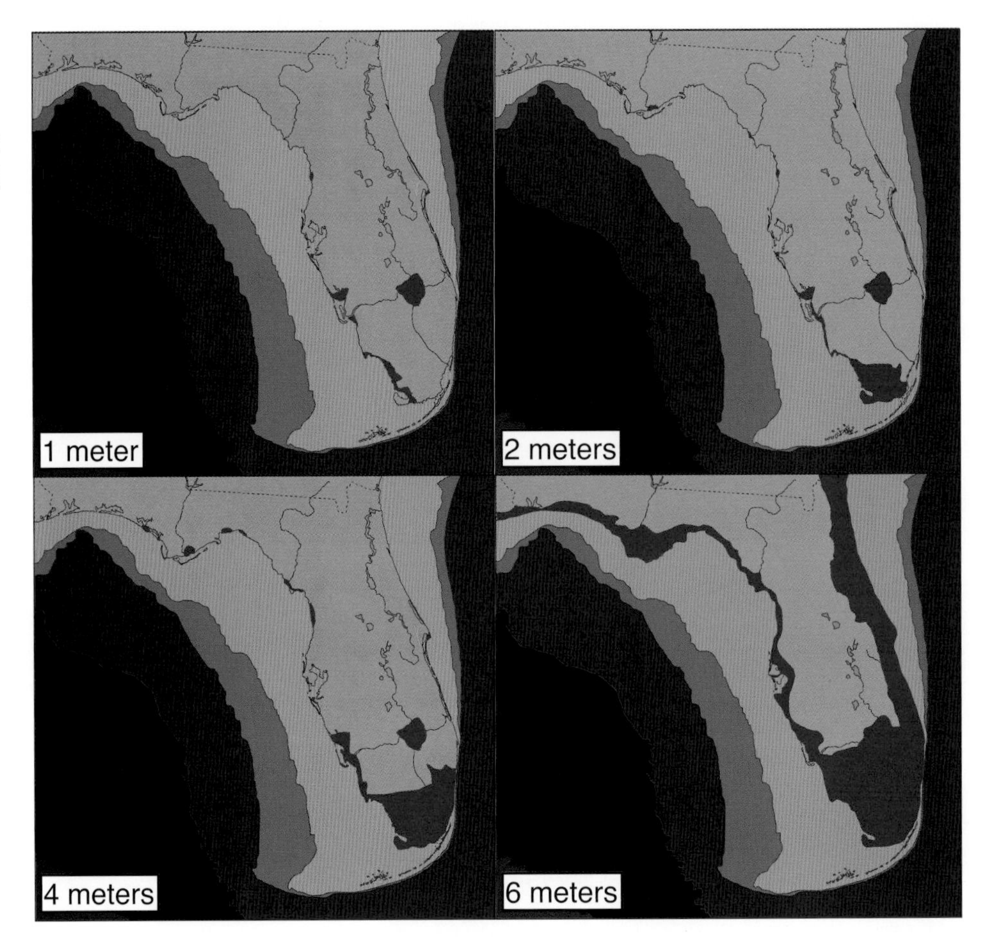

Figure 1.6. Four images of Florida illustrating the area of inundation with sea level rises of 1, 2, 4, and 6 m. The 6 m rise occurred during the last interglacial event at about 125 ka.

Florida's High Relief

Most of Florida lies underwater, and because most people do not realize this, they think that Florida has little *topographic relief* and is flat. For sure, this is an accurate conception as the highest point in the state is only ~105 m above sea level. But when considering that the base of the West Florida Escarpment lies in ~3,200 m of water in the Gulf of Mexico, the Florida Platform, indeed, is not flat and has significant relief. This underwater steep slope is nearly vertical in many places and rises about 2,000 m above the abyss (fig. 1.7).

Such relief over a short horizontal distance rivals the relief seen when viewing the Rocky Mountains. In fact, Florida has more relief (~3,400 m) than many of the 50 states. Measured this way, Alaska from the top of Denali (6,194 m) to the bottom of the Aleutian Trench (7,679 m) has the highest relief of the 50 states by having 13,873 m elevation difference. As you might expect, Hawaii is second with nearly 10 km of relief from its base on the bottom of the Pacific (~5,500 m) to the top of its highest volcano, Mauna Kea—4,205 m above sea level—making this volcano have more relief (9,705 m) than Mt. Everest (8,850 m), the highest mountain on Earth (fig. 1.8). With ~3,400 m elevation difference from top

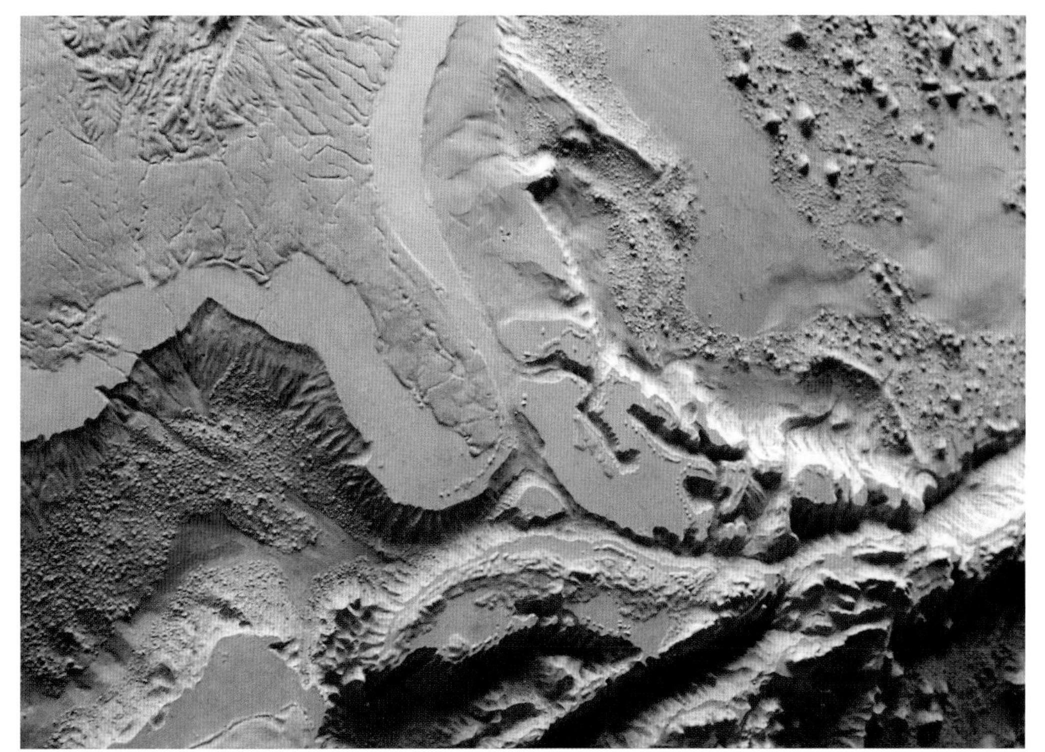

Figure 1.7. Physiographic map of general area surrounding the Florida Platform illustrating the great relief that exists beneath the sea surface. The Florida Platform has about 3.4 km of relief as compared with only ~105 m of relief seen on land. (Source: Large physiographic globe in Department of Earth Sciences at Dartmouth College; photo by A. C. Hine.)

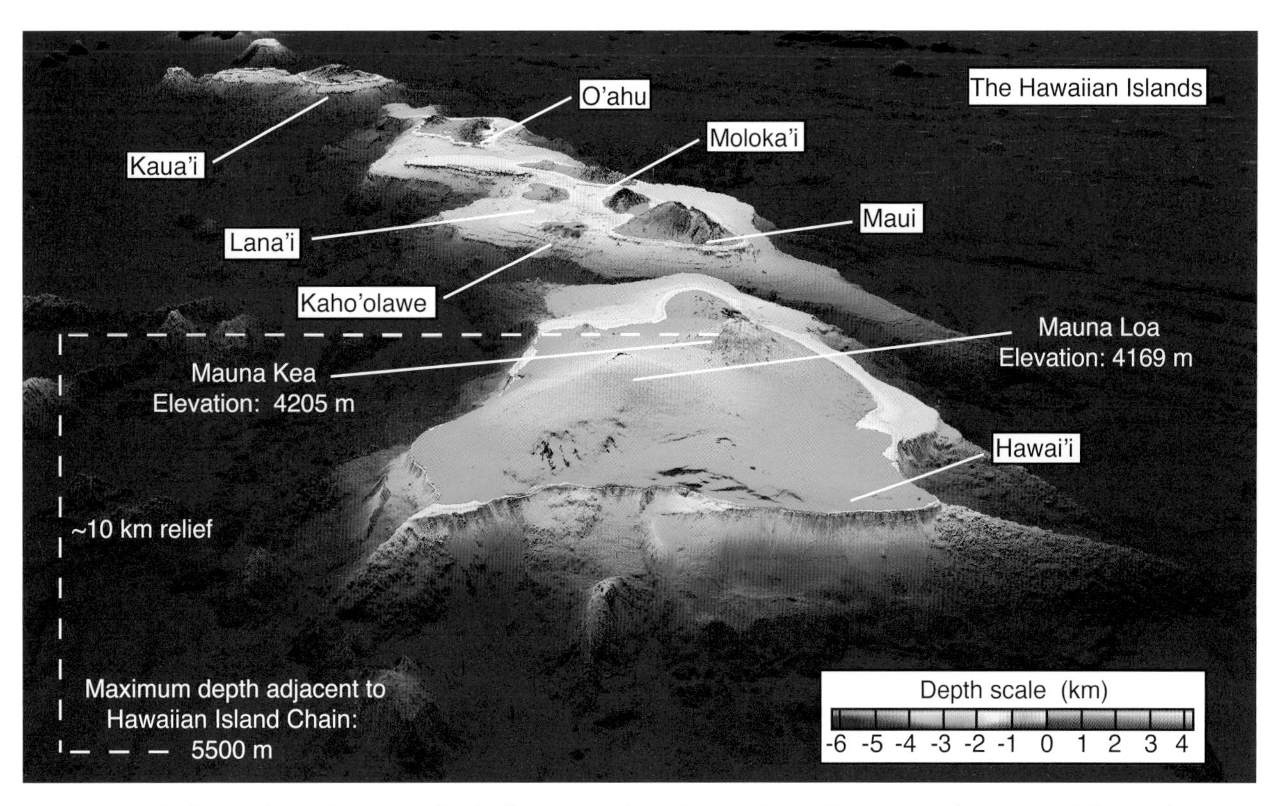

Figure 1.8. Relief map of the Hawaiian Islands illustrating the submerged as well as emerged portions of these volcanoes. Mauna Kea, rising ~10 km above the ocean floor, has the highest relief of any mountain on Earth. (Source: Image modified from original from the Hawaii Mapping Research Group at the School of Ocean and Earth Science and Technology, University of Hawaii at Manoa, http://www.soest.hawaii.edu/HMRG/Multibeam/index.php#Reference.)

to bottom, Florida is no slouch. For Floridians, such scenery lies out of sight beneath the waves. For Floridians, practically all of the geology lies beneath the land surface, extending to the seafloor.

The Future: A Brief Glimpse

Human-induced global climate change has brought with it predictions of increased sea level, not in geologic time scales but in human time scales. Some scientists predict sea level to rise 2 m or more by 2150! If these predictions turn out to be true, an American Automobile Association (AAA) road map of Florida printed in the year 2163 (only 150 years from now!) might look vastly different than the one we see today. The great-grandkids of today's kids will be alive to see the potentially new state of Florida—a much smaller piece of real estate! When driving from Naples to Ft. Lauderdale, they might cross the state on an elevated I-75 and instead of seeing the Everglades "River of Grass," they might see an open, shallow bay with no land in sight.

Because the human population has exploded over the past few hundred years and some predict that 3–5 billion more people will be added to the already ~7 billion who live today, viewing human civilization as an Earth-changing geologic agent is not a far stretch. Indeed, some geologists claim that the Holocene Epoch has now ended and that a new epoch called the *Anthropocene* (humans as a dominant geologic agent) has begun.

Back to the Present

So, in order to define Florida over geologic lengths of time, we need something a bit more stable than the highly mobile shoreline. That stability is provided by fundamental geologic structural features in the Earth's crust across which the types of rock and their ages may be fundamentally different. Many of the modern *bathymetric* changes defining the boundaries of the Florida Platform and the larger Florida-Bahama Platform are linked to these deeper structural features. We will see that the West Florida Escarpment to the west and the Bahama Escarpment to the east are geologically related to underlying basement structures associated with the opening of the Gulf of Mexico and North Atlantic Ocean, respectively (fig. 1.2). We will see that the Straits of Florida separating the Florida Platform from the Bahama Platform are tied to the tectonic collision with Cuba and past environmental distress. Finally, we will see that the modern De Soto Canyon off NW Florida actually is the remnant of a much larger rift that extended across Georgia thus defining the Florida Platform's northern boundary even though most of the rift valley lies deeply buried. The eastern side of the Florida Platform separating it from the Bahama Platform is the modern seaway called the Straits of Florida, which are in part geologically structurally bound. Therefore, the Florida Platform is a distinct geologic entity. During the course of this book, the nature and origin of these boundaries will become more apparent.

Florida's Tectonic Setting

All of these features defining the Florida Platform and the platform itself are part of the North American *tectonic plate*. Fundamental to physical geology, plate boundaries are moving laterally by each other, away from each other, or toward each other. Such movement creates features such as faults, volcanoes, earthquakes, tsunamis, and large landslides. The boundaries defining the Florida Platform were once active plate boundaries. But as fig. 1.9 illustrates, the Florida Platform is not near or adjacent to any active plate boundary. Hence it is tectonically stable as compared with Japan. (Remember the March 11, 2011, 9.0 magnitude Tohoku earthquake that generated a tsunami up to ~40 m high in some places?) Floridians do not have to worry about such drastic events because the Florida Platform lies away from such potential geologic activity. Rather, it lies in a tectonically passive setting. But that was not always the case, as we shall see.

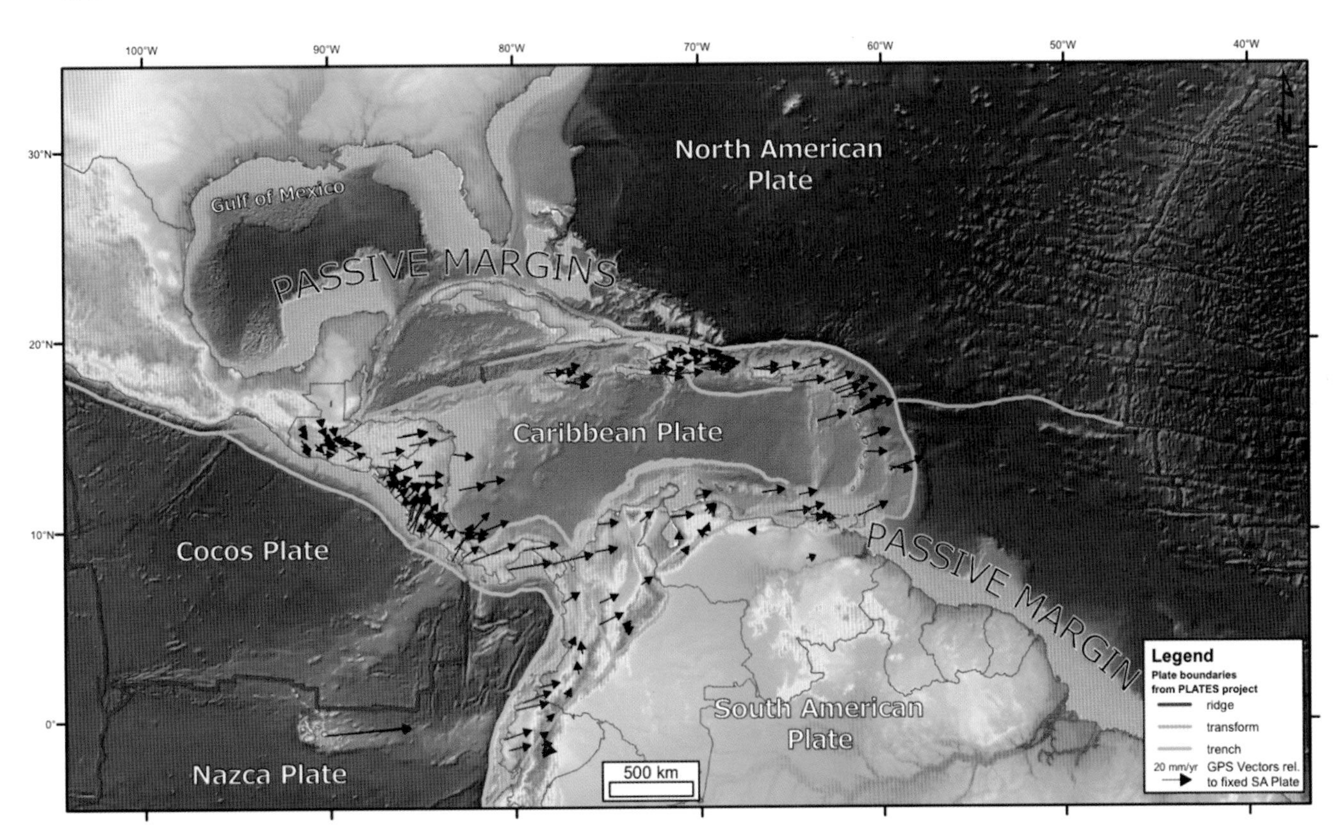

Figure 1.9. Map of various tectonic plates surrounding the Florida Platform. Note the passive margin tectonic setting within which the Florida Platform is located. (Source: Mann and Escalona 2011; used by permission from the Society of Exploration Geophysicists.)

Geology and Florida's Scenery

The underlying geology controls surface topography and therefore the scenery to a large measure. So it is in Florida. The N-S trending linear features that dominate north-central peninsular Florida are ancient coastal features formed during past elevated stands of sea level (fig. I.1). Florida's rivers and streams have been superimposed on this paleocoastal topography during sea level lowstands. Large, low-lying wetlands accompany these streams. Additionally, the thousands of circular lakes are sinkholes associated with subterranean dissolution of limestone (see chapter 7). The two large estuaries on Florida's west coast most likely resulted from dissolution of rocks at depth. The vast flat wetlands forming the Everglades rest on limestone formed in a broad shallow sea. The islands forming the upper Florida Keys are ancient coral reefs. So as we progress through geologic time—event by event, where appropriate—the surface geomorphology of the state will be tied to the geologic process responsible for creating it.

What's Next?

Now that we have defined the playing field and answered the question posed at the beginning of this chapter, let's begin our journey. For a land known as the Sunshine State, we start Florida's journey through geologic time at an improbable site—the South Pole approximately 700–650 Ma!

Essential Points to Know

1. Florida is nearly surrounded by seawater, making the ocean very important to the geologic history of the state and to the modern welfare of human society.

2. The geologic history of Florida really involves the geologic history of the Florida Platform, a much larger topographic feature than the emerged, subaerially exposed state of Florida. The Florida Platform is defined by distinct geologic boundaries.

3. The state of Florida is flat, topographically low, largely near sea level, and has minor topographic relief (~105 m). However, the Florida Platform from the base of the west Florida escarpment to the highest exposed area on land has ~3,400 m of relief, which resulted from geologic events.

4. The shape of the state of Florida, largely defined by its coastline, is a snapshot in time because of the present position of sea level. Sea level has fluctuated many times in the geologic past. During the last ice age (Last Glacial Maximum; ~26.5–19 ka), sea level was about 121 m lower than today and the subaerially exposed land mass was twice as big as it is today. Conversely, further back in the

geologic past, during sea level highstands, the entire Florida Platform (and state of Florida well before humans arrived) was completely underwater.

5. The Florida Platform is located on the North America Plate in a tectonically passive setting.

Essential Terms to Know

Anthropocene: Informal geologic time period when humans started to become a global geologic agent; considered to start in the late eighteenth century when the activities of humans first began to have a significant global impact on climate and ecosystems due to the beginning of the industrial/agricultural revolution.

continental shelf and slope: Seaward extension of a continent's exposed coastal plain having a relatively low seaward sloping gradient. The shelf is alternately flooded and exposed during multiple sea level fluctuations. Its seaward boundary is defined at the shelf/slope break (~75–200 m water depth) where the gradient becomes measurably steeper, transitioning into the continental slope. The continental slope extends into deeper water and turns into the continental rise (2–3 km water depth), where the deepest portion of the continent merges into the deep ocean basin.

drowned shorelines: Underwater paleo-shorelines formed during lower periods of sea level that have been partly preserved by undergoing insufficient erosion to fully remove their components.

escarpment: Very steep slope or cliff that has significant lateral extent and relief.

Florida Platform: Elevated, flat surface surrounded by deeper water environments whose subaerially exposed portion now forms the state of Florida. The Florida Platform has distinct geologic structural and bathymetric boundaries.

sea level highstands and lowstands: Variations in the level of the global ocean. Highstands are periods when sea level is elevated sufficiently to flood the margins of continents; lowstands are periods when sea level leaves much of the continental margin subaerially exposed.

subaerially exposed: Exposed to air as opposed to being submerged.

tectonic plate: A geologically distinct portion of the crust and mantle of the Earth that is defined by actively moving boundaries. There are eight major plates constituting the Earth's surface with many smaller plates fit in between. The Florida Platform is located on the North American Plate, one of the Earth's largest plates.

topographic relief: Difference in elevation between lowest and highest points of a geologic feature.

Keywords

Milankovitch cycles, sea level fluctuations, Florida Platform, topographic relief, Anthropocene, tectonic plates

Essential References to Know

Bryan, J. R., T. M. Scott, and G. H. Means. *Roadside Geology of Florida*. Missoula, MT: Mountain Press, 2008.

Hine, A. C., G. R. Brooks, R. A. Davis Jr., D. S. Duncan, S. D. Locker, D. C. Twichell, and G. Gelfenbaum. "The West-Central Florida Inner Shelf and Coastal System: A Geologic Conceptual Overview and Introduction to the Special Issue." *Marine Geology* 200, no. 1–4 (2003): 1–17. doi: 10.1016/s0025–3227(03)00161–0.

Mann, P., and A. Escalona. "Major Hydrocarbon Plays in the Mexican Sector of the Gulf of Mexico, the Caribbean, and Northern South America." Paper presented at the 2011 Society of Exploration Geophysicists' annual meeting, San Antonio, TX, 2011.

Press, F., R. Siever, J. Grotzinger, and T. H. Jordan. *Understanding Earth*. 4th ed. New York: W. H. Freeman, 2004.

Randazzo, A. F., and D. S. Jones, eds. *The Geology of Florida*. Gainesville: University Press of Florida, 1997.

Robbins, L. L., M. Hansen, E. Raabe, P. Knorr, and J. Browne. *Cartographic Production for the Florida Shelf Habitat (FLaSH) Map Study: Generation of Surface Grids, Contours, and KMZ Files*. Open-File Report 2007-1397. St. Petersburg, FL: U.S. Geological Survey, 2007.

Ruddiman, W. F. *Earth's Climate: Past and Future*. 2nd ed. New York: W. H. Freeman, 2008.

2

Florida Lost

Wandering the Globe and Finding Home (~700 Ma to ~200 Ma)

> We find no vestige of a beginning, no prospect of an end.
>
> James Hutton's remarks to the Royal Society of Edinburgh in 1788 concerning the age of the Earth

> Playfair later commented . . . , "the mind seemed to grow giddy by looking so far into the abyss of time."
>
> *Transactions of the Royal Society of Edinburgh*, vol. 5, pt. 3, 1805, quoted in *Natural History*, June 1999

> In the 1960s, a great revolution in thinking shook the world of geology.
>
> F. Press, R. Siever, J. Grotzinger, and T. H. Jordan, *Understanding Earth*, 2004

Very Early Beginnings

The origin of Florida is tied to the origin of the Earth, which is tied to the origin of the solar system, which is ultimately tied to the origin of the universe. The origin of the cosmos is a tale well beyond the purview of this book. But an amazing tale it is, as our astrophysicist colleagues continue to obtain new data to formulate theories on how and when it all began—some 13.88 Ga as revealed by the Hubble telescope and other instruments scanning into deep space (fig. 2.1).

However, the ignition of our Sun as a new star and the formation of the stony inner planets through intense meteorite bombardment produced external and internal heat energy essential to Earth's early development. Earth's internal heat engine is also fired by the decay of radioactive elements. Sunlight is essential to life on Earth. However, it is Earth's internal heat engine which gave us our oceans and atmosphere. Gases, including water vapor, were exhaled through volcanoes from the cooling of the young Earth.

The transfer of the energy from the planet's internal heat engine, the source of most radioactivity, from Earth's core and mantle to the Earth's surface drives plate tectonics. The recognition of these processes produced a paradigm shift in the 1960s on how we understand the motion of Earth's crust. Any modern physical geology/earth science text illustrates how the plates move in relationship to each other, thus forming our planet's primary physiographic features—the mountain ranges, ocean trenches, and the *mid-ocean ridges*. Such a text can

Figure 2.1. Close-up view of a spiral galaxy imaged by the Hubble telescope. Most likely many of these stars support solar systems containing Earth-like planets. Our solar system is in the Milky Way Galaxy. In this broad context, the Earth seems comparatively insignificant and Florida even more so. (Source: NASA; http://apod.nasa.gov/apod/ap100114.html.)

explain how the ancient *continents* formed and evolved over the eons and how the relatively young *ocean basins* come and go. Realizing that the continents were billions of years old and yet the oldest ocean basin was only <180 Myr posed a significant dilemma to the earth science community before the plate tectonics paradigm became a "game changer." Readers might want to obtain such a basic earth science text for more background on this topic.

Beginning of Florida

To pick a point in geologic time as a "golden spike," defining the dawn of Florida's geologic history, is debatable and probably without resolution. For sure, Florida's origins are tied to global plate tectonics. Peninsular Florida, or our "pistol grip" shape, did not exist before ~200 Ma. Indeed, the Atlantic Ocean, Gulf of Mexico, and Caribbean Sea did not exist at that time either. However, the *basement rocks* supporting modern Florida had formed but were embedded within the African and South American continents as part of Gondwana—*not* North America, which was part of Laurasia.

A logical but very early starting point in Florida's history is the breakup of a supercontinent called Rodinia (Russian for motherland) in the Late Precambrian around 700 Ma. The two-part breakup process is called (1) *rifting*, the initial splitting apart of a continental mass and then (2) *seafloor spreading*, the formation of a new ocean basin floored by new ocean crust. The latter formed the Iapetus Ocean (Iapetus—Greek father of Atlas), also known as the proto Atlantic Ocean, which existed before the present Atlantic Ocean was formed. The Iapetus Ocean separated continental masses such as Laurentia (ancestral North America—named after mountains in Newfoundland and Quebec Province) and Gondwana (ancestral Africa, South America, Australia, Antarctica, and Asia—named after the land of Gonds—people and the language spoken in India). The continental basement rock forming Florida was part of Gondwana. By about 650 Ma, Florida (embedded within Gondwana) was located approximately at the Earth's South Pole (fig. 2.2*A*).

Since central peninsular Florida is presently located at approximately 27° N latitude, these basement rocks have migrated, via plate motion, nearly 12,600 km during the past 650 Ma. The rifting and seafloor spreading that tore apart ancestral eastern North America created a series of embayments and promontories (Texas Promontory, Ouachita Embayment, Alabama Promontory, Tennessee Embayment, and the Virginia Promontory). This ragged, irregular *conjugate margin* prepared North America for the final collision with Gondwana that was

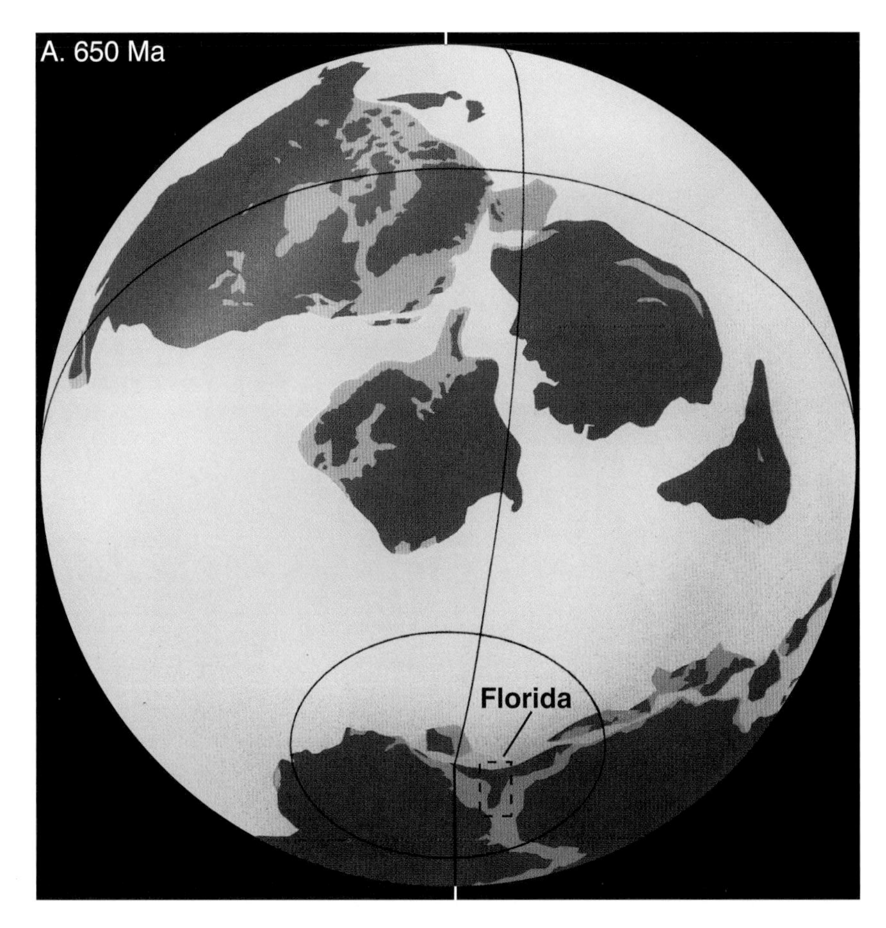

Figure 2.2*A*. Paleogeographic reconstruction illustrating distribution of continents at ~650 Ma. Note the position of Florida's basement rocks (and some for the SE United States) at the South Pole situated between the conjoined South American and African continents embedded with Gondwana. (Modified from Redfern 2001.)

Figure 2.2*B*. Paleogeographic reconstruction ~300 Ma during the extended collision between Laurentia and Gondwana forming the Appalachian Mountains in North America and the Mauritanide Mountains in Africa. (Modified from Redfern 2001.)

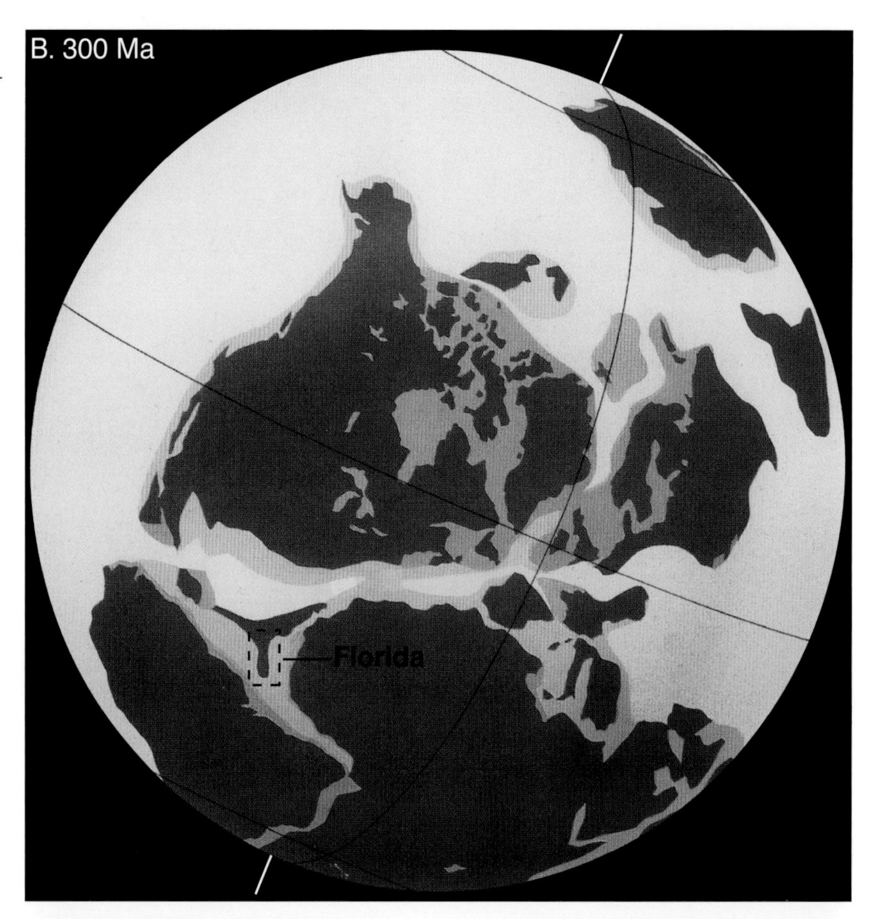

Figure 2.2*C*. Paleogeographic reconstruction of supercontinent Pangea at ~200 Ma. The basement rocks of Florida now lie north of the equator and are embedded within the new supercontinent located great distances from any ocean. The collision of Laurentia and Gondwana formed a near continuous 6,000 km long mountain chain from the Arctic to the eastern Pacific Ocean. The Florida basement rocks were sutured onto what was to become North America during the ensuing rifting and seafloor spreading forming the North Atlantic Ocean after 200 Ma. (Modified from Redfern 2001.)

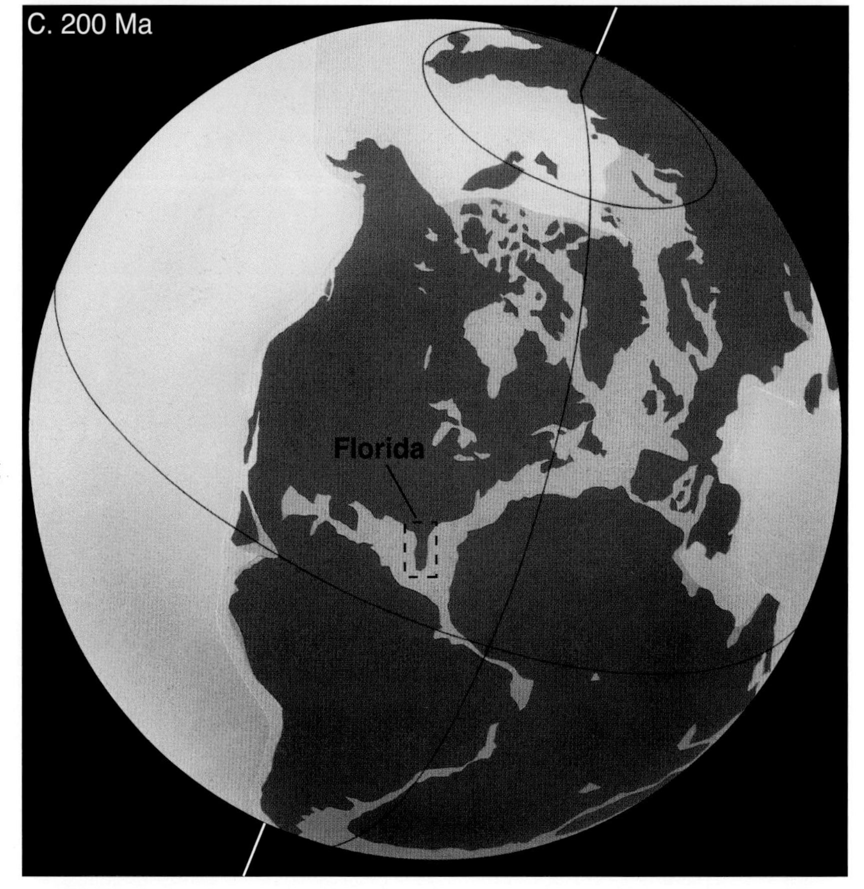

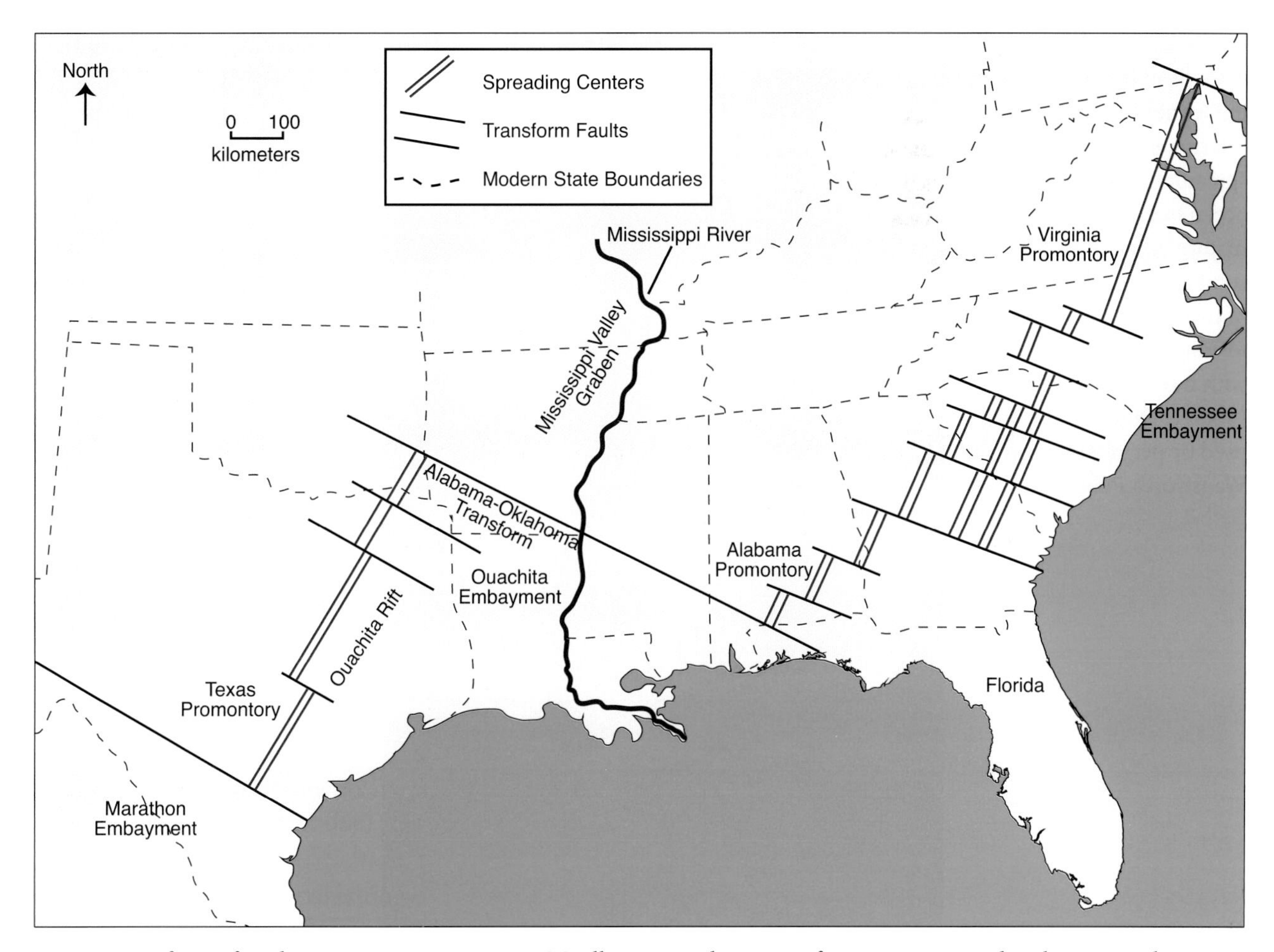

Figure 2.3. Rifting of Rodinia supercontinent ~700 Ma illustrating the series of promontories and embayments that were left behind, defining the margin of the subsequent Laurentia, or eastern North America. It was this conjugate margin that collided with Gondwana, defining where the Appalachian and Ouachita Mountains were to be created hundreds of millions of years later. And this rifting "prepared" the site where the Florida basement rocks were to be attached or sutured during that collision, the Allegheny Orogeny. (Modified from Hatcher 1989.)

to come some 350 Ma later (figs. 2.2*B*, 2.2*C*, 2.3). Depending upon the continental margin morphology and orientation, the ensuing collision could be straight on (hard hit) or oblique (soft hit). The gross morphology of this rifted margin defined where the Appalachian and Ouachita Mountains were to be located when that collision occurred.

Wilson Cycle

Much of Earth's history can be defined by various natural cycles, some lasting longer than others. There are 11 yr sun spot cycles and orbital cycles driving climate and sea level cycles. As was mentioned in chapter 1, the Milankovitch orbital cycles produce climate cycles of 23 kyr, 41 kyr, and 100 kyr. We all know that one rotation of the Earth around its axis is a daily cycle, one revolution of

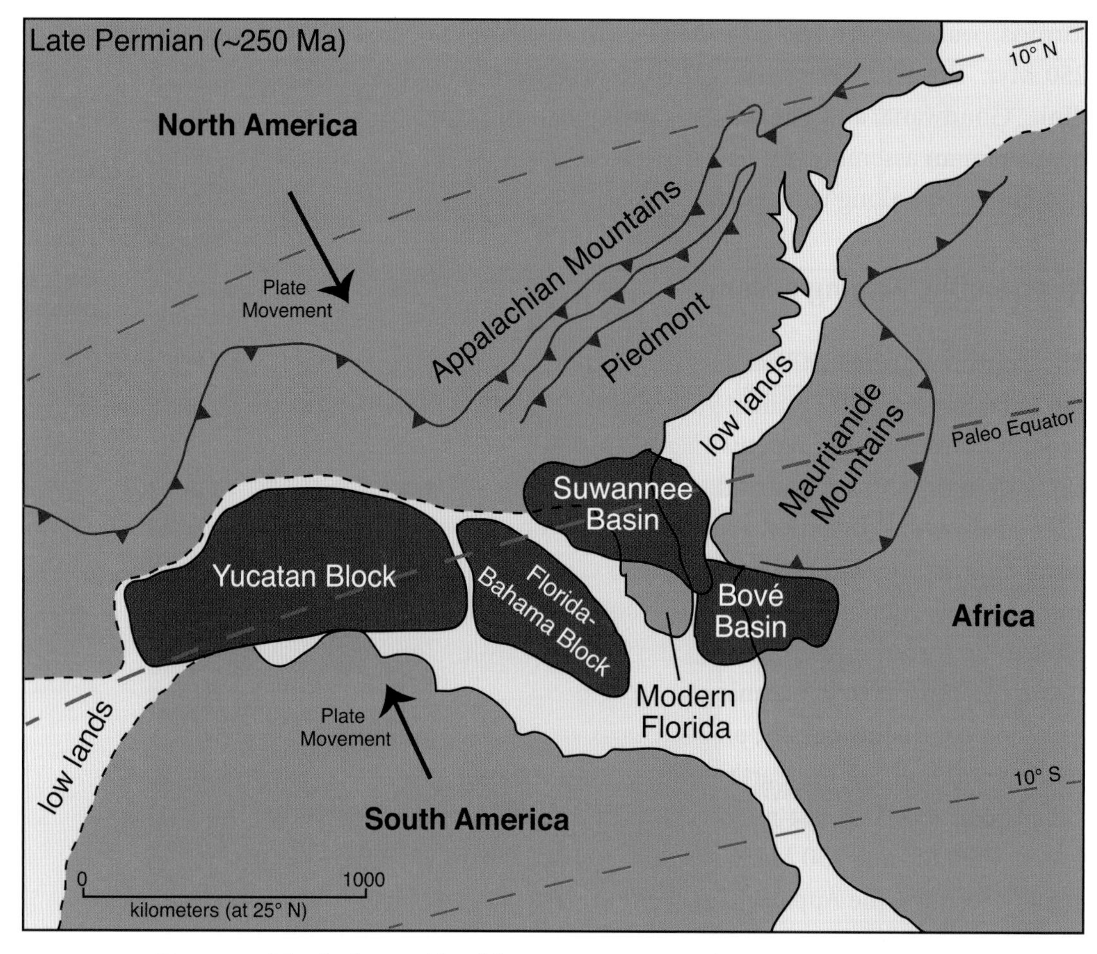

Figure 2.5. The final assembly of the Pangean supercontinent at ~250 Ma. The diagram shows the complexity of the smaller fragments and exotic terrains lying between what was to become North America, South America, and Africa upon rifting and the formation of the North Atlantic Ocean, the Gulf of Mexico, and the Caribbean Sea. Note the location of north-central peninsular Florida (NW African Suwannee Basin) and the Florida-Bahama Block, which formed the basement of south peninsular Florida. The basement rocks of Florida rested >2,000 km away from any ocean within the vast interior of this supercontinent. (Modified from Pindell and Barrett 1990.)

or accreted terrains, which formed the distinctively different geologic elements such as parts of the Valley and Ridge, the Blue Ridge Mountains and the Piedmont. The basement rocks upholding peninsular Florida form an exotic terrain in that these rocks were unrelated to the rest of the rocks of North America when they first arrived from the South American and African portions of Gondwana.

Altogether, the closing of the Iapetus Ocean and subsequent continental collision and suturing formed the active components and eroded roots of an approximately continuous mountain chain 6,000 km long from Spitsbergen to Mexico, including Greenland, Scandinavia, Scotland, northwestern Africa, and all of eastern North America.

After the Iapetus Ocean closing, the rocks that became Florida's basement were located well within equatorial Pangea >2,000 km from the global ocean called Panthalassa (*pan thalassa*—Greek for "all ocean") by 200 Ma (fig. 2.2C). Fig. 2.5 is a more detailed diagram and shows several of the key blocks of basement rock that became important to the opening of the modern Gulf of Mexico and Atlantic Ocean and the formation and attachment of Florida's basement rocks to North America.

By now (~200 Ma), Florida had migrated to within 1,600 km of today's location. At this point in time, what was to become modern Florida, now nearly surrounded by saltwater, was far removed from any ocean environment and probably was in the middle of a vast, intracontinental desert probably >2,000 km from any ocean.

The suture zone marking the contact between Laurentia and the Florida component of Gondwana now runs southwest to northeast through south Alabama, south Georgia, southern South Carolina, and eastern North Carolina.

Ultimately, Pangea did not remain a supercontinent for very long—it was intact for only ~85 Myr and then began to break up again between ~200 Ma to 160 Ma, thus progressing to the next step of the Wilson Cycle. It was this breakup that played the first essential step in forming Florida as a distinct geologic entity.

Essential Points to Know

1. Earth's internal heat left over from its origin and the continued decay of radioactive elements are the primary driving forces for plate tectonics.

2. The basement rocks beneath Florida formed many hundreds of millions (early Paleozoic era) of years to billions of years ago (late Precambrian) and were part of the Gondwana continent (ancestral South America and Africa) and not the Laurentia continent (ancestral North America). The Gondwana and Laurentia continents were split from the Rodinia supercontinent about 700 Ma to 650 Ma.

3. At one time, the Florida basement rocks were located at the South Pole, and they have migrated via plate tectonic processes to their present position—a 12,600 km journey.

4. At ~475 Ma, the new Pangean supercontinent began to assemble, forming a 6,000 km long mountain chain complex that included the Appalachian Mountains. The Florida basement rocks became sutured onto North America during multiple collisions that probably lasted <250 Myr. Final assembly of Pangea probably occurred around 250 Ma.

5. Although modern Florida is nearly surrounded by marine waters, its basement rocks were located >2,000 km inside this huge supercontinent, possibly supporting a desert environment.

6. Pangea began to split apart soon after it formed, setting the stage for Florida to be recognized with its distinctive peninsular shape.

Essential Terms to Know

basement rocks: Thick foundation of ancient metamorphic and igneous rock that forms the crust of continents, often in the form of very ancient rocks like granite, granitic gneisses, and metamorphosed igneous and sedimentary rocks. Basement rock is contrasted to younger, overlying sedimentary rocks, which were laid on top of the basement rocks after the continent was formed.

conjugate margins: Margins of continents that "fit" together like pieces of a puzzle when reconstructed—promontories fit into embayments on the opposing continents.

continents or *continental fragments*: Large landmasses consisting of thick granitic igneous and metamorphic basement rocks rich in silicon and aluminum, hence the term *sial* or *sialic*. Continental fragments are small, distinct pieces of continental crust, such as Madagascar, but are not elevated to full continent status by convention.

exotic terrain: Fragment of crustal material formed on, or broken off from, one tectonic plate and accreted—"sutured"—to crust lying on another plate. Piece of crust that has been transported laterally, usually as part of a larger plate, and is relatively buoyant due to thickness or low density. When the plate of which it was a part subducted under another plate, the buoyant terrain failed to subduct, detached from its transporting plate, and accreted onto the overriding plate. Therefore, the terrain transferred from one plate to the other. Typically, accreting terrains are portions of continental crust that rifted off another continental mass and were transported surrounded by oceanic crust.

microbial, microbe: An organism that is microscopic. The study of microorganisms is called microbiology.

mid-ocean ridge: An underwater mountain range, typically having a valley known as a rift running along its spine, formed by plate tectonics. This type of oceanic ridge is characteristic of what is known as an oceanic spreading center, which is responsible for seafloor spreading and widening of the ocean basin.

ocean basin: Structural basin underlain by ocean crust types of rocks—basalt, gabbro (rocks containing Mg, Fe rich silicate minerals—denser than continental crust rocks, which are rich in Al and Si).

orogenic belt/orogens/orogeny: Related to mountain building. They result from plate tectonic movement. Orogens are usually long, arcuate tracts of rock that have a pronounced linear structure resulting in terrains or blocks of deformed rocks.

plate tectonics: Describes the large-scale motions of Earth's lithosphere. The outermost part of Earth's interior is made up of two layers: the lithosphere and the asthenosphere.

Above is the lithosphere, consisting of the crust and the relatively rigid uppermost part of the mantle.

Below the lithosphere lies the asthenosphere. Although solid, the asthenosphere has relatively low viscosity and shear strength and can flow like a liquid on geological time scales. The deeper mantle below the asthenosphere is more rigid, again, due to the higher pressure.

The lithosphere is broken up into what are called tectonic plates. In the case of Earth, there are currently eight major and many minor plates. The lithospheric plates ride on the asthenosphere.

rifting: Tectonic process where the Earth's crust and lithosphere are being pulled apart. This is the initial process of continental breakup. Seafloor spreading immediately follows rifting, creating a new ocean basin. The East African Rift valleys partially filled with lakes (Albert, Rudolph) are good examples of ongoing rifting.

seafloor spreading: Occurs at mid-ocean ridges, where new oceanic crust forms through volcanic activity and then gradually moves laterally.

subduction: The process that takes place at convergent boundaries by which one tectonic plate moves under another tectonic plate, sinking into the Earth's mantle, as the plates converge. A subduction zone is an area where two tectonic plates move toward one another and subduction occurs.

suturing: Process of crust from another area (exotic terrain) accreting onto another continent. The boundary of accretion is a suture zone.

volcanic arc: A chain of volcanic islands or mountains formed by plate tectonics along a continental margin (i.e., Cascades in Oregon and Washington) as an oceanic tectonic plate subducts under another tectonic plate and produces magma. There are two types of volcanic arcs: oceanic arcs (commonly called island arcs, a type of archipelago) and continental arcs. In the former, oceanic crust subducts beneath other oceanic crust on an adjacent plate, while in the latter case, the oceanic crust subducts beneath continental crust.

Wilson Cycle or *supercontinent cycle:* Describes the quasi-periodic aggregation and dispersal of Earth's continental crust. There are varying opinions as to whether Earth's budget of continental crust is increasing, decreasing, or remaining about constant, but it is agreed that this inventory is constantly being reconfigured. One supercontinent cycle is said to take 250 to 500 Myr to complete.

Keywords

Seafloor spreading, mid-ocean ridge, ocean basin, basement rocks, orogeny, exotic terrain, Laurentia, Gondwana, Pangea

Essential References to Know

Dutch, S. I., J. S. Monroe, and J. M. Moran. *Earth Science.* Wadsworth Earth Science and Astronomy Series. Boston: Brooks/Cole, 1998.

Hatcher, R. D., Jr. "Tectonic Synthesis of the U.S. Appalachians." In *The Appalachian-Ouachita Orogen in the United States*, ed. R. D. Hatcher Jr., W. A. Thomas, and G. W. Viele, 767. Boulder: Geological Society of America, 1989.

Klitgord, K. D., P. Popenoe, and H. Schouten. "Florida: A Jurassic Transform Plate Boundary." *Journal of Geophysical Research* 89, no. B9 (1984): 7753–72. doi: 10.1029/JB089iB09p07753.

Pindell, J. L., and S. F. Barrett. "Geological Evolution of the Caribbean Region: A Plate Tectonic Perspective." In *The Caribbean Region*, ed. G. Dengo and J. E. Case, The Geology of North America, 405–32. Boulder: Geological Society of America, 1990.

Press, F., R. Siever, J. Grotzinger, and T. H. Jordan. *Understanding Earth.* 4th ed. New York: W. H. Freeman, 2004.

Redfern, R. *Origins: The Evolution of Continents, Oceans, and Life.* Norman: University of Oklahoma Press by special arrangement with Cassell and Co., UK, 2001.

Sawyer, D. S., R. T. Buffler, and R. H. Pilger Jr. "The Crust underneath the Gulf of Mexico Basin." In *The Gulf of Mexico Basin*, ed. A. Salvador, The Geology of North America, 52–72. Boulder: Geological Society of America, 1991.

Shallow, J. "Geologic Map of the Blue Ridge Parkway." Paper presented at the 2004 annual meeting of the Geological Society of America, Denver, 2004.

Sheridan, R. E., H. T. Mullins, J. A. Austin, M. M. Ball Jr., and J. W. Ladd. "Geology and Geophysics of the Bahamas." In *The Atlantic Continental Margin: U.S.*, ed. R. E. Sheridan and J. A. Grow, The Geology of North America, 329–64. Boulder: Geological Society of America, 1988.

3

The Big Split

Formation of Three Oceans and the Establishment of the Florida Basement (~225 Ma to ~140 Ma)

"Breaking Up Is Hard to Do"
Neil Sedaka, Hit #1 *Billboard* Hot 100, August 11, 1962

Breakup Basics

Why *supercontinents* break up into smaller pieces creating new *ocean basins* between them is poorly understood, particularly when the supercontinent has been assembled for only geologically short periods. They take a relatively long time to form and only a short time to break apart. It seems like there must be some fundamental instability at work denying supercontinents a long-term existence.

One theory is that a huge continental mass traps heat beneath the continent, thus leading to enhanced *thermal uplift* and rifting. Regardless, the Pangean supercontinent proceeded to break up about 225 Ma after having been fully assembled by about 300–250 Ma prior—it took several hundred million years to assemble, but began to split apart only ~75 Myr after final assembly. The origin of Florida, the Gulf of Mexico, the Caribbean, and the North Atlantic Ocean is intimately tied to this breakup (fig. 3.1).

Another theory about the mechanics of continental breakup requires that specific zones of increased upwelling of hot material from the Earth's mantle beneath the crust form *hot spots*. These are areas of enhanced outpourings of lava from concentrated volcanic activity—the Hawaiian Island chain, Iceland, and the Galapagos Islands are excellent modern examples of this process. Additionally, this idea suggests that *rifted zones* forming within continents link up or tie together multiple hot spots, thus leading to an irregular continental split creating *conjugate continental margins* facing a new ocean basin. Thus, a continent can break apart along promontories and embayments (fig. 2.3). The promontory on one side of an ocean neatly nests into an embayment on the other side when fitted back together. The North Carolina Outer Banks extend further eastward into the Atlantic Ocean than the shoreline to the north or south because it rests

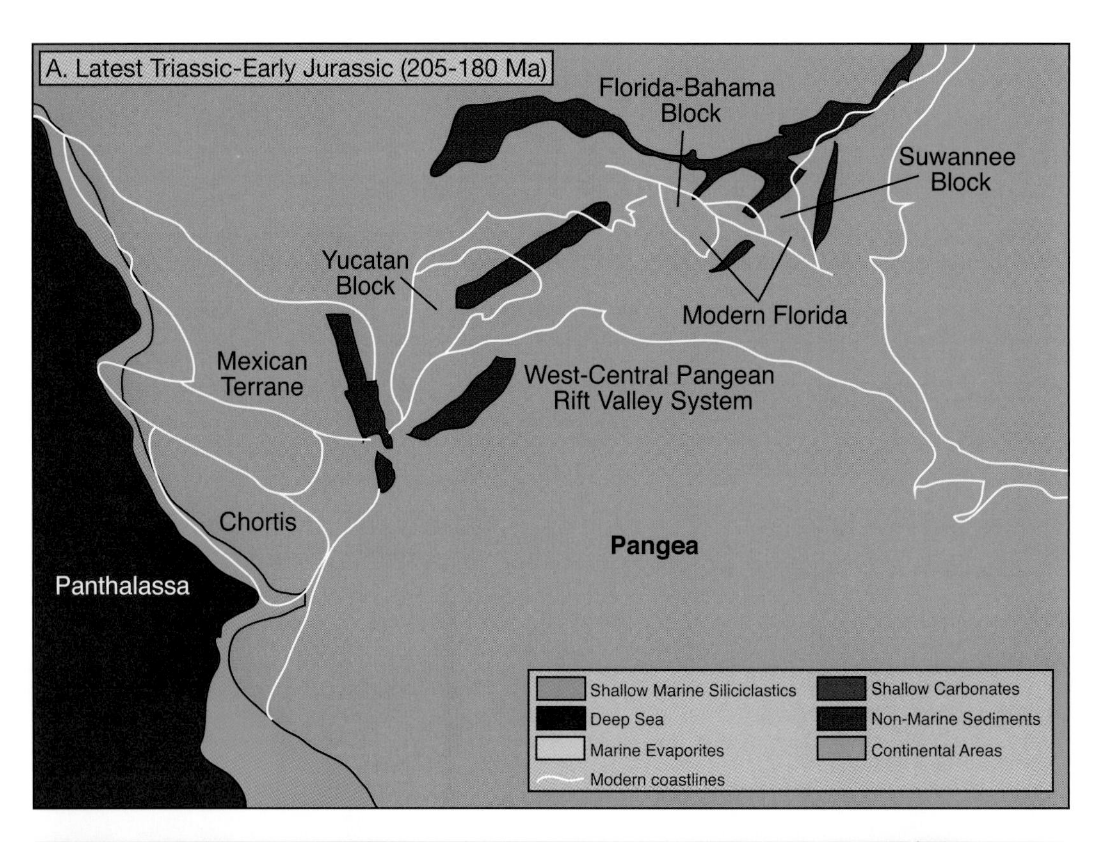

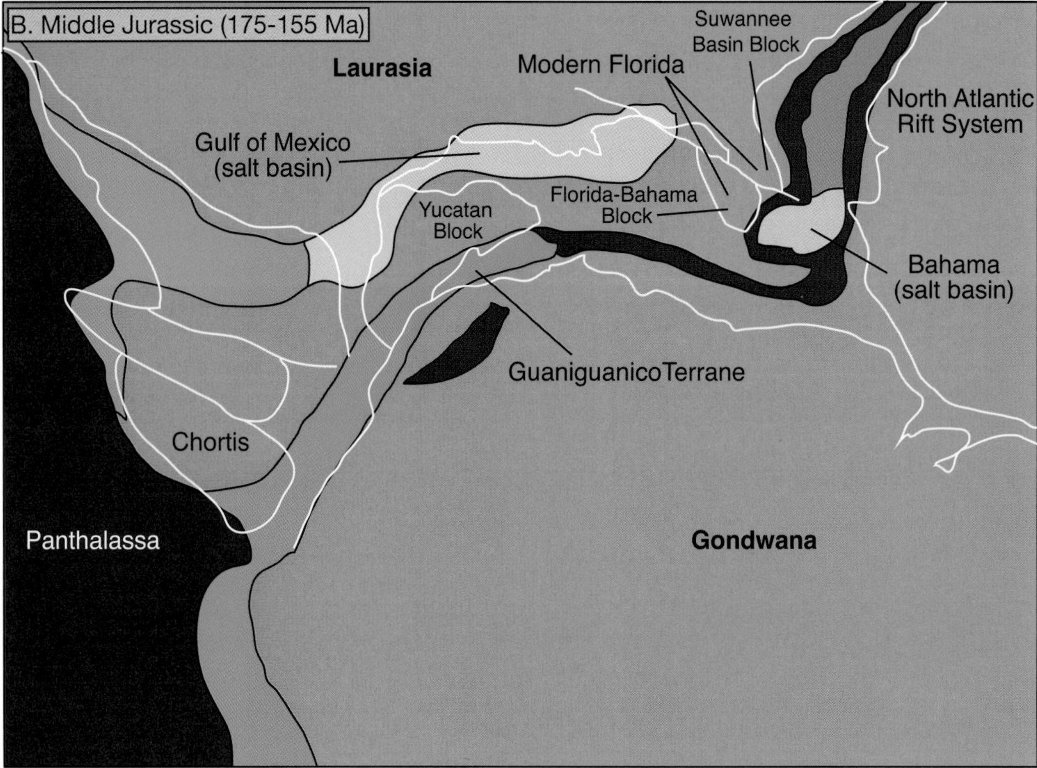

Figure 3.1 A–D. The breakup of Pangea. Note that during this time (205 Ma to 136 Ma) the North Atlantic Ocean, Gulf of Mexico, and Caribbean Sea all formed. Note the two components of the basement rocks forming the basement of peninsular Florida (Suwannee Basin and Florida-Bahama Blocks). These ocean basins opened up through a complex system of seafloor spreading systems. (Modified from Iturralde-Vinent 2003.)

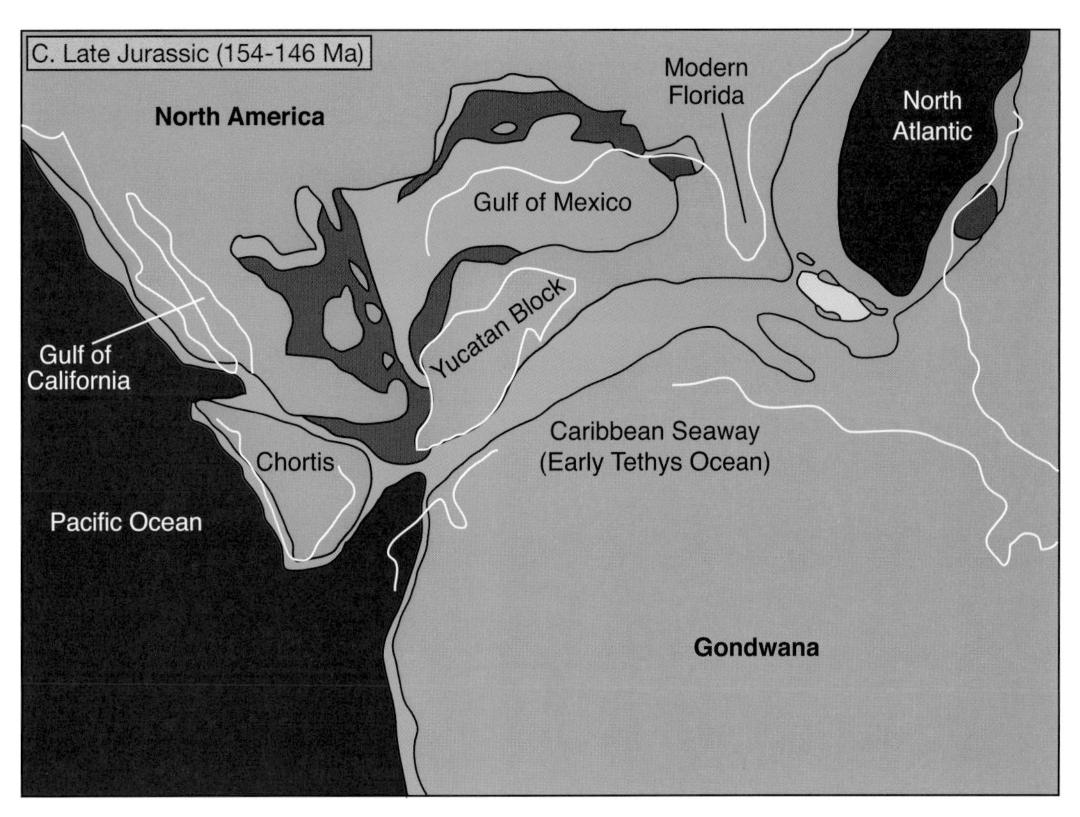
C. Late Jurassic (154-146 Ma)
North America
Modern Florida
North Atlantic
Gulf of Mexico
Yucatan Block
Gulf of California
Chortis
Caribbean Seaway (Early Tethys Ocean)
Pacific Ocean
Gondwana

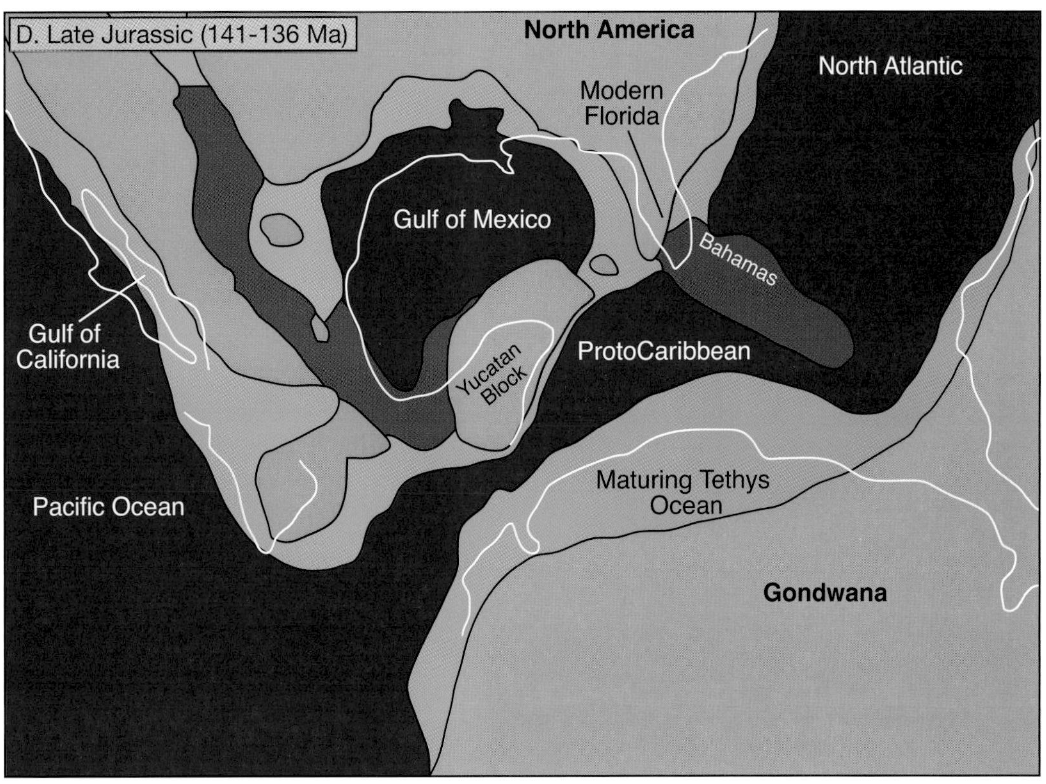
D. Late Jurassic (141-136 Ma)
North America
North Atlantic
Modern Florida
Gulf of Mexico
Bahamas
Gulf of California
ProtoCaribbean
Yucatan Block
Maturing Tethys Ocean
Pacific Ocean
Gondwana

on one of these deep crustal promontories separated by two large embayments whose conjugate counterparts can be found along the NW Africa margin from whence they rifted long ago.

Furthermore, this hot spot continental breakup theory indicates that three major rifted arms extend away from the hot spot center at ~120°. It is along these arms that the new ocean basins may form. Following this, some geologists have theorized that a large hot spot formed in central Pangea near what was to become peninsular Florida and the Bahamas. One rifted arm led to the opening of the North Atlantic Ocean, a second rifted arm led to the opening of the Gulf of Mexico and the *proto Caribbean Sea*, and the third arm led to the much later opening of the South Atlantic Ocean separating South America from Africa. This, of course, is highly simplified. But an important fact about the formation of Florida is that it required the opening of three ocean basins (North Atlantic, Gulf of Mexico, and the Caribbean Sea) and the destruction/reformation of one of them (proto Caribbean Sea crust destroyed and replaced by a "new" Caribbean Sea—see chapter 6) to leave behind the distinctive peninsular shape of the platform. The opening of the South Atlantic Ocean, starting ~125 Ma, occurred much later than the opening of the North Atlantic Ocean, the Gulf of Mexico, and the Caribbean Sea—most of which occurred in the 180 Ma to 140 Ma interval.

The First Signs of Breakup

Prior to rifting, a broad transition between Laurasia (North America) and Gondwana (South America and Africa) consisted of a mosaic of crustal blocks and old plate *tectonic boundaries* (fig. 2.3). It was within and through this area that complex rifting processes propagated. Fig. 2.5 illustrates the various components prior to breakup. These are like pieces of a puzzle already assembled that become disassembled. Geologists have become very good at figuring out where and when fragments of the Earth's crust have moved in the past using various geophysical, sedimentological, and paleontological techniques. In figs. 2.5 and 3.1, note how the Florida-Bahama Block, the Yucatan Block, and the Suwannee Basin Block are nested together connecting South America and Africa as part of Gondwana (southern portion on Pangea). The Florida-Bahama Block and the Suwannee Basin Block would eventually form the basement rocks of peninsular Florida (fig. 3.2).

Natural events never occur as cleanly or as neatly as the cartoon diagrams in science books illustrate. Rifting of continents and ensuing seafloor spreading forming new ocean basins are no exceptions. Associated with initial continental pull-apart are a series of ragged, tensional tears in the crustal fabric forming linear basins, which become filled with sediments mostly from rivers and lakes. These are called *rift basins,* and the long, narrow, and deep lakes

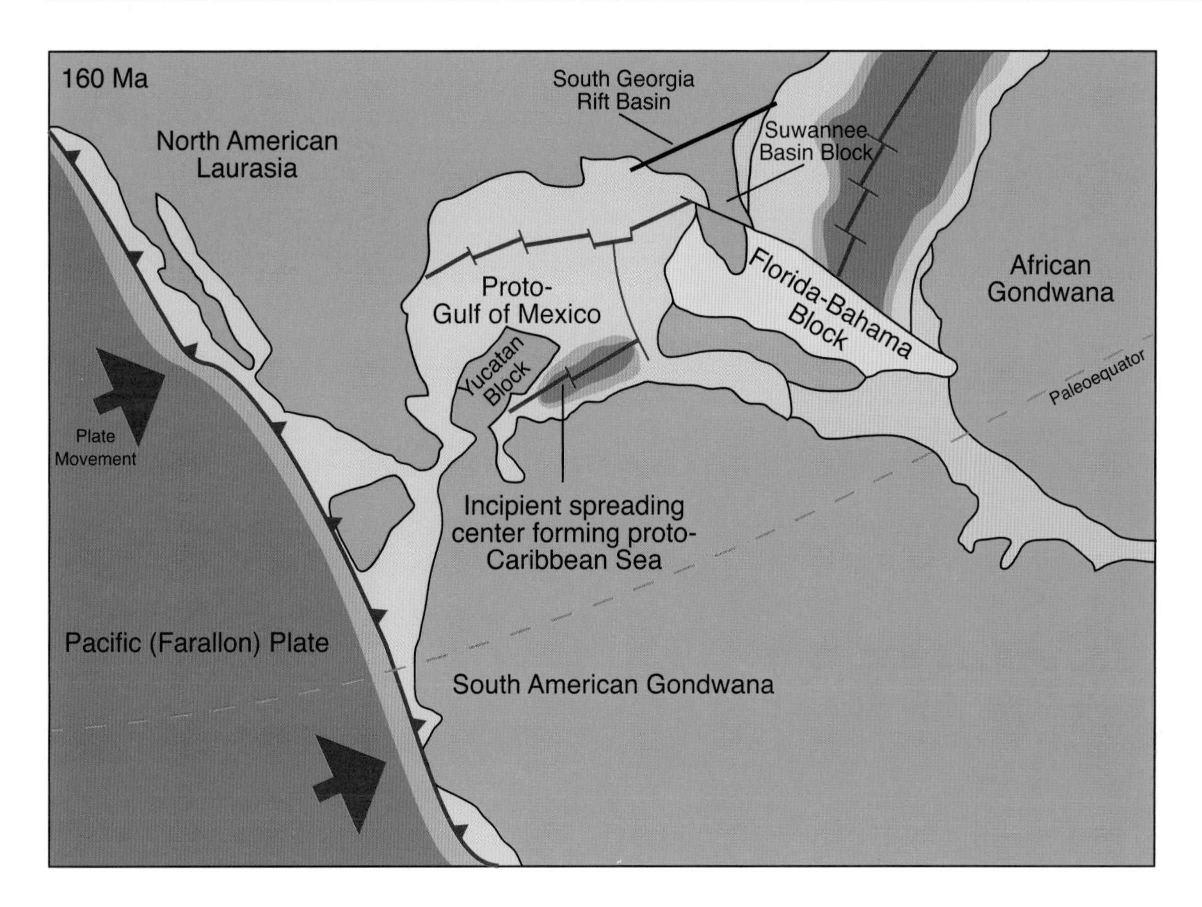

Figure 3.2. Paleogeographic map at ~160 Ma showing opening of Gulf of Mexico by movement of the Yucatan and Florida-Bahama Blocks. Note also new spreading center beginning to open up the Caribbean Sea. (Modified from Redfern 2001.)

of eastern Africa, Tanganyika, Rudolph, and Albert, are outstanding modern examples. The Gulf of California is a more mature rift system in that a *seafloor spreading center* forming *ocean crust* is present. However, all along the eastern United States were early rift basins that formed during the initial pull-apart that eventually led to the North Atlantic Ocean basin. These rift valleys formed as the continent was stretched (fig. 3.3). These basins filled mostly with sediment formed in nonmarine sedimentary environments such as alluvial fans, playa lakes, flood plains, and lavas and ash from volcanic activity associated with regional continental breakup. The famous three-toed dinosaur footprints seen in many museums were formed in shallow-water mudflats surrounding lakes that formed in these rift valleys.

The earliest part of this rifting in the Triassic (~200 Ma) created a line of basins from north to south along the eastern North Atlantic margin. Some of these basins can be seen today with their volcanic rocks (Palisades Sill on the west side of the Hudson River) and red sedimentary rocks (central Connecticut Valley extending north into Massachusetts). Most of these rift valleys eventually were buried, including one that formed across south Georgia—called the South Georgia Rift Basin near the old suture zone marking the boundary between Gondwana and Laurasia—NW Africa and SE North America. This basin is of great importance to the development of the Florida Platform later in geologic time as it eventually filled with seawater to become the ~100 km wide

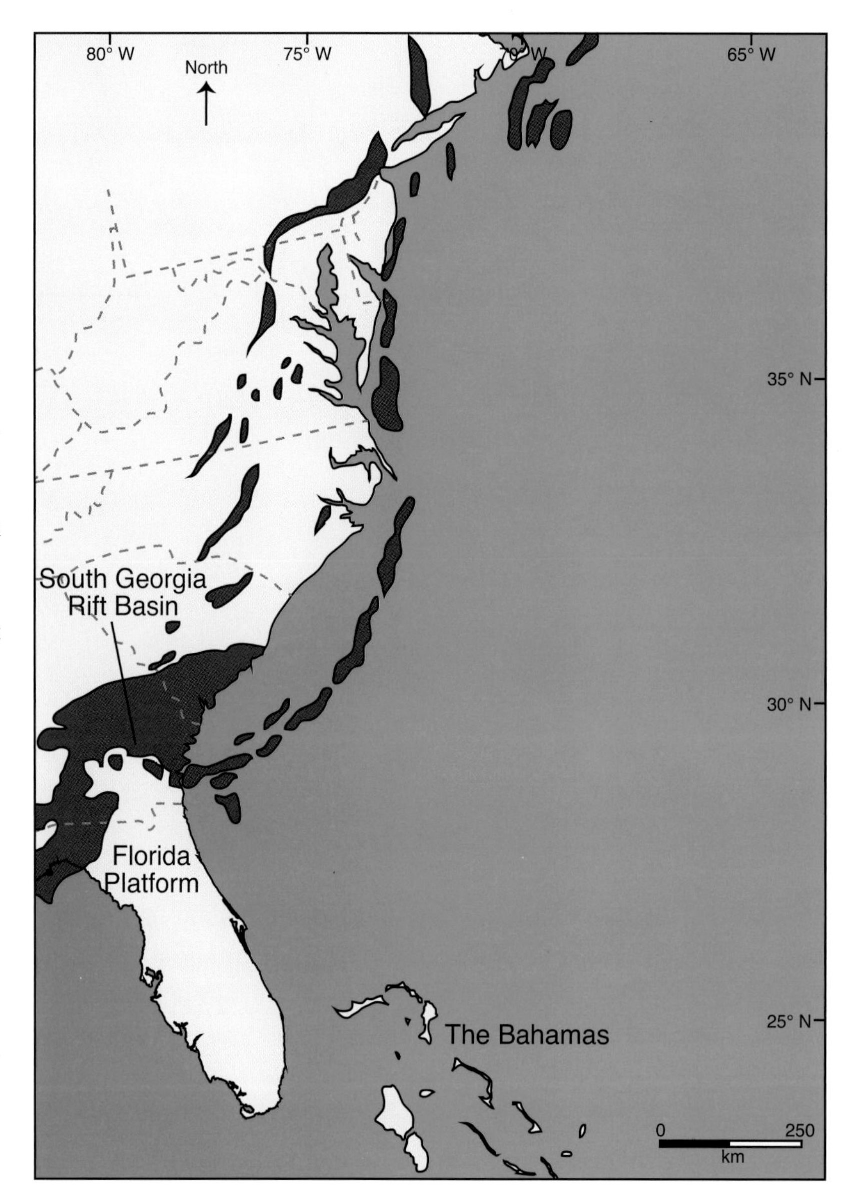

Figure 3.3. Map of eastern North American margin showing distribution of rift valleys (shown in dark) formed during the initial pull-apart of North America from NW Africa during the Triassic. One of the largest is the South Georgia Rift Basin, which separated the Florida Platform basement rocks from the rest of North America. The South Georgia Rift Basin eventually filled with seawater to become the Georgia Channel Seaway. (Modified from Klitgord et al. 1988.)

and ~300 m deep Georgia Seaway Channel. Eventually the Georgia Seaway Channel completely filled in and now lies buried beneath southern Georgia with no surficial expression. The offshore De Soto Canyon is the Georgia Seaway Channel's last surficial (bathymetric) manifestation (fig. 1.3*A*).

Rifting commonly occurs along zones of structural weakness in the Earth's crust. It seems likely that the suture defining the collision between NW Africa and SE North America could have been such a zone of weakness. Therefore, it would have made sense for the new Atlantic Ocean to have opened up where the "old" (proto) Atlantic or Iapetus Ocean once closed. For reasons not well understood, the "new" Atlantic Ocean opened up by abandoning the South Georgia Rift Basin and leaving behind a section of NW Africa still attached to North America. This piece, the Suwannee Basin Block (fig. 3.4), became the northern section of peninsular Florida's basement. Both the Suwannee Basin

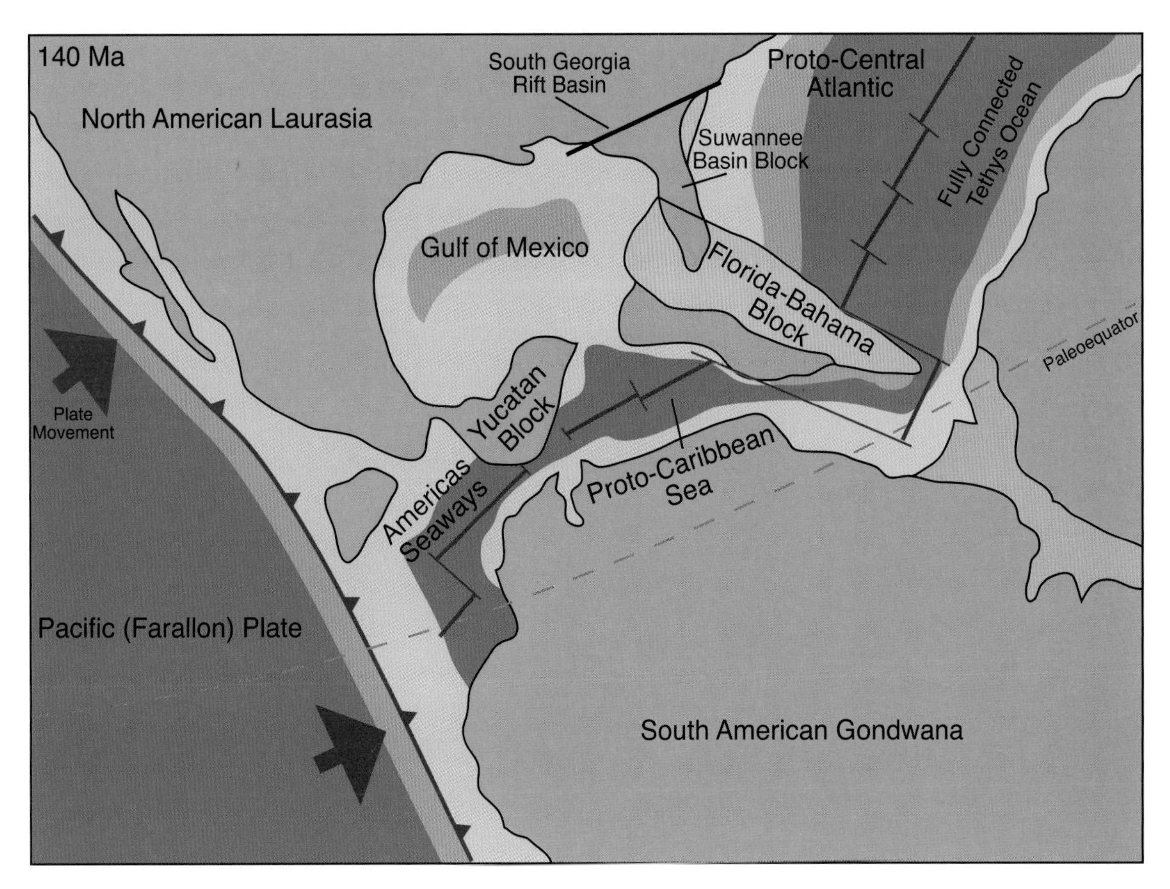

Figure 3.4. Paleogeographic map ~140 Ma. Seafloor spreading has ceased in the Gulf of Mexico. Yucatan and Florida-Bahama Blocks have ceased moving. The Atlantic Ocean and Caribbean Sea together have formed part of the Tethys Ocean. The assembly of the Florida basement rocks is now complete. (Modified from Redfern 2001.)

and Florida-Bahama Blocks, being part of Gondwana, became attached to Laurasia, thus forming *exotic terrains* or fragments of crust on one tectonic plate later attached to another tectonic plate.

Creating Three New Oceans

A three-ocean-basin sequence was initiated by the separation of North America from Africa, creating the North Atlantic Ocean, the Gulf of Mexico, and the Caribbean Sea. A seafloor spreading system propagated from north to south approaching the Florida-Bahamas area by the end of Early Jurassic ~180 Ma. During this time, there was great stretching, thinning, and extension as the Yucatan Block started to move south and the Florida-Bahama Block started to move east along *transform faults,* creating a basin that was to become the Gulf of Mexico. This early phase of the Gulf of Mexico subsided and aperiodically flooded with seawater entering gaps between these two blocks and elsewhere. The invading saltwater would repeatedly become trapped and evaporated, leaving behind brines that precipitated ever-thickening deposits of *evaporite minerals*—halite, gypsum, and anhydrite. Upwards of 4 km of evaporite minerals (mostly salt—NaCl) was deposited in this expanding Gulf of Mexico basin.

Between ~160 Ma and 140 Ma, a seafloor spreading center had created the Gulf of Mexico as a true ocean basin, floored by ocean crust and filled with normal salinity seawater, and it had divided the salt deposits into northern and southern provinces. Additionally a new seafloor spreading center located south of the Gulf of Mexico began to separate North America from South America, forming the very early (proto) Caribbean Sea.

By 140 Ma the seafloor spreading center forming the new North Atlantic Ocean had migrated around the Florida-Bahama Block via transform faults joined up with the spreading center separating North and South America. By now, there was a complete oceanic connection to the Pacific Ocean, however narrow. This was the new Tethys Ocean or Seaway. The Block had moved south of the Suwannee Basin Block, completing the basement of peninsular Florida and the Bahamas.

There has been little to no seafloor spreading in the Gulf of Mexico since ~140 Ma. Since that time, enormous amounts of sediment have been shed off North America, being deposited upon the initial evaporite formations known as the Louann Salt (northern province) and Challenger Salt (southern province). Seafloor spreading continued to the south for some time, thus widening the gap between North and South America and widening the Caribbean Sea.

While the Florida-Bahama Block was migrating to the southeast, it was stretched by *extensional forces*. This deformation caused a series of basins and arches to form by *block faulting* on these basement rocks. These deep-seated features controlled carbonate sedimentation that was to follow (see chapter 4). The key point is that these structures may control potential hydrocarbon accumulation—a potential that has yet to be fully realized by energy industry exploration.

Importance of the Gulf of Mexico

The Gulf of Mexico has become one of three mega-provinces on Earth, producing oil and gas. The other two are the western Siberia and the Arabia-Iran provinces—each having produced well over 100 billion barrels of oil. The Gulf of Mexico provided the five key ingredients required to build such a hydrocarbon mega-province: (1) organic source to create the hydrocarbons in the first place, (2) the transformation from organic matter to oil and gas, (3) permeable and porous rocks for the oil and gas to migrate and reservoir rocks to store these fluids, (4) a trap so they accumulate, and (5) impermeable seals so these fluids could not escape. The huge volume of hydrocarbons in this small ocean basin is largely due to the early salt deposits, followed by deposition of organic-rich sediments (formed by the Oceanic Anoxic Events—OAEs, see chapter 5) and finally a very thick sedimentary cover (maybe up to 20 km!) introduced by rivers carrying sediments off North America. The organic-rich sediments ultimately yielded the hydrocarbons. The weight of the thick sedimentary cover forced the salt to deform and move vertically and horizontally, forming *diapirs* and many other structures trapping migrating oil and gas.

The carbonate Florida Platform, it seems, sat off to the side and appears not to have the same potent mix of these five ingredients as the western and northern Gulf of Mexico. However, the full extent of oil and gas in Florida has yet to be determined, and hydrocarbon exploration and exploitation will remain controversial.

The Gulf of Mexico also has enormous importance in that it provides a source of moisture for weather systems passing across the North American continent. The west-to-east parade of low-pressure and high-pressure systems across North America draws enormous volumes of water from the Gulf of Mexico and provides rainfall for much of the central and eastern United States. Had the Gulf of Mexico never formed, the huge interior drainage system of rivers in the central United States, including the Mississippi River, one the world's largest rivers, never would have formed. Life would be quite different and probably would have resembled the arid and desolate Australian Outback. The American breadbasket of the Midwest would not exist, and our food would have to be grown elsewhere. Florida would be much more arid, probably dominated by unvegetated sand dunes on land rather than the heavily vegetated countryside we now see in undeveloped areas. Certainly there would be no wetlands, no Everglades, and no springs.

Connecting the Rocks across the Ocean

What are these basement rocks forming peninsular Florida, now that we know how and when they were emplaced? We have disassembled the original pieces of crust that existed between Laurasia and Gondwana, but what kinds of rocks composed these pieces? Do we see the same rocks beneath peninsular Florida as we see in NW Africa? If so, we could be reasonably certain that they were once physically connected.

About 140 deep borings (deepest ~3 km) in Florida and a handful more in the Bahamas (deepest ~5.5 km) retrieving rocks from great depth as well as a geophysical remote sensing survey detecting *gravity/magnetic anomalies* and *seismic reflection* and *refraction data* provide us with an understanding of what lies below. In Florida, the shallowest basement rocks are ~1 km below the surface, and the deepest basement rocks are ~6 km down. These basement rocks form the structural backbone of peninsular Florida. This structure is appropriately called the Peninsular Arch (fig. 3.5).

In the Bahamas, it has been estimated that the top of the basement rocks are as deep as 14 km (fig. 1.3*B*). Thus geologists working in Florida or the Bahama Banks do not have the luxury of seeing the basement rocks in natural outcrops, road cuts, or rock quarries that geologists can examine in mountainous terrains, for example.

The rocks forming the Suwannee Basin Block consist of granite, sandstones, and shales. Geologists provide names to specifically defined rock formations whether they are igneous, metamorphic, or sedimentary in origin. The oldest rocks are the granites, >500 Ma, both beneath central peninsular Florida and NW Africa. In Florida, the granite body is called the Osceola Granite, and the presumed same rock, torn apart and separated by rifting and seafloor spreading, is called the Coya Granite in NW Africa. Perhaps 100 Myr later, sedimentary rocks derived from the eroding granites formed sandstones and shales of the Suwannee Basin lying beneath north-central peninsular Florida. In NW Africa rocks of the same age formed the Bové Basin—indicating that they were once connected, forming one larger sedimentary basin (fig. 3.6).

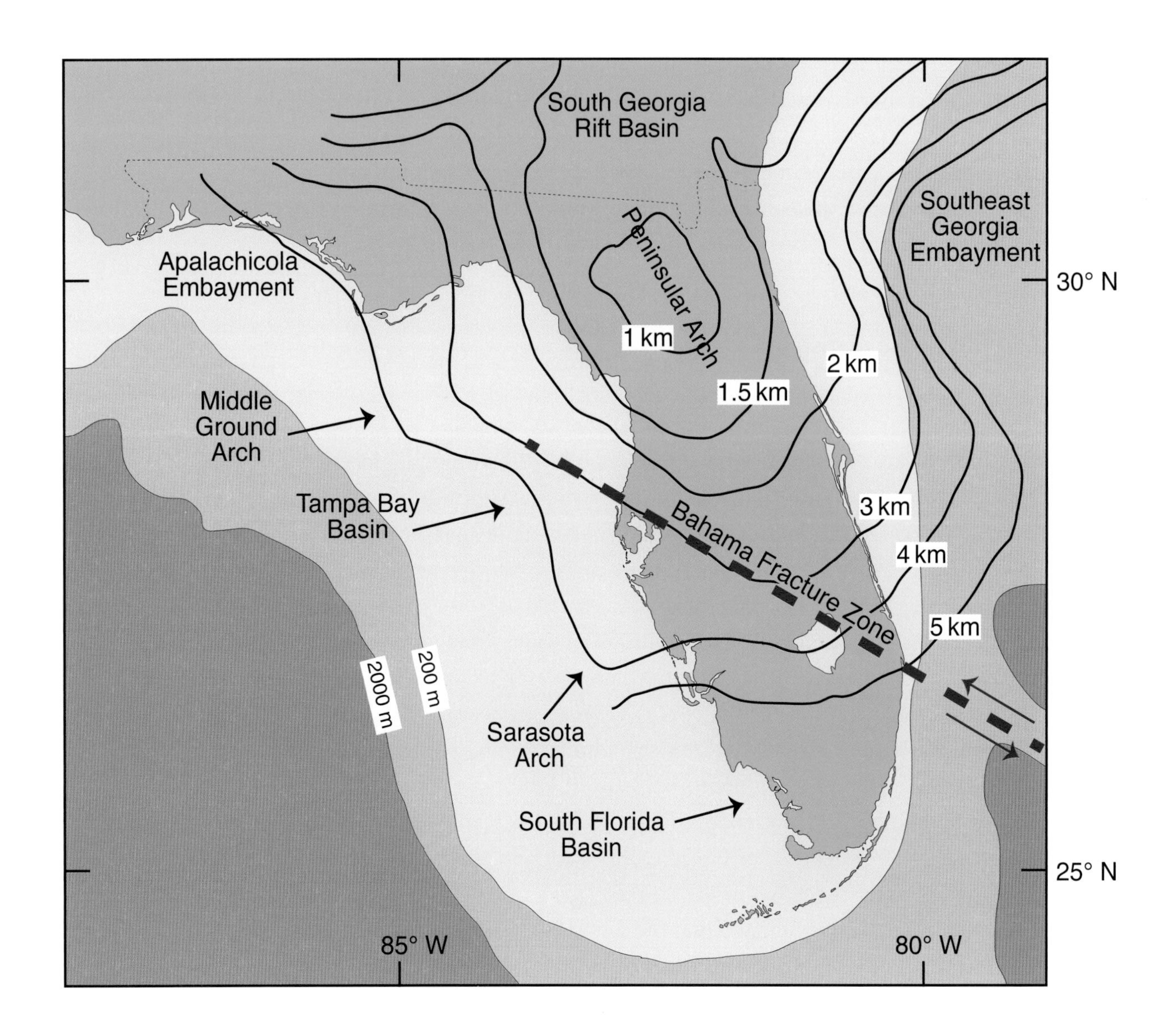

Figure 3.5. Depth to basement map (in km) illustrating the shape of the Peninsular Arch and the location of smaller structural features, particularly along the western margin of the Florida Platform. These features formed during the opening of the eastern Gulf of Mexico. (Modified from Klitgord et al. 1988.)

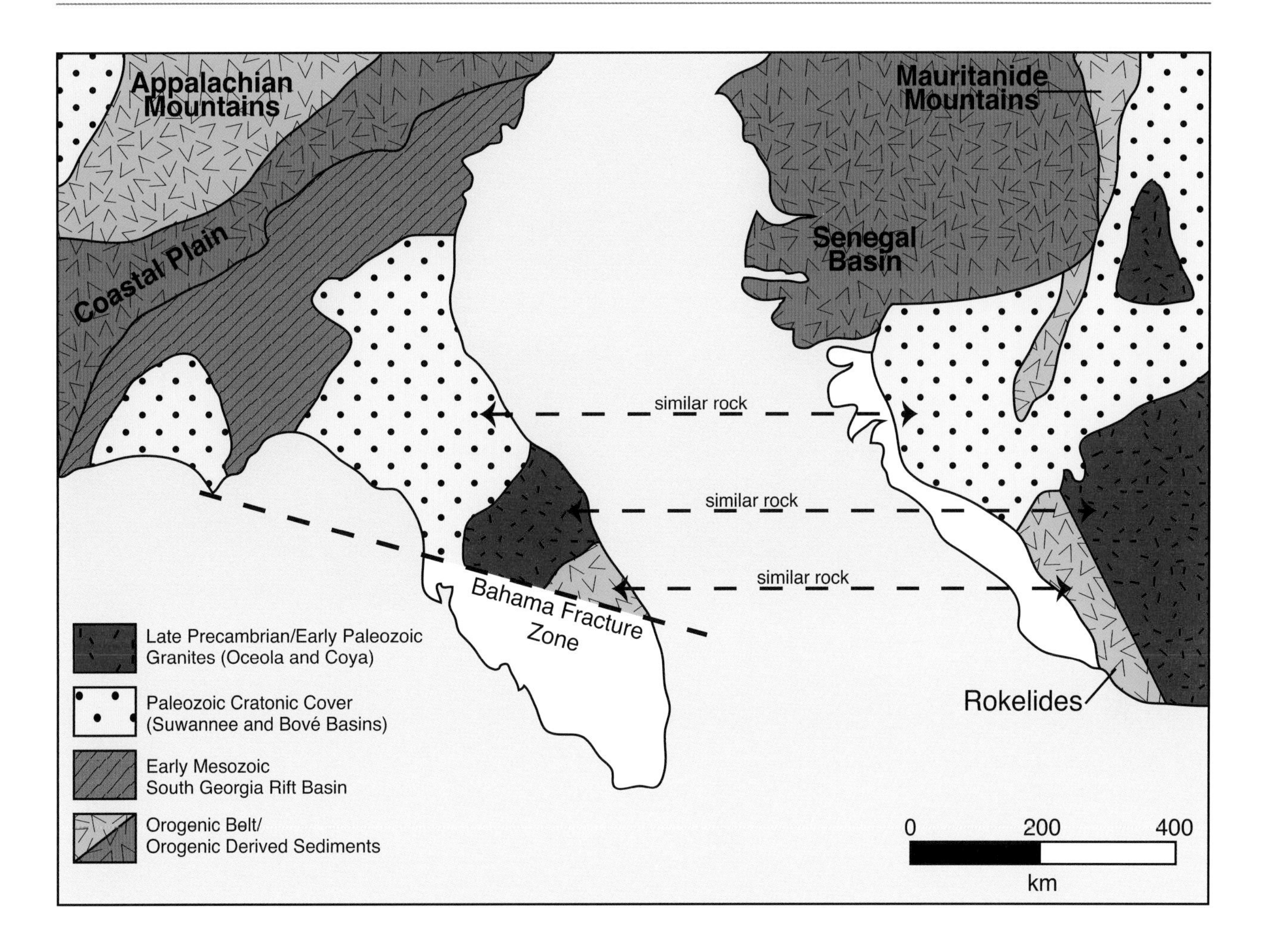

Figure 3.6. Paleogeographic map illustrating similarity and connectivity of rock beneath peninsular Florida and NW Africa. (Modified from Dallmeyer et al. 1987.)

The Structural Boundaries of the Florida-Bahama Platform

In chapter 1, we asked the question "What is Florida?" and presented fig. 1.2 illustrating the key structural boundaries that make Florida a distinct geologic entity. Rather than just viewing Florida as defined by its coastline and *political* boundaries, we pointed out that Florida should be viewed as a distinct geologic feature defined by *geologic* boundaries—actual structural features in the crust of the Earth.

The Georgia Rift Valley (or Basin) defines the northern boundary—a rift that failed to form an ocean basin floored by ocean-crust rocks. Nevertheless, this elongate basin was flooded with seawater forming the Georgia Channel Seaway—a bathymetric barrier that separated Florida and the Bahamas from North America for many tens of millions of years. The western margin of peninsular Florida facing the Gulf of Mexico, consisting of the Suwannee Basin and Florida-Bahama Blocks, is a complex rifted/transform margin where continental rocks transition into oceanic crust underlying this new ocean basin—a *passive margin* (fig. 3.7). The southern margin is also a complex rifted (passive) margin that faced the newly forming proto Caribbean Sea. Some 100 Myr later,

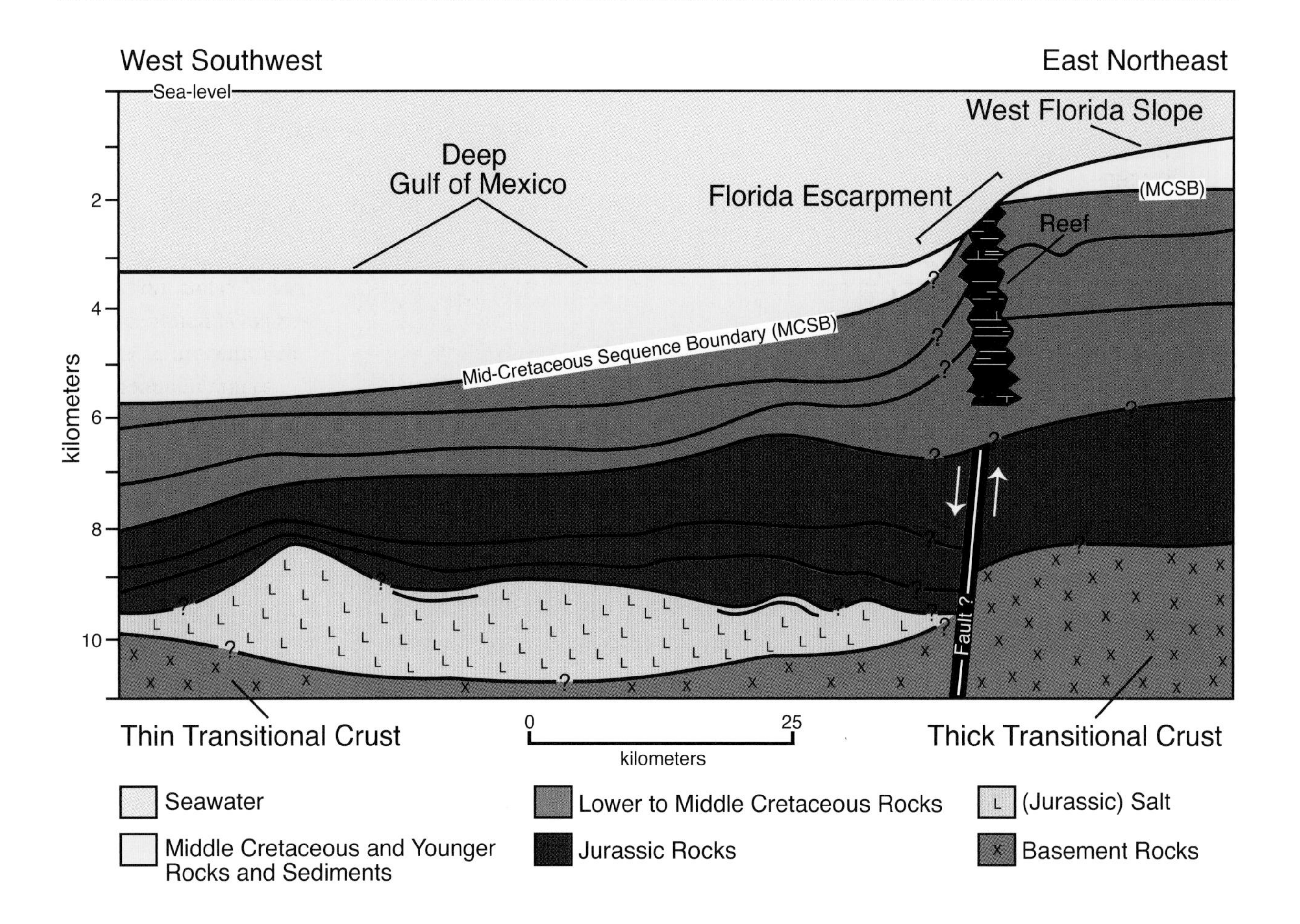

Figure 3.7. Transect across outer margin of west Florida illustrating faulted basement structural boundary. Note the Middle Cretaceous Sequence Boundary. (Modified from Hine 1997; originally from DeBalko and Buffler 1992.)

this passive margin was to be radically changed into a *collision margin* with the NE migrating Caribbean Plate (chapter 6). Finally, the eastern margin of the Florida-Bahama Platform is defined by a complex series of transform faults. The most famous is the Bahama Fracture Zone along which the Florida-Bahama Block moved to the southeast. The NW–SE linear trend of the modern eastern Bahamas resulted from this now deeply buried fault trend—a *transcurrent margin*. Eventually, as will be seen in chapter 5, the Florida Platform became a separate geographic entity from the Bahama Platform with the formation of the Straits of Florida providing a deep, bathymetric barrier between the two carbonate platform systems.

So the stage is now set for Florida to build its ~1–6 km thick cover of carbonate rock on top of the basement rock. With the formation of the North Atlantic Ocean, Gulf of Mexico, and the Caribbean Sea, the exposed surface of the basement became flooded by seawater forming a huge shallow ocean, ideal for the production of carbonate sediments.

Essential Points to Know

1. The Pangean supercontinent proceeded to break up 225 Ma after having been fully assembled by 300–250 Ma. It took several hundred million years to assemble, but began to split apart only ~25–75 Myr after final assembly. The origin of Florida, the Gulf of Mexico, the Caribbean, and the North Atlantic Ocean is intimately tied to this breakup.

2. The formation of Florida required the opening of three ocean basins (North Atlantic, Gulf of Mexico, and the Caribbean Sea) and the destruction/reformation of one of them (proto Caribbean Sea crust destroyed and replaced) to leave behind the distinctive peninsular shape of the platform.

3. The Florida-Bahama Block, the Yucatan Block, and the Suwannee Basin Block were nested together connecting South America and Africa as part of Gondwana (southern portion on Pangea). The Florida-Bahama Block and the Suwannee Basin Block eventually formed the basement rocks of peninsular Florida.

4. Most of the rift valleys eventually were buried, including the South Georgia Rift Basin, near the old suture zone marking the boundary between Gondwana and Laurasia—NW Africa and SE North America. This basin was of great importance to the development of the Florida Platform later in geologic time as it eventually filled with seawater to become the Georgia Channel Seaway System.

5. The "new" Atlantic Ocean opened up by leaving behind a section of NW Africa still attached to North America. This piece, the Suwannee Basin Block, became the northern section of peninsular Florida's basement. Both the Suwannee and Florida-Bahama Blocks, being part of Gondwana, became attached to Laurasia, forming exotic terrains or fragments of crust on one tectonic plate later attached to another tectonic plate.

6. The early phase of the Gulf of Mexico subsided and repeatedly flooded with seawater. The invading saltwater became trapped and eventually evaporated, leaving behind ever thickening deposits of evaporite minerals, mostly halite.

7. While the Florida-Bahama Block migrated to the southeast, it was stretched by extensional forces. This deformation caused a series of basins and arches to form by block faulting on these basement rocks. These deep-seated features controlled carbonate sedimentation that was to follow.

8. The proto Caribbean ocean crust formed by seafloor spreading between North and South America was destroyed and replaced by a small, eastward-migrating tectonic plate that was to become the modern Caribbean Plate. A key point is that geology south of Florida is some of the most intricate on Earth, and Florida was eventually influenced by this complexity.

9. The Suwannee Basin was formed beneath north-central peninsular Florida sandstone and shales of early Paleozoic age. In NW Africa similar rocks of

the same age form the Bové Basin, indicating that they were once connected, forming one larger sedimentary basin.

Essential Terms to Know

basalt: Crustal portions of oceanic tectonic plates are composed predominantly of basalt, produced from upwelling mantle below ocean ridges. Basalt underlies the ocean basin and is a type of dark volcanic rock that is extruded onto or near the seafloor or land surface. It cools and crystallizes rapidly so that the constituent minerals do not form large crystals. It is composed of Mg and Fe rich silicate minerals.

block faulting: Faults creating large vertical displacements of large pieces of continental crust. Vertical motion of the resulting blocks, sometimes accompanied by tilting, can then lead to high escarpments even forming small mountains. These mountains are formed by the Earth's crust being stretched and extended by tensional forces. Fault block mountains commonly accompany rifting, another indicator of tensional tectonic forces.

collision margin: Convergent margins of plates that move toward each other, causing collision. Mountain ranges and volcanic island arcs are products of such plate-margin collisions.

conjugate continental margins: Margins of continents that "fit" together like pieces of a puzzle when reconstructed, as promontories fit into embayments.

diapirs: Generally salt intrusions in which a more mobile and ductile, deformable material is forced into brittle overlying rocks. Diapirs commonly intrude vertically upward along fractures or zones of structural weakness through denser overlying rocks because of density contrast between a less dense lower rock mass and overlying denser rocks. The density contrast manifests as a force of buoyancy. The resulting structures are also referred to as *piercement structures*. This term is used to describe vertical salt movement upward into overlying sediments, forming salt domes.

evaporite minerals: Minerals such as halite (NaCl, salt), gypsum ($CaSO_4 \cdot 2H_2O$), and anhydrite ($CaSO_4$) that form when seawater evaporates.

exotic terrain: Fragment of crustal material formed on, or broken off from, one tectonic plate and accreted—"sutured"—to crust lying on another plate; piece of crust that has been transported laterally, usually as part of a larger plate, and is relatively buoyant due to thickness or low density. When the plate of which it was a part subducted under another plate, the terrain failed to subduct, detached from its transporting plate, and accreted onto the overriding plate. Therefore, the terrain is transferred from one plate to the other. Typically, accreting terrains are portions of continental crust that have rifted off another continental mass and been transported surrounded by oceanic crust or old island arcs formed at some distant subduction zone.

extension forces: Tectonic horizontal forces in Earth's crust that stretch or extend the crust, commonly thinning it until it breaks.

gravity/magnetic anomalies: A gravity anomaly is the difference between the observed gravity and a value predicted from a standard model—a slight variation in the Earth's gravitational field to a change in local mass within the Earth. A magnetic anomaly is a change in the Earth's magnetic field as a result of the presence or absence of Fe rich rocks and the depth and configuration of such rocks.

hot spot: A persistent source of magma produced by partially melting the overriding plate. The magma, which is lighter than the surrounding solid rock, then rises through the mantle and crust to erupt onto the seafloor, forming an active seamount.

ocean basin: Structural basin underlain by ocean crust types of rocks, such as basalt and gabbro, containing Mg and Fe rich silicate minerals. These rocks are denser than granitic continental crust rocks, which are rich in Al and Si.

ocean crust: Part of Earth's lithosphere that underlies the ocean basins. Oceanic crust is primarily composed of Mg and Fe rich rocks. It is thinner and denser than continental crust, which consists of Si and Al rich rocks. Ocean crust is generally less than 10 km thick and is denser than continental crust. Basalt is the common igneous rock of the upper ocean crust, which is extruded from seafloor spreading centers.

passive margin: A continental margin that has been rifted and faces a seafloor spreading center. Mature passive continental margins are low relief. They have subsided slowly for long periods of time and have accumulated a thick sedimentary cover.

rift basin/rift zone: A place where the Earth's crust and lithosphere are being pulled apart—an example of extensional tectonics. Typical rift features are a central elongate down-dropped fault segment, called a graben, with parallel extensional faulting and uplifts on either side forming a rift valley. The axis of the rift area commonly contains volcanic rocks, and active volcanism is a part of many but not all active rift systems.

seafloor spreading: Occurs at mid-ocean ridges, where new oceanic crust is formed through extrusive igneous activity and that gradually moves away from the ridge. Seafloor spreading helps explain continental drift in the theory of plate tectonics. It starts as a rift in a continental land mass, similar to the Red Sea–East Africa Rift System today. The process starts with heating at the base of the continental crust, which causes it to become more plastic and less dense. Because less dense objects rise in relation to denser objects, the area being heated becomes a broad dome. As the crust bows upward,

The Carbonate Factory Cranks Up

Florida Being Born from the Sea (~160 Ma to Present)

> I think you will now allow that I did not overstate my case when I asserted that we have as strong grounds for believing that all the vast area of dry land, at present occupied by the chalk, was once at the bottom of the sea, as we have for any matter of history whatever; while there is no justification for any other belief.
>
> T. H. Huxley, "On a Piece of Chalk," *Macmillan's Magazine* 18 (1868)

The Basic Ingredients of Carbonate Platform Construction

Cross sections illustrating the vertical distribution of rocks in the Earth's crust are fundamental tools of geology allowing geologists to peer back in time. Cross sections represent reconstructions of events and processes that must have occurred to create the stratigraphic succession depicted in such diagrams. A simplified cross section from the West Florida Escarpment extending eastward ~1,000 km to the Florida-Bahama Escarpment reveals an enormously thick carbonate rock section covering much older basement rocks (fig. 1.3). Indeed, the carbonate rock cover is nearly 14 km thick beneath the Bahamas and nearly 6 km thick beneath south peninsular Florida. Most of the rocks were formed from carbonate sediments originally deposited in shallow water.

Through geophysical remote sensing techniques, geologists can begin to construct such cross sections. Drilling long holes and extracting rock cores from them allow geologists to "ground-truth" the remotely sensed data and to determine the rock type and age. From the remote sensing and the ground-truth from drilling, geologists can determine past climates, past sea level behavior, past life, and the rates of change of Earth processes. The cross sections shown in figs. 1.3*B* and 4.1 beg some fundamental questions: (1) how did carbonate sediments form over such a wide area and for such a long time, and (2) how is it possible for carbonate sediments that mostly formed in water a few meters deep to have built a platform that is 1,000 km wide and up to 14 km thick? Somehow we have to explain the geology.

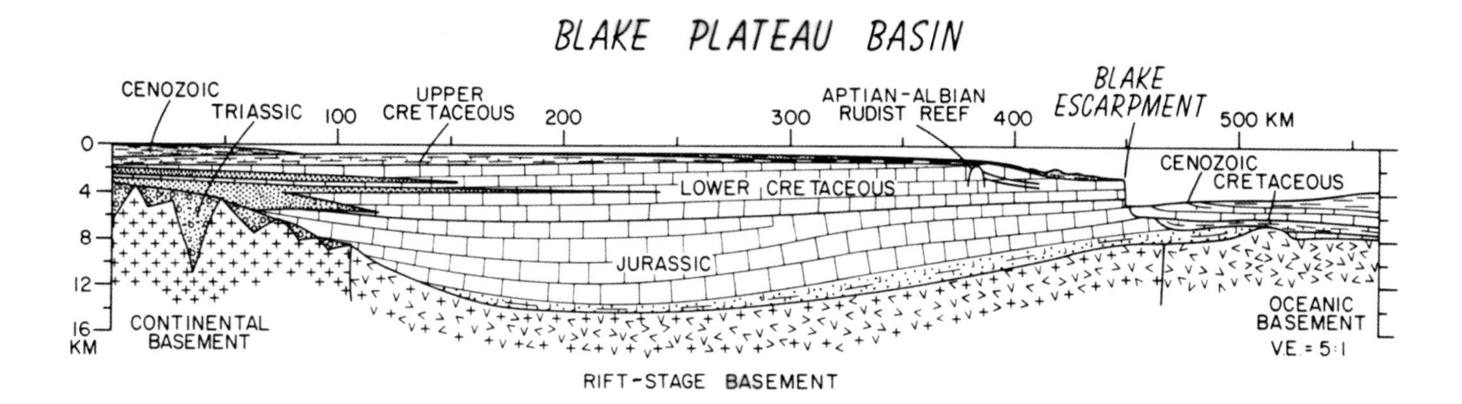

Figure 4.1. Idealized section across the northern Bahama Platform including the Blake Plateau, which is essentially the eastern half (~500 km wide) of the combined Florida-Bahama Platform complex—probably >1,000 km wide at its peak. Note that the maximum thickness of carbonate rock is ~14 km. The eastern portion of this platform abruptly ends at the Blake Escarpment (northern extension of the Bahama Escarpment), which drops ~2 km vertically in places into the deep western Atlantic Ocean. Most of these rocks formed in shallow water, and this great thickness has formed through long-term subsidence. (Source: Dillon and Popenoe 1988; used by permission from the Geological Society of America.)

How Do Carbonate Platforms Get Started?

As was seen in chapter 3, the breakup of Pangea and the rifting of North America from Gondwana (combined South American and African continents) created the Florida basement—a broad, elevated surface underlain by igneous and metamorphic (and some sedimentary) basement rocks (pieces of South American and African basement) that were exposed to air causing weathering and flooded by marine waters in a warm, low-latitude setting. This initial flooding eroded weathered basement rocks enriched in *silicate minerals* and first deposited siliciclastic rich sedimentary rocks—shales and sandstones. These initial sedimentary rocks on top of the Florida basement were similar to the Late Jurassic (~160–145 Ma) Cotton Valley Formation and the Norphlet Formation (both sandstones, mostly siliciclastic, quartz rich) found elsewhere in the deep subsurface around the Gulf of Mexico where they now produce gas and oil. To what extent these formations might trap hydrocarbons on the Florida Platform is unknown, but they are deeply buried—at least 4 km down into the subsurface of west-central peninsular Florida, probably beneath most of the modern west Florida outer shelf.

However, shortly after these siliciclastic sedimentary units formed, environmental conditions changed to favor carbonate sediment deposition. Perhaps, with continued sea level rise in the Late Jurassic/Early Cretaceous, the Florida Platform became sufficiently separated from North America by open seawater (see global sea level curve from Mesozoic to Cenozoic on page ii and fig. 4.2).

This would have prevented muddy and nutrient-rich waters from reaching Florida (more about this point in chapter 8). As a result, shallow-water carbonate deposition, which requires clear, warm waters for light to penetrate to stimulate *photosynthesis*, began to dominate.

Additionally, these marine waters would have contained dissolved calcium (Ca^{+2}) and bicarbonate (HCO^{-3}), the essentials to make calcium carbonate ($CaCO_3$) or limestone. Two important chemical reactions are:

(1) $CO_2 + H_2O \rightarrow CH_2O + O_2$; *photosynthesis*—makes organic matter (CH_2O—simplified chemical notation for organic matter). Some light-dependent plants such as algae secrete large amounts of calcium carbonate (calcification).

(2) $Ca^{+2} + 2(HCO_3)^{-} \rightarrow CaCO_3 + 2H^{+} + CO_2$; *calcification*—makes the carbonate skeletons ($CaCO_3$) of plants and animals.

Given these environmental factors, the early seas rising over and flooding this Jurassic rocky basement surface deposited the first carbonate sediments, providing the first carbonate rock layer for the Florida-Bahama Platform—just as the new North Atlantic Ocean and Gulf of Mexico were beginning to form.

In the Late Jurassic, as Pangea continued to break apart by rifting and seafloor spreading, a great new circumglobal tropical ocean was opening (figs. 4.3, 4.4). This created a continuous east-west ocean in a low-latitude setting—called the

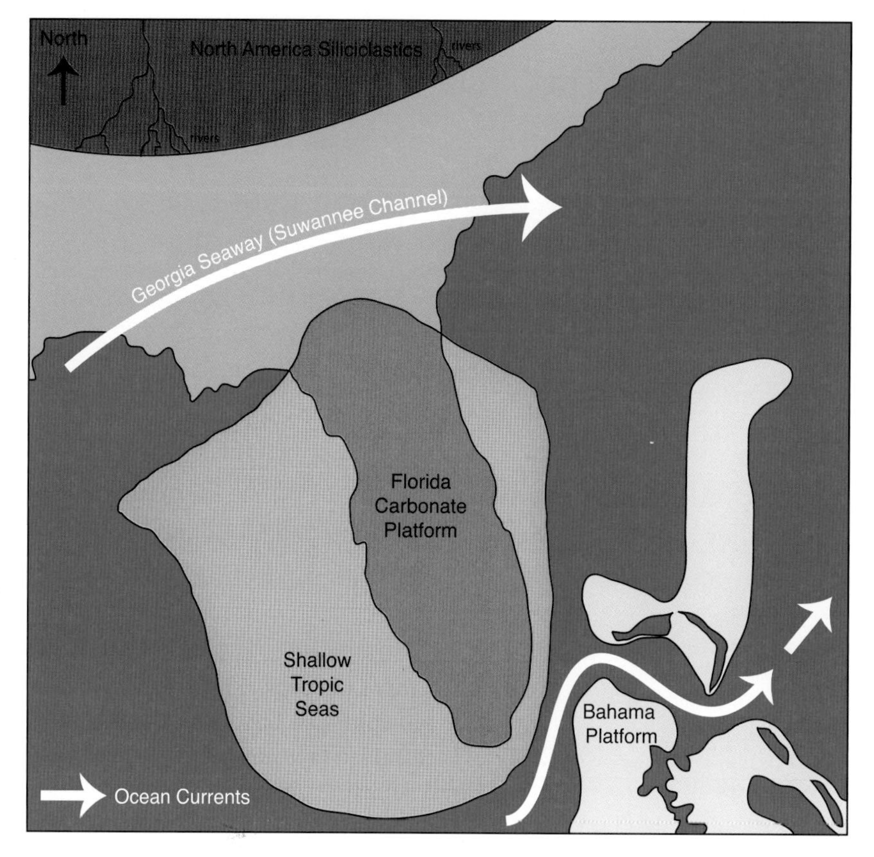

Figure 4.2. Paleogeographic map showing the shallow-water Florida Platform separated from North America by the flooded rift valley called the Georgia Seaway Channel (also known as the Suwannee Channel). The ancestral Loop Current passed between siliciclastic sediment (quartz rich) shedding from North America (southern Appalachian Mountains) and the warm, shallow clear waters covering the Florida Platform (accumulating carbonates). (Modified from Redfern 2001.)

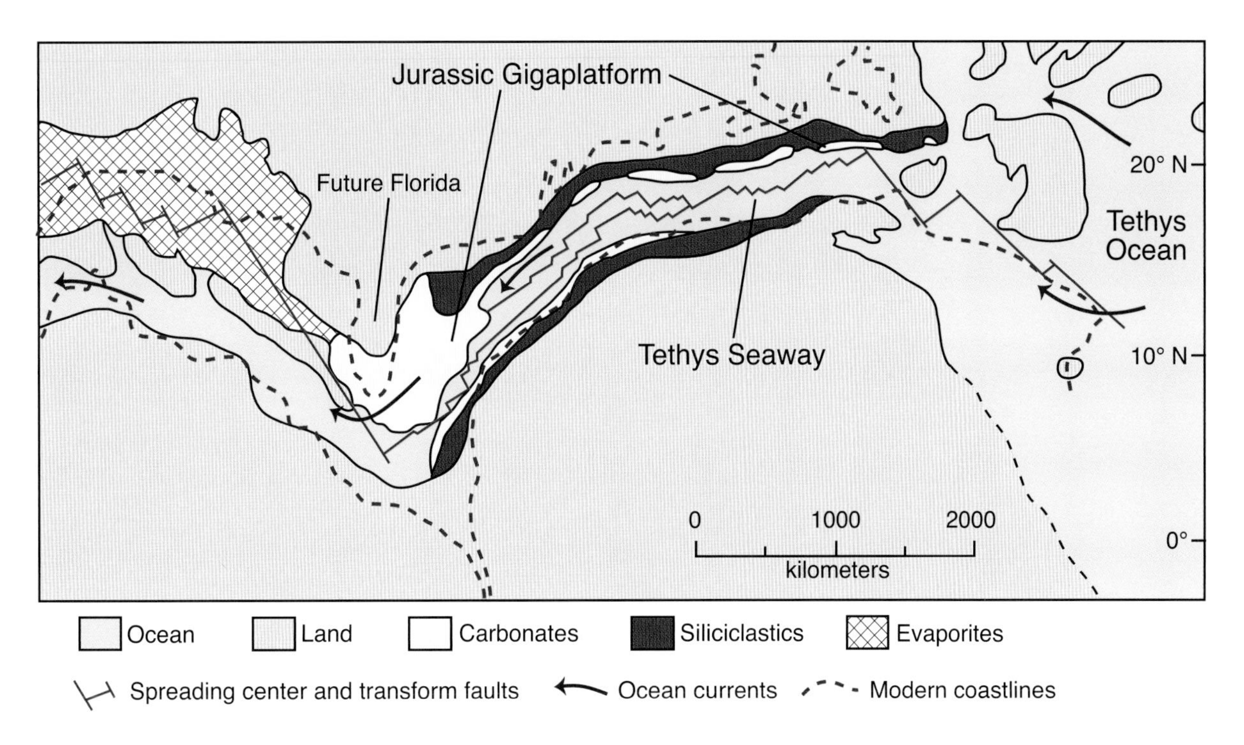

Figure 4.3. Map of the early opening of the Tethys Ocean (or Seaway) forming the initial North Atlantic Ocean at ~160 Ma (Late Jurassic). Carbonate platforms developed on both the North American and African sides of this new, low latitude ocean. The Florida-Bahama Platform area is the largest part of the widespread carbonate system. Note extensive evaporites forming in the early Gulf of Mexico basin. (Source: Poag 1991; used by permission from Elsevier.)

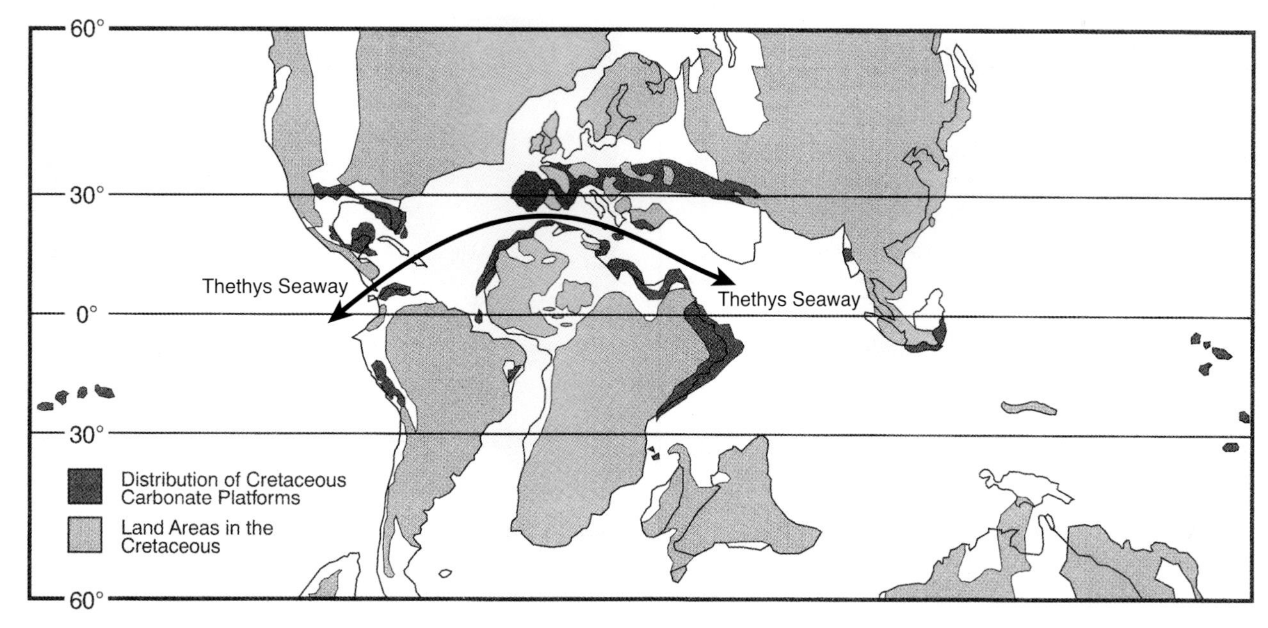

Figure 4.4. Tethys Ocean in the mid-Cretaceous (~100 Ma), when sea level was at its highest point. The Tethys formed a circumglobal tropical ocean surrounded by the Florida-Bahama and other carbonate platforms. (Source: Randazzo and Jones 1997.)

Tethys Ocean. The margins of this new ocean became dominated by carbonate sedimentation forming nearly ubiquitous limestone platforms by the mid- to Late Cretaceous. The Florida, the Bahamas, and the Yucatan all formed one huge mega-platform. At the same time many carbonate platforms extended along this ancient tropical ocean. Eventually Florida, the Bahamas, and Yucatan were separated by deep seaways to form the three carbonate platform systems we see today. What are these carbonate sedimentary building blocks that form these platforms in the first place? What are the components, what are the pieces, and how are they produced? Let us take a closer look.

Carbonate Sedimentary Environments on Platforms: The Carbonate Factory

Carbonate sediments are enormously diverse because they form from a huge range of organisms (microbes, plants, and animals) that produce skeletal material. There are even nonskeletal carbonate grains such as *fecal pellets* and coated grains called *ooids* (figs. 4.5*A*, 4.5*B*). Biologic reefs (solid, wave-resistant rock framework—not sediments) are very important products of the carbonate factory (fig. 4.5*C*). Carbonate muds (fine-grained sediments) are perhaps the most important part of the carbonate sediment factory because of the high volume of material generated and the relative ease with which they can be transported great distances (fig. 4.5*D*). Carbonate mud formation has been a very controversial subject.

Biological processes from a variety of organisms (commonly corals, encrusting organisms) build a rock framework that may produce enormous structures—just think of the Great Barrier Reef! However, as spectacular as they appear in the modern ocean, reefs built by organisms may be volumetrically quite small as compared with the total mass of carbonate rock and sediment in a large platform such as the Bahamas, which are largely built from lime mud (fig. 4.6).

Coral reefs and other types of carbonate reefs are critically important environment indicators. In lower Cretaceous rocks, instead of coral reefs, massive accumulations of bivalves (clams) known as rudists formed the reefs along the outer margin of the west Florida Platform (fig. 4.5*C*).

Once created, this diverse array of grains may be physically distributed, dispersed, concentrated, and sorted by waves and currents. Carbonate sediments can be described by their grain size (microscopic clays to building-sized boulders) and by the type of skeletal-producing organisms that made them (e.g., mollusks, algae, bryozoans, foraminifera). Depending upon the community of skeletal-producing organisms found in a particular environment and the level of physical energy capable of moving these sedimentary particles around, spe-

Figure 4.5*A*. Ooid sand grains under normal light. These are nonskeletal carbonate (formed by the mineral aragonite) particles, formed in high-energy tidal channels and along beaches where there is constant water agitation and movement. A fecal pellet or a small skeletal fragment forms the nucleus, which is rolled around constantly in a high-energy environment and coated by fine layers of precipitating carbonate, forming a cortex, thus giving these grains their well-rounded, shiny appearance. These grains are from a tidal channel south of Shroud Cay, Exumas, Bahamas. (Courtesy of Don McNeill.)

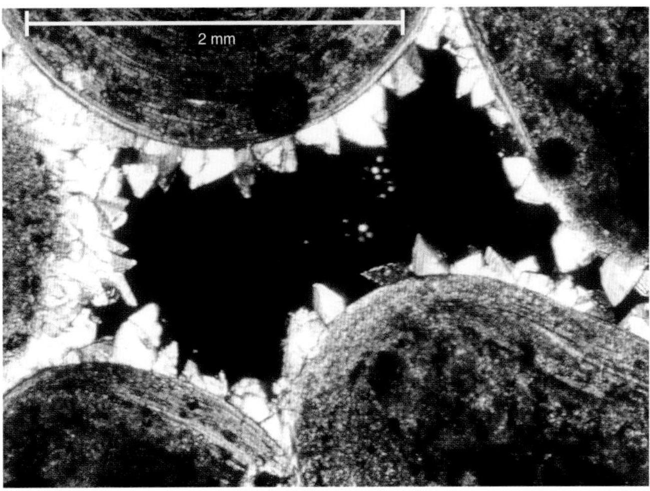

Figure 4.5*B*. Photomicrograph of ooid sand grains whereby polarized light passes through very thin, cut slices of the grains, making them translucent. In this manner, the internal structure can be seen. Both the nuclei and cortices (zone where fine layers exist) can be seen. Note the jagged white crystals that are made of calcite (not aragonite), which bind the ooid grains together, forming a carbonate rock. These calcite crystals are natural cement that formed in freshwater, thus revealing that these marine-created sediments were cemented in a fresh groundwater environment such as an island. (Courtesy of S. D. Locker.)

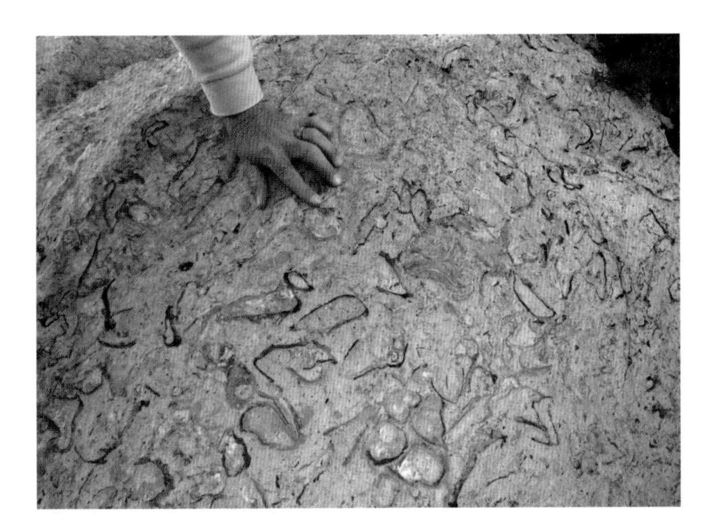

Figure 4.5*C*. Outcrop of requiinid rudist bivalves from the mid-Cretaceous (Albian) in Sonora, Mexico. These shallow-water colonies may have dominated the margins (instead of coral reefs) of the Florida Platform during the Early to mid-Cretaceous. They are now deeply buried or perhaps even exposed to deep water along the eroded West Florida Escarpment. (Courtesy of Charles Kerans.)

Figure 4.5*D*. Aerial photo of suspended carbonate muds being transported off the Great Bahama Bank into the deep Straits of Florida to the west (left). The muds are produced on the shallow carbonate bank (right), resuspended by storm waves and currents, and exported offbank by storm currents. (Courtesy of A. Conrad Neumann.)

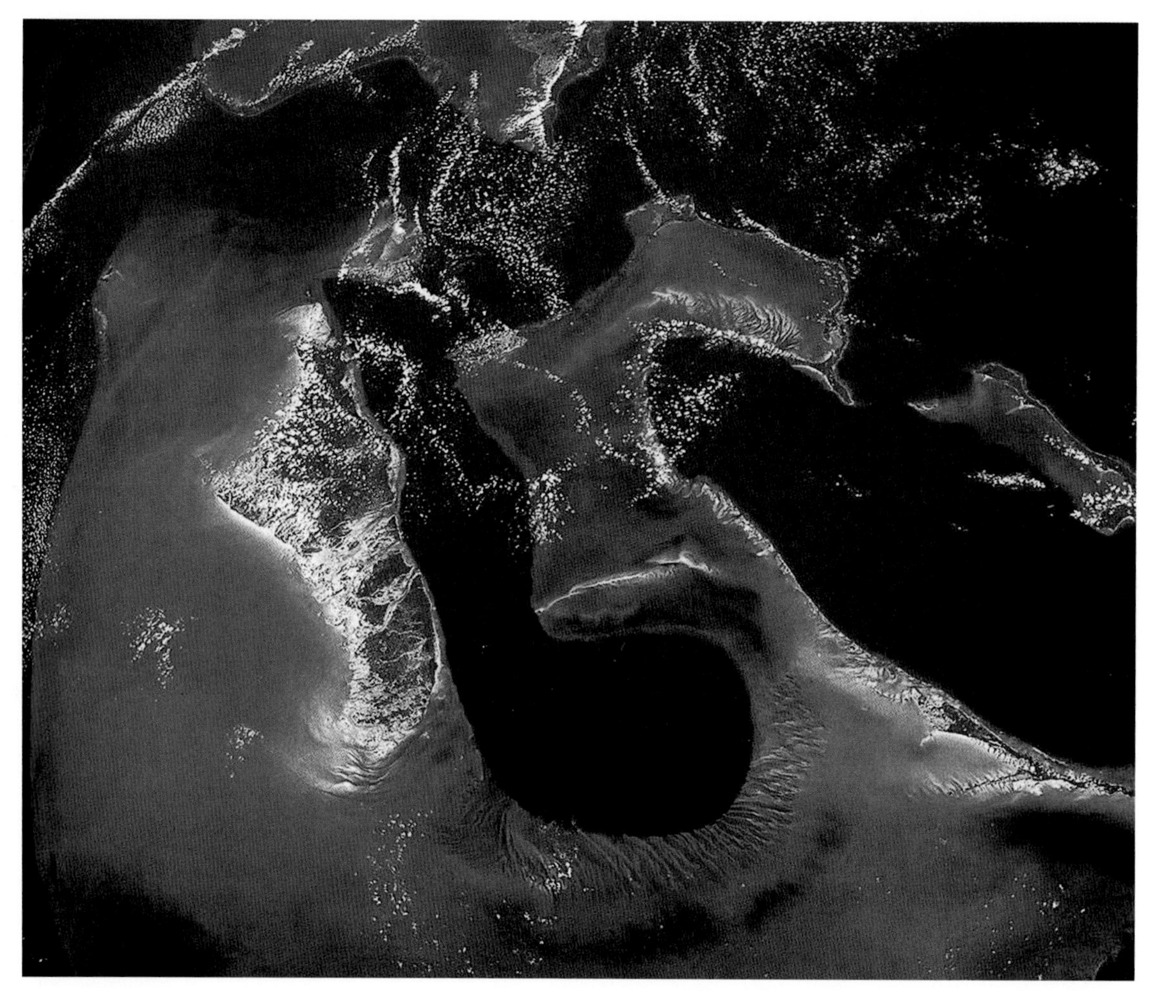

Figure 4.6*A*. Space image of the Bahamas illustrating one of the modern Earth's great carbonate platform provinces. (Source: NASA.)

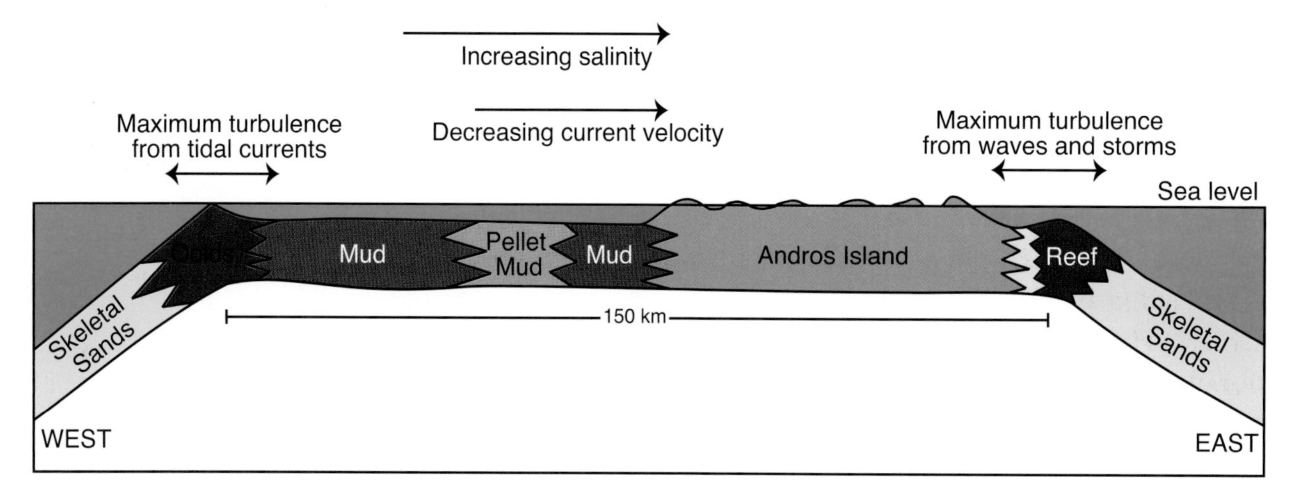

Figure 4.6*B*. Transect across the modern Great Bahama Bank showing the distribution of sedimentary environments, changes in water properties (salinity, currents), and resulting in different sedimentary facies. (Source: Jones 2010—their modification of Bosence and Wilson 2003; used by permission from the Geological Association of Canada.)

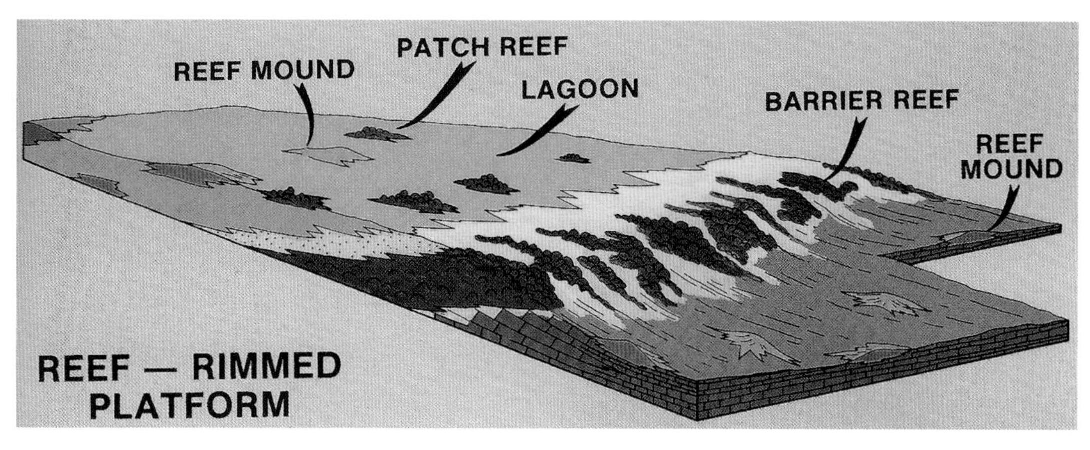

Figure 4.7. Three-dimensional diagram across a reef-dominated carbonate platform margin. Note the lateral distribution of different sedimentary environments and the vertical distribution as these sedimentary facies accumulate with time. In this case, the margin is building seaward out into the deeper water. (Source: James 1983; reprinted by permission from the American Association of Petroleum Geologists, whose permission is required for further use.)

cific groupings of sediments will form and define that environment. Geologists call these groupings *sedimentary facies* (figs. 4.6*B*, 4.7).

From shallow to deep or from high energy to low energy, different sedimentary facies are formed adjacent to each other, sometimes with distinct and sometimes with diffuse boundaries between them. A "layer" of carbonate sediments that eventually becomes limestone may contain a laterally diverse array of carbonate sediments. Each array or facies formed in distinct environments is defined by distinct biological and physical processes. It makes sense that carbonate muds would form in a lagoon far from large waves and that coral reefs requiring high energy would form along the margins. A carbonate platform accumulates a complex lateral suite of facies depending upon how much the processes change across it. Even though they might all be carbonate sediments, they are not all the same.

Fig. 4.6*B* illustrates a two-dimensional transect across a portion of the modern Bahama Banks showing the different environments—each defined by a specific suite of sediments, each forming a distinct sedimentary facies. Fig. 4.7 is a three-dimensional depiction, illustrating different sedimentary environments on a carbonate bank. Starting at the platform margin, a steep slope or escarpment basically plunges off into as much as 5 km of water after platform maturity is reached. Here, at the base of the escarpment or slope, sediments shed off the platform top accumulate. They may be fine-grained sediments or much coarser material, sometimes huge blocks that have slumped downslope due to gravity-induced instability (see chapter 5).

To make matters more complex, these sedimentary environments shift laterally or appear and disappear as a function of sea level fluctuations and climate change. With time and with sufficient accumulation of sediments, the lateral and vertical distribution of the carbonate sedimentary facies may become quite complex—such is the nature of *stratigraphy*.

Upon burial, the grains, regardless of size, become cemented together by dissolved material in the water that fills the spaces between the grains (fig. 4.5*B*). These natural cements are also composed of calcium carbonate minerals. The *cementation* process can be very fast and may begin soon after the sediments become deposited. In fact, geologists have discovered aluminum cans and glass bottles encased in limestone (albeit somewhat soft), proving that rock formation may occur within a few decades.

Carbonate rocks may be affected by groundwater anywhere from decades to hundreds of millions of years. Just look at the underground caves and sinkholes that form as a result of the dissolution of carbonate rock (see chapter 7). Indeed, the chemistry of the rock itself may change. A common transformation is limestone ($CaCO_3$) to dolomite ($CaMg[CO_3]_2$) whereby magnesium has become incorporated into the crystal structure. As a result, some of the carbonate in the Florida Platform has been altered from limestone to dolomite—but that is another story.

The Evaporites

Not everything deposited on a carbonate platform is necessarily carbonate. With the Florida-Bahama Platform being ~1,000 km wide, seawater circulation at mid-platform became very sluggish, evaporation began to dominate, salinities became elevated, and eventually *evaporite minerals* formed, such as anhydrite ($CaSO_4$) and gypsum ($CaSO_4 \cdot 2H_2O$). These are not carbonates but *sulfates*. Since they formed by evaporation, causing the seawater to become concentrated in all *cations* and *anions* found in normal seawater, these minerals are classified as evaporites. Such minerals are found by deep drilling into the subsurface of the Florida Platform, which appears not to have any halite (NaCl, a chloride or plain table salt)—just the sulfates. The halite dominates the deep Gulf of Mexico.

Volumetrically, the evaporites form only a tiny portion of the total mass of rock *on the platform*. As we shall see in chapter 5, the dissolution of these evaporites played a key role in shaping the modern Florida Platform and in forming a strange and unusual benthic community in very deep water adjacent to the base of the platform. Fig. 4.8 illustrates what modern and ancient evaporites look like.

Top: Figure 4.8*A*. Halite (NaCl) being precipitated in Salt Lake in the Coorong region of south Australia. This is an excellent example of evaporite minerals being deposited. Such environments were common in restricted areas on the Florida Platform during the Mesozoic, where there was an aperiodic influx of saltwater that underwent evaporation. (Courtesy of Noel James.)

Middle: Figure 4.8*B*. Outcrop of gypsum from southern Spain. The gypsum was formed in the Late Miocene in the Mediterranean Sea when narrow gateways (like the modern Straits of Gibraltar) leading to the eastern Atlantic Ocean repeatedly opened and closed, allowing saltwater to enter the deep basin when open and evaporate, leaving behind evaporite minerals—primarily gypsum (not NaCl—halite) in this case when the gateway closed. Due to active tectonic activity, these evaporite deposits (actually rocks) were uplifted and now cap the higher elevation in southern Spain due to their resistance to erosion. Such rocks occur at depth beneath the Florida Platform. (Photo by A. C. Hine.)

Bottom: Figure 4.8*C*. Impermeable gypsum crystals "locked together" in a "swallow tail" structure. (Photo by A. C. Hine.)

Building the Florida Carbonate Platform: Going Vertical

Now that we know how carbonate (and some evaporite) sediments form and are distributed *laterally* across a platform, how did the great thickness of these shallow-water sediments accumulate to form carbonate rock up to 6 km thick beneath Florida and 14 km in places beneath the Bahamas?

As carbonate sediments were deposited on the basement surface, the entire margin of the newly formed North American continental margin was undergoing rapid *tectonic subsidence*. Newly rifted continental margins (10s Myr) subside much more quickly than mature, rifted margins (>100s Myr) as a result of early rapid cooling and thermal contraction of the crust (fig. 4.9). As the new ocean basin widens due to seafloor spreading, the margins move further and further away from the heat source at the seafloor-spreading center—the mid-ocean ridge. With the source heat diminishing, the margins subside less rapidly due to heat loss and subsequent contraction.

Subsidence creates space for sediments to accumulate, providing an increased *lithostatic load* (weight from sediments) on the margin. With sea level

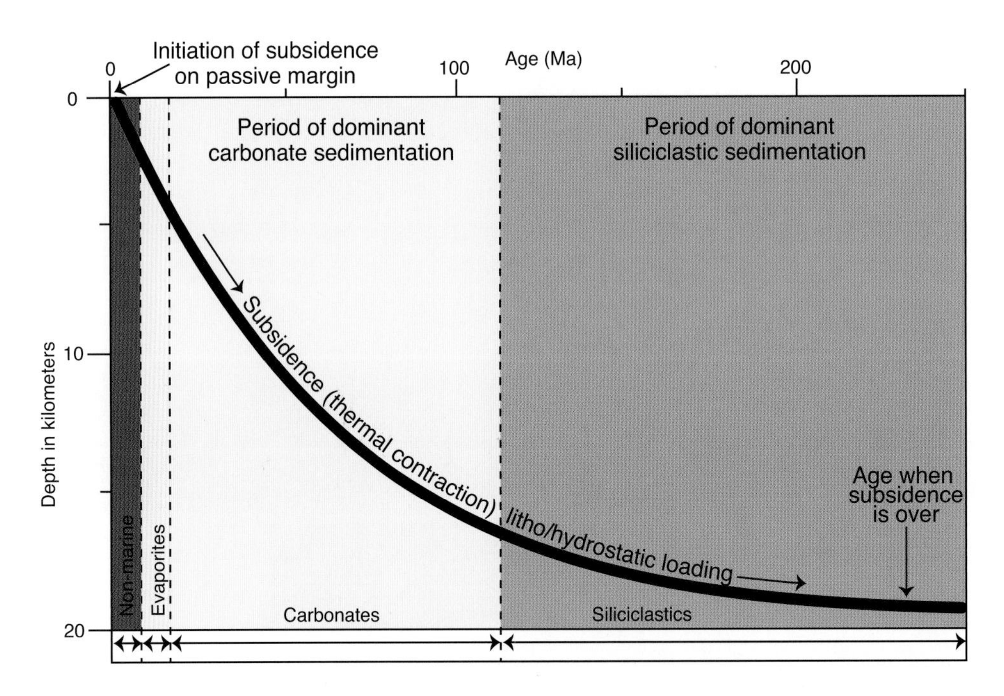

Figure 4.9. The exponentially slowing rate of subsidence of a rifted continental margin occurs as an ocean basin widens. As the source of heat associated with the spreading center becomes more distant, the margin contracts and subsides as a result. During the early period of ocean basin formation, the most rapid subsidence occurs, allowing thick carbonate sediments to accumulate, causing further subsidence due to the added weight of rocks and seawater—lithostatic and hydrostatic loading. Note the sequence of nonmarine sediments, evaporites, thick carbonates, and finally siliciclastic sedimentation. Total thickness beneath the Mississippi River or the Ganges-Brahmaputra River is >20 km, indicating that such subsidence is possible. (Modified from Bally et al. 1981.)

continuing to rise during this period of time (Late Jurassic/Early Cretaceous), increased *hydrostatic loading* (weight from water) occurred as well. The lithostatic and hydrostatic loading, in addition to the rapid heat loss from the widening ocean basin, enhanced subsidence, resulting in a thick, early phase of carbonate sediment accumulation given a healthy carbonate factory. With time, the rate of subsidence decreased exponentially. As a result, the Jurassic and lower Cretaceous rocks beneath the eastern portion of the Florida-Bahama Platform are much thicker than the younger, overlying rocks because of this early, rapid, exponentially high rate of subsidence (figs. 1.3*B*, 4.1, 4.9).

Much of the Florida Platform formed very early and very quickly in its geologic development. Once subsidence began to slow and the rate decreased exponentially with time, the rate of sediment (ultimately rock) accumulation slowed as well. The tectonic subsidence, where Earth's crust moved downward, created space so that carbonate sediments could continue to accumulate. This is how great thicknesses of carbonate rocks could form with the water never having been more than a few dozen meters deep. As will be shown in chapter 6, if combined subsidence and sea level rise are too rapid, the shallow-water carbonate factory is drowned and sediment production and accumulation slow considerably.

Added to the rapid tectonic subsidence is an extended period of sea level rise and highstand. The extended sea level highstand during the Cretaceous greenhouse Earth and its prolonged warmth were essential to the early formation of a carbonate "gigaplatform," a huge carbonate platform (figs. 4.10, 4.11). Tectonic subsidence and long-term rising sea level (see page ii, sea level curve from ~200 Ma to ~100 Ma) allowed for a huge carbonate platform to form in the Yucatan, Florida, Bahamas area. This platform system extended all the way up to present-day Nova Scotia, forming a carbonate platform 6,000 km long, up to 1,000 km wide, and eventually up to 14 km thick—probably the largest carbonate platform complex ever on Earth. Fig. 4.11 shows the extent of the Late Jurassic/Early Cretaceous carbonate gigaplatform as compared with the modern Great Barrier Reef off Australia. By Late Cretaceous, however, most of this gigaplatform had been buried by siliciclastic sediments washed off North America through rivers, leaving only Florida, the Bahamas, and the Yucatan as active carbonate sediment–accumulating areas—still a very large system, but not nearly as large as the gigabank during its maximum extent earlier in geologic time.

This platform-building zenith appeared about ~140 Ma at the end of the Late Jurassic and into the Early Cretaceous. Eventually, the platforms were terminated by environmental deterioration and by burial resulting from sediments carried off the continents by rivers to the ever-widening North Atlantic Ocean. These Mesozoic carbonate platforms now lie several kilometers in the subsurface beneath the present-day siliciclastic-dominated continental shelves

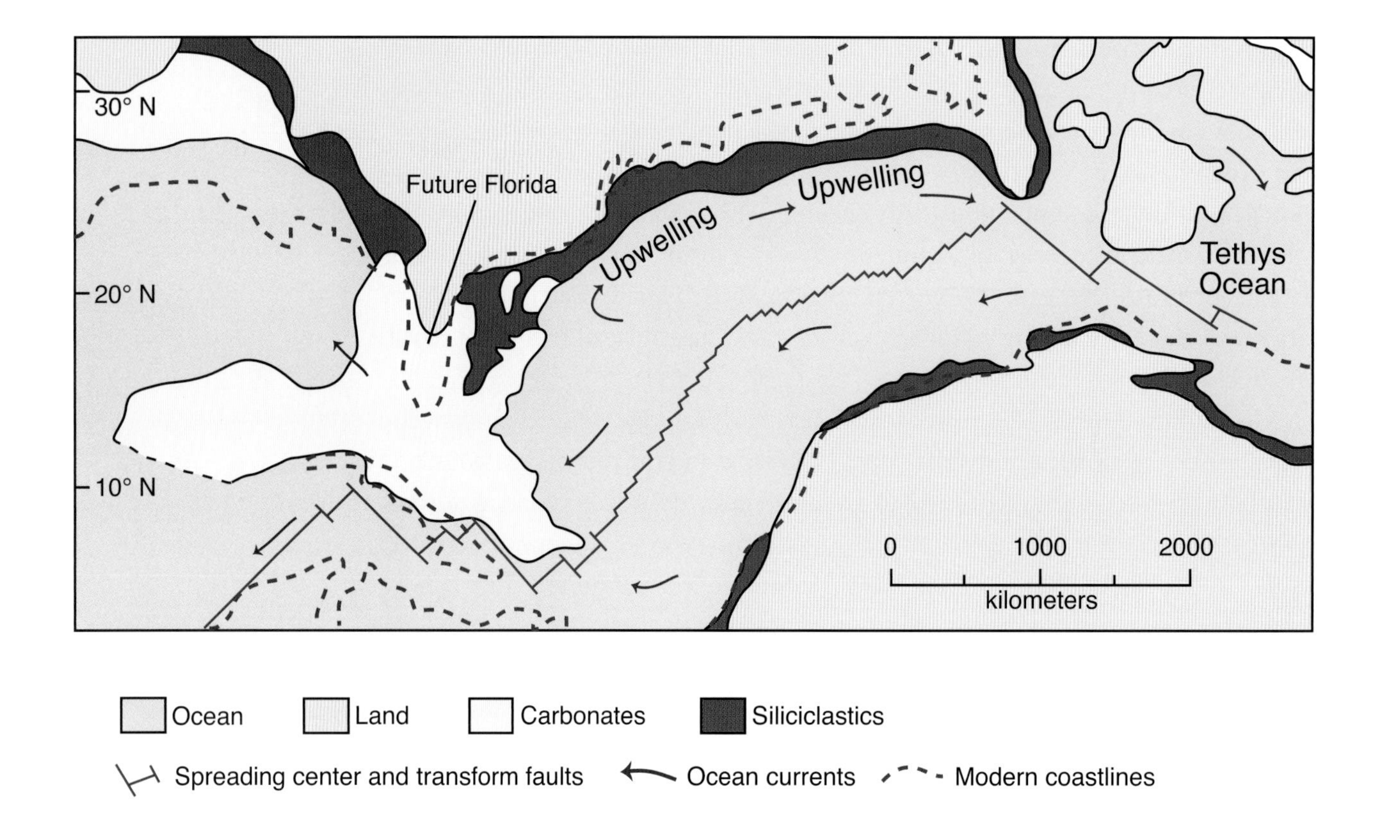

Above: Figure 4.10. The carbonate platform complex along the east coast and Gulf of Mexico coast during the combined Jurassic and Cretaceous was one of the largest carbonate accumulation areas on Earth. Note that the Florida-Bahama Platform was the widest and most extensive single component of this complex. The center of this platform was ~1,000 km from the open ocean, preventing vigorous seawater exchange and allowing for evaporites to form occasionally. (Source: Poag 1991; used by permission from Elsevier.)

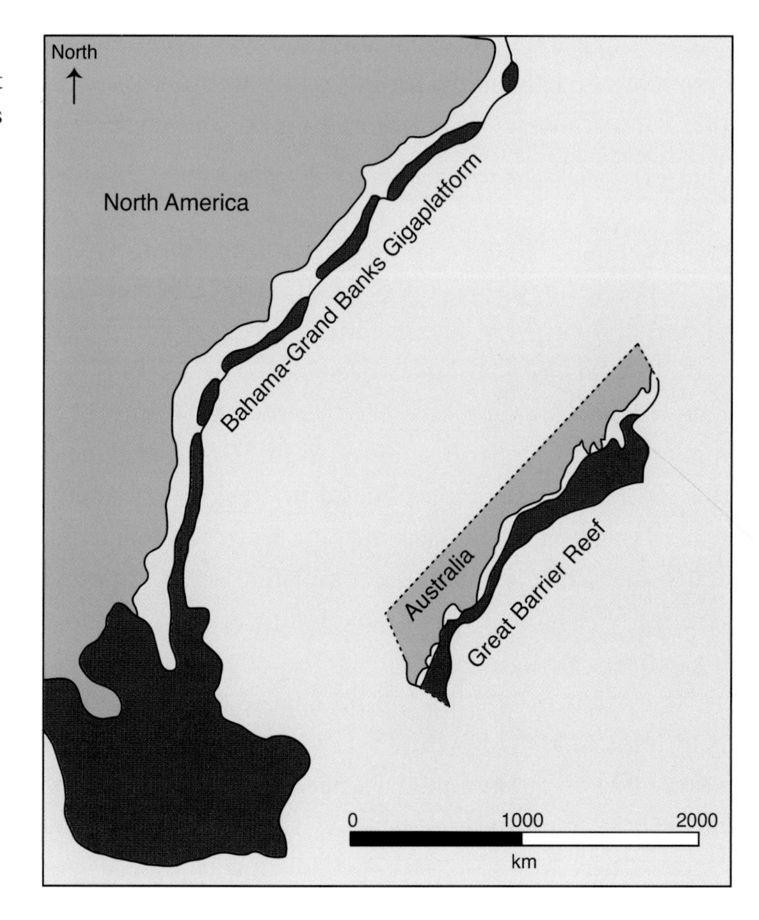

Right: Figure 4.11. The extent of the Late Jurassic/Early Cretaceous carbonate gigaplatform of Florida and the Bahamas extending up the U.S. East Coast as compared with the modern Great Barrier Reef off Australia. (Modified from Poag 1991; previously used in Hine 1997.)

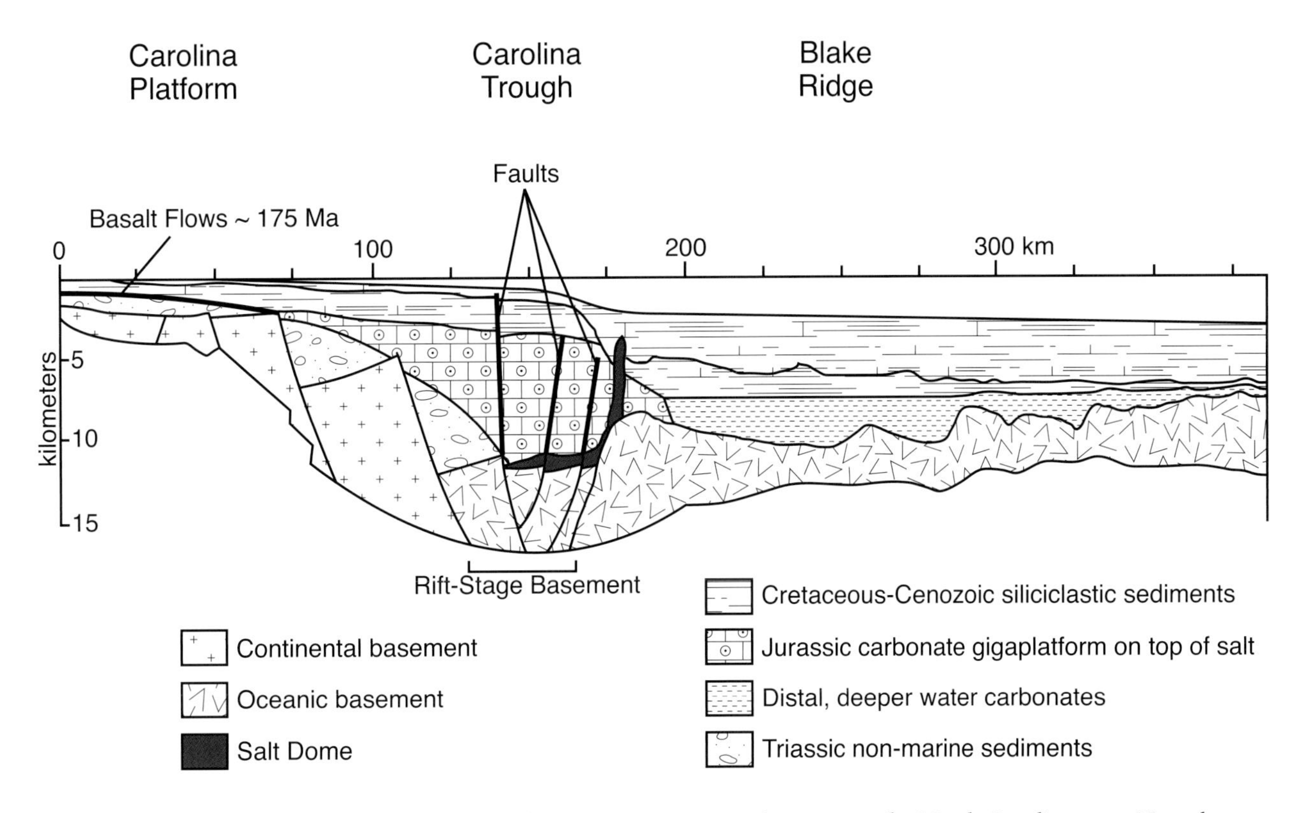

Figure 4.12. Cross section over the eastern North American continental margin in the North Carolina area. Note the buried carbonate platform that was once part of the contiguous carbonate gigaplatform that extended from the Gulf of Mexico to Nova Scotia (identified by "brick" pattern). Most of this huge carbonate platform complex was buried by rivers transporting sediments off North America in the middle to late Mesozoic. This left the Yucatan, Florida, and the Bahamas as the remaining active portion of this once huge carbonate sediment accumulating system. (Source: Dillon and Popenoe 1988; used by permission from the Geological Society of America.)

(fig. 4.12). But the south Florida-Bahama-Yucatan carbonate platform system has persisted to the present day.

In the end, the combination of (1) a laterally extensive, warm, clear-water marine environment, (2) persistent subsidence, and (3) long-term sea level rise during the period of rapid subsidence formed a thick stratigraphic succession of carbonate sediments, which eventually became carbonate rock through the process of cementation. A complex history of multiple sea level highstands and lowstands has created different types of carbonate rock formations lying on top of one another—commonly being separated by erosional surfaces known as unconformities. The many geologic formations within the Florida Platform define this three-dimensional, lithologic variability and heterogeneity. So, instead of being one solid mass of homogeneous carbonate rock, the stratigraphy of the Florida Platform consists of many formations, each identifiable and defined by different carbonate sedimentary facies and less common evaporites.

Finally, the development of the south Florida carbonate stratigraphic succession extended into the Cenozoic and, of course, to modern times as seen in the Florida Keys and the outer portion of the west Florida margin. Because the Cenozoic was a period of slower overall regional subsidence and actually falling sea level, the carbonate formations created during this time were much thinner than their Mesozoic counterparts. And, as will be shown in chapter 8, much of Florida became buried by the quartz-rich influx from the southeastern United States, eventually restricting carbonate sedimentation further to the south along peninsular Florida. As mentioned earlier, carbonate sedimentation today is restricted to south and southwestern Florida.

As this thick "carbonate cake" was being formed, a number of geologic events significantly altered the carbonate sedimentary regime—drowning, tectonic collision, widespread dissolution and erosion, climate change, sediment influx from the noncarbonate terrain of North America, and changing oceanographic regimes. Such changes altering the Florida Platform will be presented in the coming chapters.

Essential Points to Know

1. To build a carbonate platform, a relatively shallow-water substratum bathed by warm, clear tropical waters is needed. Such an environment cannot be dominated by continental river runoff introducing suspended sediments and nutrients that decrease water clarity, thus reducing sunlight penetration and photosynthesis. The mostly Paleozoic/Mesozoic igneous and metamorphic rocks constituting the Florida basement provided such a substrate to initiate the Florida Platform.

2. The carbonate factory is a suite of warm, clear, shallow-water environments in which microbes, plants, and animals produced a variety of carbonate sediments. Distinct mixtures or suites of sediments are called facies and form in distinct sedimentary or depositional environments such as lagoons, beaches, and tidal flats.

3. As sea level fluctuates, these depositional environments move laterally and vertically depending on how they respond to changing water depth. In this manner, sediments from different facies may be deposited on top of each other, producing variability in a vertical section.

4. As soon as carbonate sediments are deposited, water enriched in dissolved calcium carbonate may migrate through the pore spaces or interstices between the grains. This stimulates the precipitation of carbonate cements (minerals), which bind the sedimentary grains together, forming carbonate rock or limestone. Later, the limestone may become dolomite due to the addition of magnesium into the carbonate crystal structure from groundwater or evaporated seawater.

5. Over millions of years, thick accumulations of carbonates may form due to tectonic subsidence—a sinking of the Earth's crust due to cooling and the weight added to it by both the sediment and water loading. Some carbonate platforms may be as much as 14 km thick, consisting of carbonate rocks formed from sediments originally deposited in shallow water.

6. As Pangea split apart and new seafloor spreading commenced, a circum-global tropical ocean (or seaway) called Tethys formed, reaching its peak during the Cretaceous. A carbonate gigaplatform formed along the eastern margin of North America that included the Yucatan, Florida, and the Bahamas. This giant carbonate platform was part of an even larger complex of platforms that rimmed the Tethys Ocean.

7. Eventually most of these platforms surrounding the Tethys Ocean became buried (or uplifted by plate tectonics). The gigaplatform off the east coast of North America and rimming the Gulf of Mexico was buried. Now only the Yucatan (Campeche Bank), southern Florida, and the Bahamas remain as active carbonate sediment producers.

8. The Florida Platform has undergone significant altering events almost from its initiation. The remaining chapters in this book address these major platform-changing events.

Essential Terms to Know

anion: An ion is an atom or molecule where the total number of electrons is not equal to the total number of protons, giving it a net positive or negative electrical charge. An anion is an ion with more electrons than protons, giving it a net negative charge (since electrons are negatively charged and protons are positively charged).

cation: An ion with more protons than electrons.

cementation: Process by which chemically enriched interstitial water (fresh or salt) precipitates minerals such as calcite or aragonite in spaces between sedimentary grains; the resulting cements bind the grains together, forming a sedimentary rock.

cyanobacteria: A phylum of bacteria that obtain their energy through photosynthesis. They are a significant component of the marine nitrogen cycle and an important primary producer in many areas of the ocean. Stromatolites of fossilized oxygen-producing cyanobacteria have been found beginning at ~2.8 Ga. Cyanobacteria conduct oxygenic photosynthesis, converting the early reducing atmosphere into an oxidizing one, which dramatically changed the composition of life forms on Earth by allowing an explosion of biodiversity.

evaporites: Sedimentary rocks consisting of water-soluble minerals that result from the partial or complete evaporation of bodies of water. As evaporation occurs, the remaining water is enriched in salts, and minerals precipitate when the water becomes oversaturated. Evaporite minerals include the chlorides, sulfates, nitrates, some carbonates, and other unusual minerals.

facies or *sedimentary facies*: A body of rock or sediments with specific characteristics that form under certain conditions of sedimentation, reflecting particular processes or environments.

fecal pellet: Excrement of invertebrates occurring in marine deposits and as fossils in sedimentary rocks. They commonly lithify and form carbonate sand grains.

microbe: An organism that is microscopic. The study of microorganisms is called microbiology; microbial is the adjective form.

ooids: Small (<2 mm), spheroidal, "coated" (layered) sedimentary grains, usually composed of calcium carbonate, but sometimes made up of iron- or phosphate-based minerals. Ooids usually form on the sea floor, most commonly in shallow tropical seas. These ooid grains can be cemented together to form a sedimentary rock called an oolitic (containing ~<25 percent ooids) or ooid grainstone (>25 percent ooids).

photosynthesis: A process that converts carbon dioxide and water into organic compounds using the energy from sunlight. Photosynthesis occurs in plants, algae, and cyanobacteria. Photosynthetic organisms create their own food. In plants, algae and cyanobacteria photosynthesis uses carbon dioxide and water, releasing oxygen as a waste product. Photosynthesis is vital for life on Earth. It maintains the normal level of oxygen in the atmosphere, and nearly all life depends on it either directly as a source of energy or indirectly as the ultimate source of the energy in their food.

rudists: A group of generally cone-shaped marine heterodont bivalves that arose during the Late Jurassic and became so diverse during the Cretaceous that they were the major reef-building organisms.

stratigraphy/stratigraphic unit: A layer of sediments or sedimentary rock that can be defined by distinctive features, thus distinguishing it from sedimentary layers above, below, or laterally.

stromatolites: Layered, accretionary structures formed in shallow water by the trapping, binding, and cementation of sedimentary grains by biofilms of microorganisms, especially cyanobacteria (commonly known as blue-green algae). They include some of the most ancient records of life on Earth.

sulfates: Minerals containing the sulfate anion (SO_4^{2-}), most commonly gypsum ($CaSO_4 \cdot 2H_2O$) and anhydrite ($CaSO_4$).

tectonic subsidence: Sinking of Earth's crust due to forces transmitted within it. The space created by the subsidence allows sediments and water to accumulate, providing lithostatic and hydrostatic loading and adding to the rate and magnitude of the subsidence.

unconformity: A buried surface (commonly erosional) separating two rock masses or strata of different ages, indicating that sediment deposition was not continuous. In general, the older layer was exposed to erosion for an interval of time before deposition of the younger, but the term is used to describe recognizable significant breaks in the sedimentary geologic record.

Keywords

Carbonate factory, sedimentary facies, stratigraphy, photosynthesis, tectonic subsidence, evaporites, rudists, gigabank

Essential References to Know

Bally, A. W., A. B. Watts, J. A. Grow, W. Manspeizer, D. Bernoulli, C. Schreiber, and J. M. Hunt. *Geology of Passive Continental Margins: History, Structure, and Sedimentologic Record (with Special Emphasis on the Atlantic Margin) for the AAPG Eastern Section Meeting and Atlantic Margin Energy Conference Education Course Note Series #19.* Tulsa: American Association of Petroleum Geologists, 1981.

Bathurst, R.G.C. *Carbonate Sediments and Their Diagenesis.* 2nd ed. Amsterdam: Elsevier Scientific, 1975.

Bosence, D. W. J., and R.C.L. Wilson. "Carbonate Depositional Systems." In *The Sedimentary Record of Sea Level Change*, ed. A. L. Coe, 209–33. Cambridge: Cambridge University Press, 2003.

Dillon, W. T., and P. Popenoe. "The Blake Plateau Basin and Carolina Trough." In *The Atlantic Continental Margin: U.S.*, ed. R. E. Sheridan and J. A. Grow, The Geology of North America, 291–328. Boulder: Geological Society of America, 1988.

Halley, R. B., P. M. Harris, and A. C. Hine. "Bank Margin." In *Carbonate Depositional Environments*, ed. P. A. Scholle, D. G. Bebout, and C. H. Moore, AAPG Memoir, 463–506. Tulsa: American Association of Petroleum Geologists, 1983.

James, N. P. "Reef Environment." In *Carbonate Depositional Environments*, ed. P. A. Scholle, D. G. Bebout, and C. H. Moore, AAPG Memoir, 346–453. Tulsa: American Association of Petroleum Geologists, 1983.

Jones, B. "Warm-Water Neritic Carbonates." In *Facies Models 4*, ed. N. P. James and R. W. Dalrymple. St. Johns, Newfoundland: Geological Association of Canada, 2010.

Klitgord, K. D., and D. R. Hutchinson. "U.S. Atlantic Continental Margin: Structural and Tectonic Framework." In *The Atlantic Continental Margin, U.S.*, ed. R. E. Sheridan and J. A. Grow, The Geology of North America, 19–55. Boulder: Geological Society of America, 1988.

Poag, W. C. "Rise and Demise of the Bahama–Grand Banks Gigaplatform, Northern Margin of the Jurassic Proto-Atlantic Seaway." *Marine Geology* 102, no. 1–4 (1991): 63–130. doi: 10.1016/0025-3227(91)90006-p.

Randazzo, A. F., and D. S. Jones, eds. *The Geology of Florida.* Gainesville: University Press of Florida, 1997.

Redfern, R. *Origins: The Evolution of Continents, Oceans, and Life.* Norman: University of Oklahoma Press by special arrangement with Cassell and Co., UK, 2001.

Schopf, J. W. *Cradle of Life: The Discovery of Earth's Earliest Fossils.* Princeton, NJ: Princeton University Press, 1999.

Scoffin, T. P. *An Introduction to Carbonate Sediments and Rocks.* New York: Chapman and Hall and Methuen, 1987.

Tucker, M. E., and P. V. Wright. *Carbonate Sedimentology.* Oxford: Blackwell Science, 1990.

5

An Environmental Crisis

Drowning of the West Florida Margin and Development of the West Florida Escarpment (~100 Ma to ~80 Ma)

> We must commence by separating from the various other changes which affect the level of the stand, those which take place at an approximately equal height, whether in a positive or negative direction, over the whole globe; this group we will distinguish as eustatic movements.
>
> Austrian geologist Eduard Suess in 1888 coining the term *eustatic* meaning worldwide changes in sea level

Things Constantly Change

As the Florida Platform was being constructed (see chapter 4) through extensive and long-term carbonate sediment production, other geologic processes began to alter it almost immediately. Such is the nature of geology—as soon as features start to form, other forces begin to change them—a never-ending process. Mountains are elevated by various *internal* processes. As they are being built, *surface* processes, such as weathering and glacial ice movement, begin to alter and shape them. So it is with carbonate platforms, except that the altering processes are quite different.

The first of these altering events for the Florida Platform was the failure of the west Florida margin to keep up with subsidence of the crust and sea level rise starting in the Early to mid-Cretaceous (100 Ma–80 Ma). The second event, occurring simultaneously, was the erosion of the west-facing margin of the Florida Platform, forming a steep escarpment and, in some places, a vertical underwater cliff extending ~1.5–2 km from top to bottom. It remains uncertain how or even if the two events are connected—but they significantly and permanently changed the western portion of the Florida Platform.

The Seascape

A distinctive character of the modern Florida Platform is the very wide, gently westward-sloping ramp forming the west Florida shelf and upper slope (see figs. 1.3, 5.1*A*).

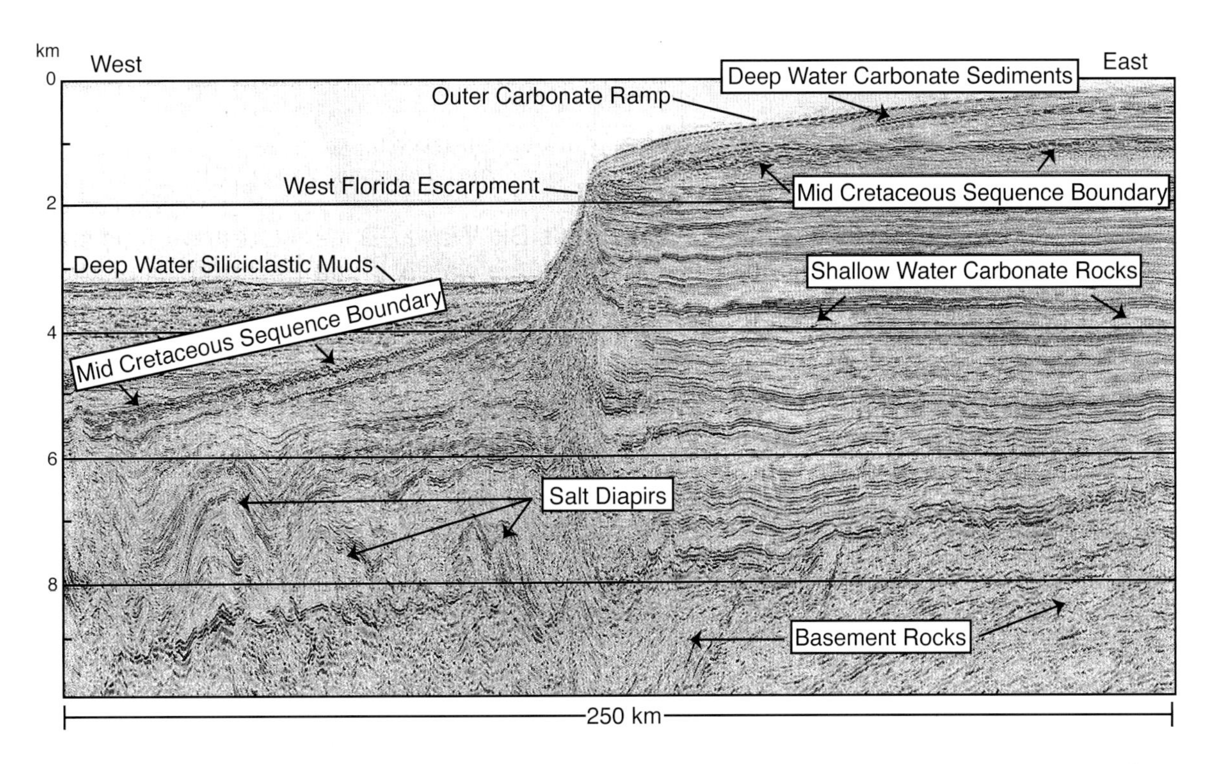

Figure 5.1*A*. The seaward portion of the ramp forming the west margin of the Florida Platform. Note that the steep West Florida Escarpment is about 1 km high at this location. This is a seismic reflection profile obtained by industry geophysical ships at sea and later processed by computers. The image shows deeper structure, such as the presence of salt diapirs, that may be of interest to the oil industry. The deeper sediments that onlap the base of the escarpment came from the Mississippi River. The flat-lying structure of the platform is evident as is the prominent unconformity separating lower Cretaceous rocks from upper Cretaceous rocks—or the Middle Cretaceous Sequence Boundary (MCSB)—a mid-Cretaceous unconformity that is sometimes called the Middle Cretaceous Unconformity (MCU). (Courtesy of Spectrum, Inc., from public promotional literature.)

There is a subtle, break-in-slope at about 75 m water depth, dimly defining the lower gradient shelf (0.2–4 m/km) landward of that depth and a slightly higher gradient slope (6–9 m/km) seaward of that depth. However, this shelf/slope break is not as distinct as that defining the shelf/slope boundary along most other continental margins. The west Florida shelf/slope system extending further seaward terminates at the top of the West Florida Escarpment, having a very steep gradient (500 m/km—45°, and in places nearly vertical) that plunges to ~3.2 km water depth into the deep Gulf of Mexico (figs. 1.9, 5.1*B*). The entire system (shelf/slope/escarpment) is called by geologists a *distally inclined ramp*. This ramp is huge, as it extends nearly 900 km long by 250 km wide and now supports the modern coastal ocean—the shallow-water portion of the eastern Gulf of Mexico lying immediately adjacent to the west coast of Florida. This modern coastal ocean is critically important to the state as it supports a large recreational and commercial fishing industry, features many recreational diving sites, and contains critically important habitats such as coral reefs and rocky hardbottoms.

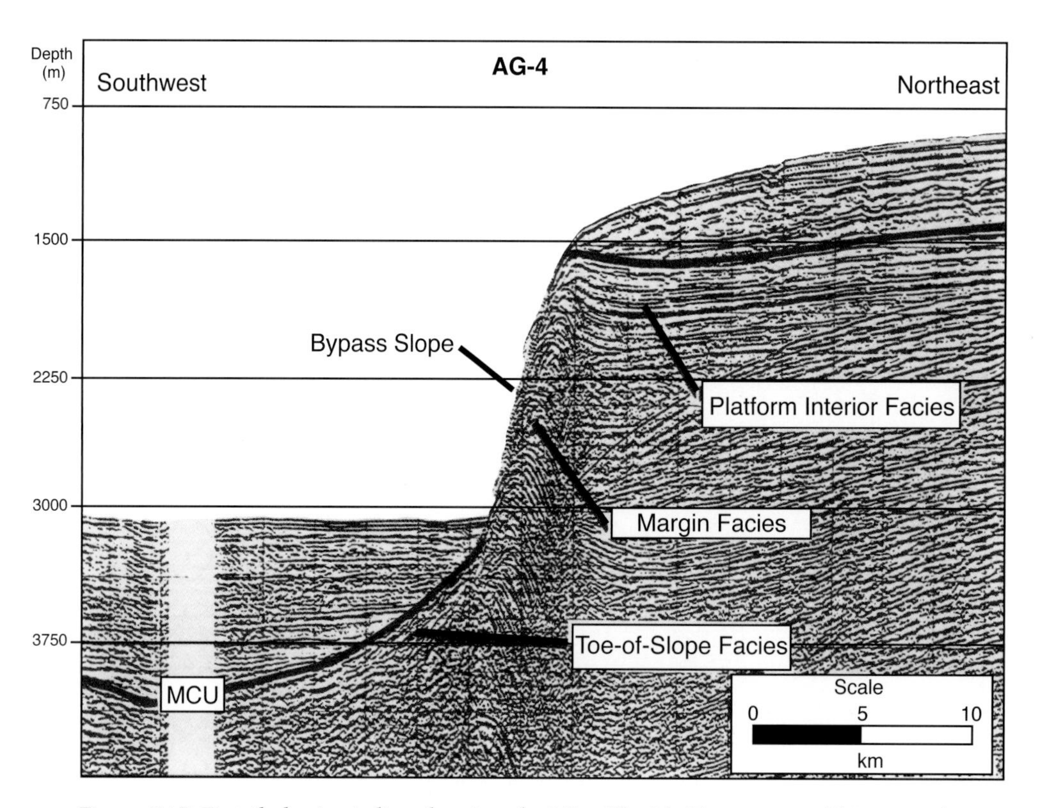

Figure 5.1*B*. Detailed seismic line showing the West Florida Escarpment. This steep slope is vertical in places and may approach ~2 km in relief. The MCU is the same as the MCSB in fig. 5.1*A*. (Source: Corso et al. 1988; reprinted by permission from the American Association of Petroleum Geologists, whose permission is required for further use.)

How did this seascape of a seaward-sloping ramp form, terminating abruptly at a huge drop-off? Why is it not a broad, shallow, flat platform like the Bahamas?

A Special Boundary

Shallow-water carbonate sedimentation began on top of Florida's basement rocks ~160 Ma in the Late Jurassic and lasted up until ~100 Ma in the Early Cretaceous on the western portion of the Florida Platform. Near the end of Early Cretaceous, a wide and very distinctive surface called an *unconformity* formed. Many scientists refer to it as the MCU or MCSB (Mid-Cretaceous Unconformity or mid-Cretaceous *Sequence Boundary*) because of its prominence, extent, and importance. It extends across the Florida-Bahamas Platform and separates the rocks of the lower Cretaceous from those of the upper Cretaceous. This surface, seen clearly in geophysical data, marks an abrupt change from shallow-water carbonate sedimentation to deep-water carbonate sedimentation. Some submarine erosion probably occurred during this abrupt change in depositional systems as well.

The shallow-water depositional environments described in chapter 4 were replaced by a different type of carbonate sedimentation. Instead of the shallow-water *rudistid* reefs and other light-dependent organisms growing at the margin, much finer grained sediments and skeletal debris from deeper-water, pelagic organisms began to accumulate, such as coccolithophorids, *foraminifera,* and *pteropods.*

Why was a highly sediment-productive, shallow-water environment (a few meters deep) changed to one that is now up to 1,800 m deep? How did this happen?

Blame It on Enhanced Seafloor Spreading and Submarine Volcanoes

During much of the Early Cretaceous, very high rates of seafloor spreading worldwide (92 mm/yr; average rates today are ~20–40 mm/yr) produced significant amounts of new ocean crust (fig. 5.2). Additionally, deep *mantle plumes* reached the Earth's surface (seafloor), creating numerous volcanoes from which huge volumes of lava formed large *oceanic plateaus.* Both processes decreased the volume of the global ocean basins—the mid-ocean ridges that accommodated seafloor spreading swelled and enlarged as did the new, large oceanic plateaus, making the global ocean basin volume smaller and displacing seawater higher up onto the continents. Both processes, through volcanic degassing, also enriched the atmosphere with CO_2, a potent greenhouse gas. This elevated CO_2 concentration (maybe ~10 times greater than today's ~390 ppm) created conditions known as *greenhouse Earth.*

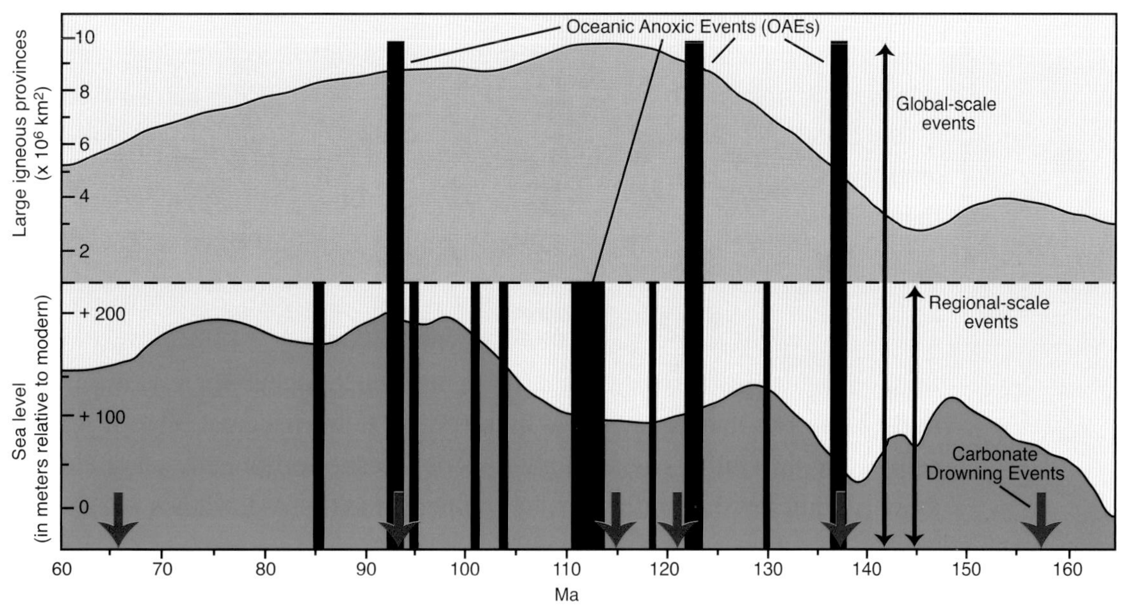

Figure 5.2. The correlation between the high sea level and the expanded ocean crust production in the mid-Cretaceous. The added CO_2 from enhanced submarine volcanic activity created the greenhouse Earth—an extended period of time of unusual warmth, high sea level, and sluggish atmospheric and ocean circulation leading to ocean anoxic events (OAEs). The import of low oxygen water onto shallow carbonate platforms during a period of higher sea level probably retarded or terminated the shallow-water carbonate sediment production factory. (Modified from Ocean Drilling Program planning document.)

suggested that these mid-Cretaceous environmental crises might have caused selective drowning within them. The deep seaways such as Straits of Florida, the Tongue of the Ocean, Exuma Sound, and the NE/NW Providence channels (fig. 4.6*A*) that separate Florida from the Bahamas and segment the Bahama Platform may have started as a result of this environmental stress. As will be shown in the next chapter, significant tectonic activity associated with the collision between the Greater Antilles (Antillean Orogeny) and the passive carbonate margin of the Yucatan-Florida-Bahama Platform was not fully developed until ~56 Ma. But it is possible that early tectonic activity associated with this event could have played a role in early seaway (Straits of Florida) development. Additionally, other scientists have suggested that the deep seaways within the Florida-Bahama Platform were faulted depressions within the basement rock and thus have been deep water environments since the beginning of platform development.

Whatever the process of seaway formation, platform segmentation was an important event. By the mid-Cretaceous, Florida and the northern Bahamas had become separated into distinctly different geologic provinces. Before, they were essentially the same feature—one huge, mega-carbonate platform (fig. 4.10). Afterwards, the Bahamas became and remained isolated from the effects of the North American continent, and the islands are still one of the most important carbonate-producing environments in the world today (chapter 4). Florida, on the other hand, remained attached to North America and began to be influenced by siliciclastic runoff and ground water, particularly starting in the early Cenozoic (see chapters 7 and 8).

Reshaping the West Florida Escarpment

By the mid-Cretaceous, the western margin of the Florida Platform had built vertically, creating several kilometers of relief. This margin probably did not significantly prograde seaward into the Gulf of Mexico if at all during its growth stage (fig. 3.7). As mentioned earlier, the margin supported shallow-water rudistid reefs, not coral reefs, at this point in geologic time. Rudists were bivalves—clam-like mollusks that lived in massive colonies trapping mud and forming a rock framework. Coral reefs forming rigid, rocky framework were not dominant until later in geologic time.

As the margin became higher and higher, it became over-steepened and unstable, and it began to erode back by removing as much as 6 km of rock from its original position (fig. 5.5). Erosion probably removed much of the original rudistid reef that may have dominated the margin construction phase. As a result, huge scalloped embayments or re-entrants now dominate the morphology of the central portion of the West Florida Escarpment (fig. 5.6). These embayments resulted from enormous, destabilized slabs of carbonate rock and sediment,

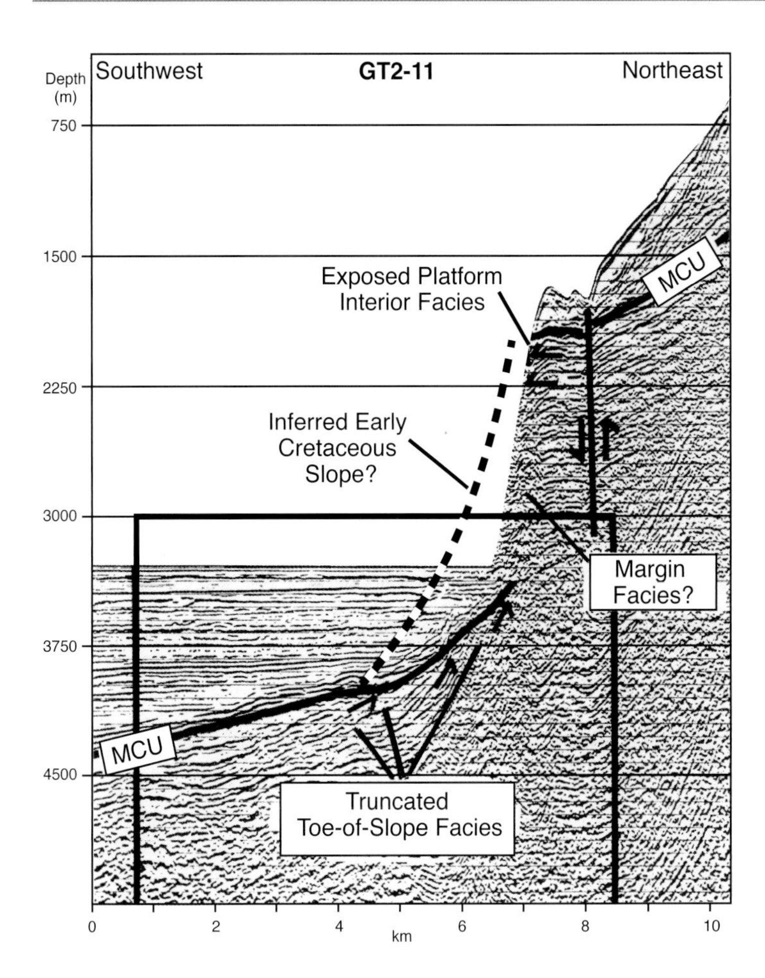

Figure 5.5. Seismic line indicating as much as 6 km of erosion that may have occurred along the West Florida Escarpment, making it much steeper than during its construction phase. The MCU marks the drowning event. (Source: Corso et al. 1988; reprinted by permission from the American Association of Petroleum Geologists, whose permission is required for further use.)

some 30 km long, 15 km wide, and 300 m thick, which slid off into the adjacent Gulf of Mexico forming *sediment gravity flow deposits* (fig. 5.7). These deposits may have extended many tens if not hundreds of kilometers out into the deep Gulf of Mexico. However, there is no direct evidence of massive, thick mass-wasting deposits seen in seismic data nor have any been drilled in the adjacent deep Gulf of Mexico. Quite possibly, these massive slides disintegrated into very broad, very thin sedimentologically chaotic units. They would now be buried beneath finer-grained terrigenous sediments from the Mississippi and the many other rivers that now flow to the Gulf of Mexico (fig. 5.1). Further to the north off Panhandle Florida, the West Florida Escarpment becomes more slope-like (no escarpment), resulting from reduced erosion and increased deposition from the Mississippi River, eventually burying the carbonate platform.

Erosion probably was always part of platform-margin development as a result of over-steepening, causing instability. With sufficient sediments being produced by the shallow-water carbonate factory and exported off into the deeper water, deposition might have balanced sediment removal. However, when the shallow-water carbonate sediment factory disappeared due to drowning, erosional processes probably began to dominate and the margin began to retreat as

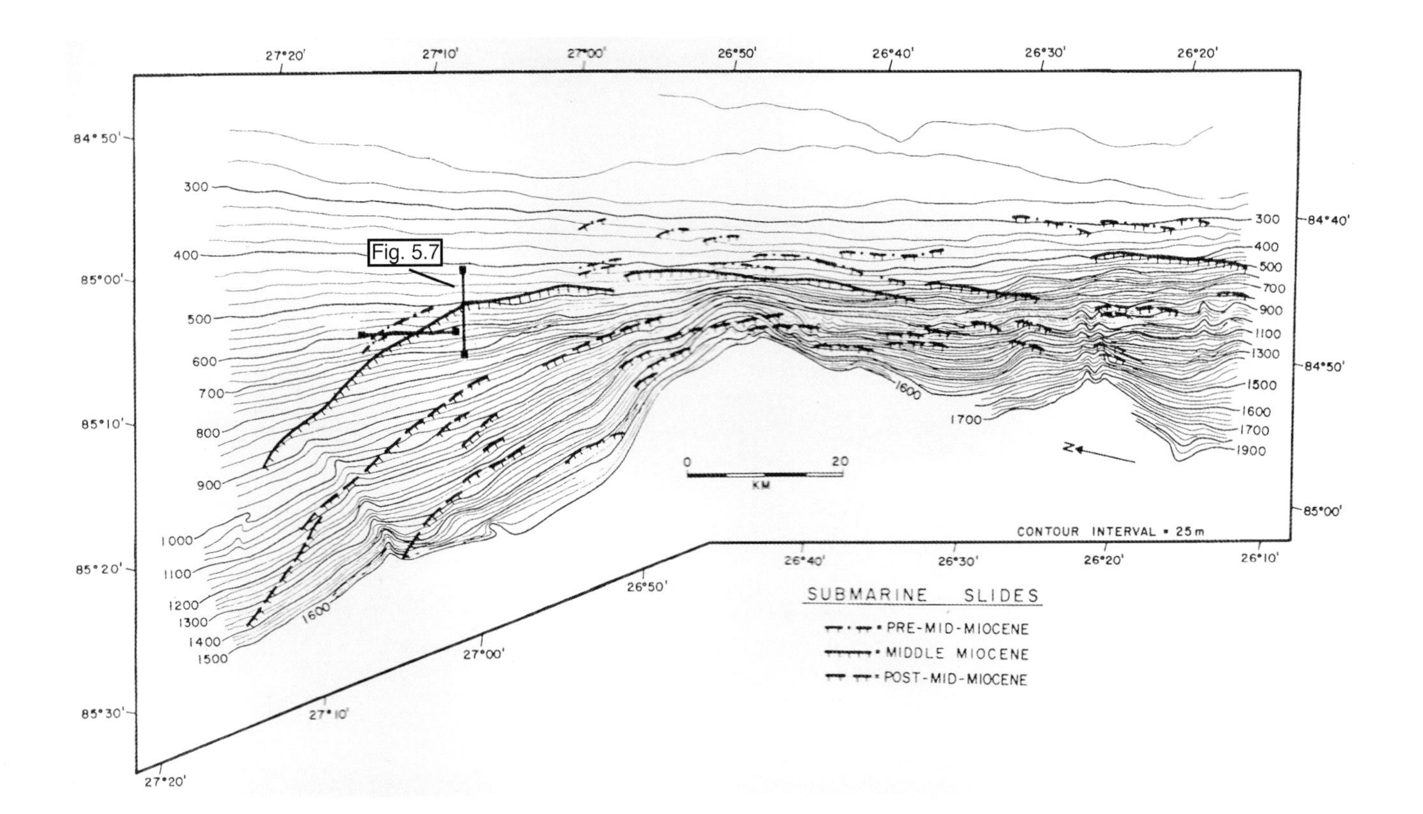

Above: Figure 5.6*A*. Bathymetric map showing large, scalloped embayments along the edge of the West Florida Escarpment due to huge slabs of rock spalling off aperiodically. Map also illustrates locations of buried detachment surfaces indicative of past submarine slides. (Source: Mullins 1986; courtesy of the Colorado School of Mines Press.)

Right: Figure 5.6*B*. Multibeam image of upper portion (down to 1,020 m) of the West Florida Escarpment and west Florida margin. Note steep and highly eroded nature of the escarpment and large scalloped features cut within the margin indicating massive slumping due to sediment gravity processes. (Source: Data collected, processed, and visualized by the Center for Coastal and Ocean Mapping/ Joint Hydrographic Center, University of New Hampshire as part of the U.S. Extended Continental Shelf Mapping Program; courtesy of Dr. Larry Mayer.)

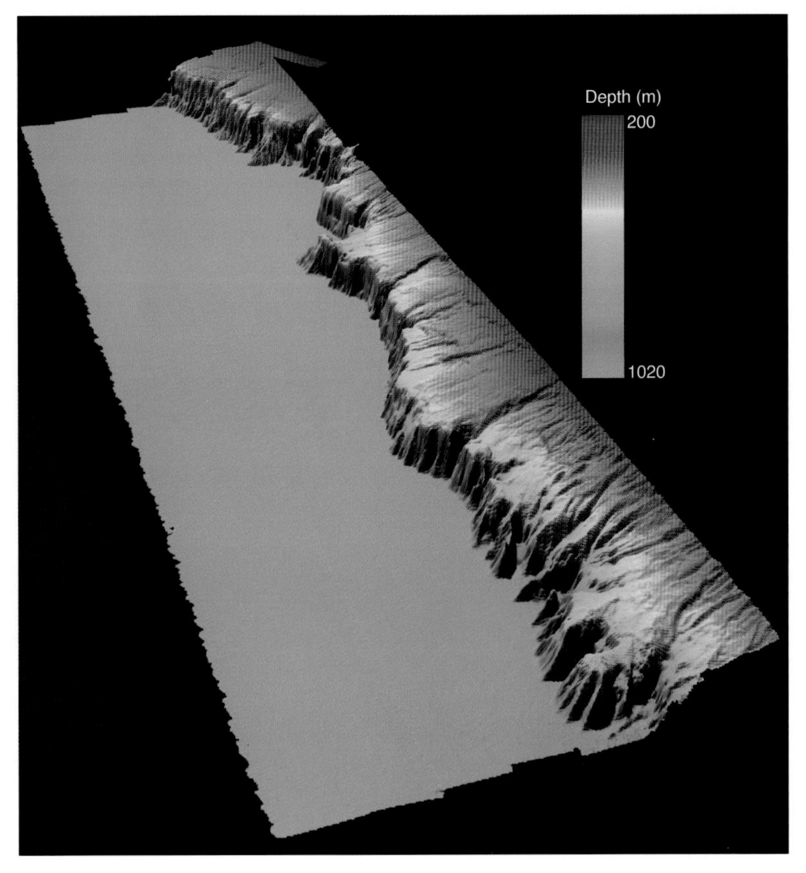

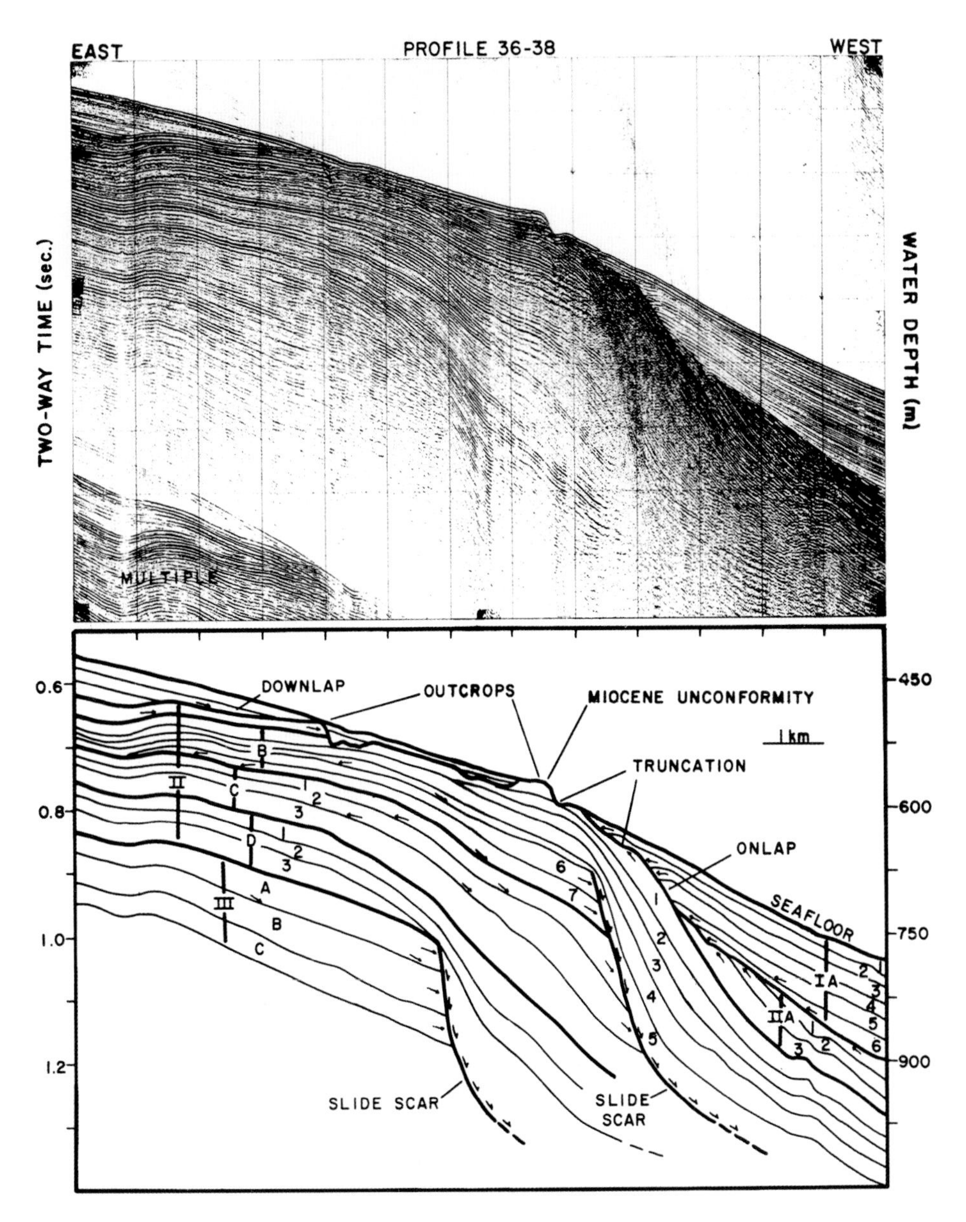

Figure 5.7. Seismic line and line-drawing interpretation of buried detachment surface (slide scar) indicating removal of units IB and II. Note location of line on fig. 5.6*A*. (Source: Mullins 1986; courtesy of Colorado School of Mines Press.)

a result. So perhaps platform-margin drowning and erosional retreat are linked in this manner.

Further to the south, these broad embayments turn into enormous submarine canyons up to 25 km long incised into the margin (fig. 5.8) that are dissimilar to classic submarine canyons found elsewhere such as the Hudson Canyon off New York. The embayments probably only occur along the top of the escarpments whereas the canyons cut all the way to base in some 3.2 km of water. These SW Florida canyons have steep sides but flat floors (not V-shaped), and they do not appear to be exporting sedimentary debris from above and to be cut by physically transporting large amounts of rocky and sedimentary material down through them, forming submarine fans. There are no submarine fans associated with them as seen with canyons, which support vigorous downslope sediment transport.

Chemically enriched brines dissolved from lower Cretaceous gypsum ($CaSO_4 \cdot 2H_2O$) evaporite deposits (see chapter 4) within the platform emanate from the base of the escarpment along the deep portion of the SW Florida Platform margin (fig. 5.9). These fluids carry dissolved sulfides that combine with oxygen in the seawater to become acidic. This acidic seawater dissolves limestone, thus undercutting and over-steepening the escarpment. This over-steepening causes blocks to fall, allowing the margin to erode back even more. There is little evidence of large blocks at the base of the escarpment, indicating that they, too, are dissolved eventually. Where these seeps are concentrated, the dissolution, over-steepening, and erosion are concentrated and a submarine canyon begins to erode back into the margin.

A strange and unusual benthic community exists at the base of the canyon walls and the escarpment. These organisms are similar to *vent-type communities* (bacterial mats, tube worms, clams, shrimp, and other organisms) seen on the mid-ocean ridge, seafloor spreading centers. These communities derive their energy from the same source—chemosynthesis, even though these base-of-escarpment and mid-ocean ridge vent environments could not be more different.

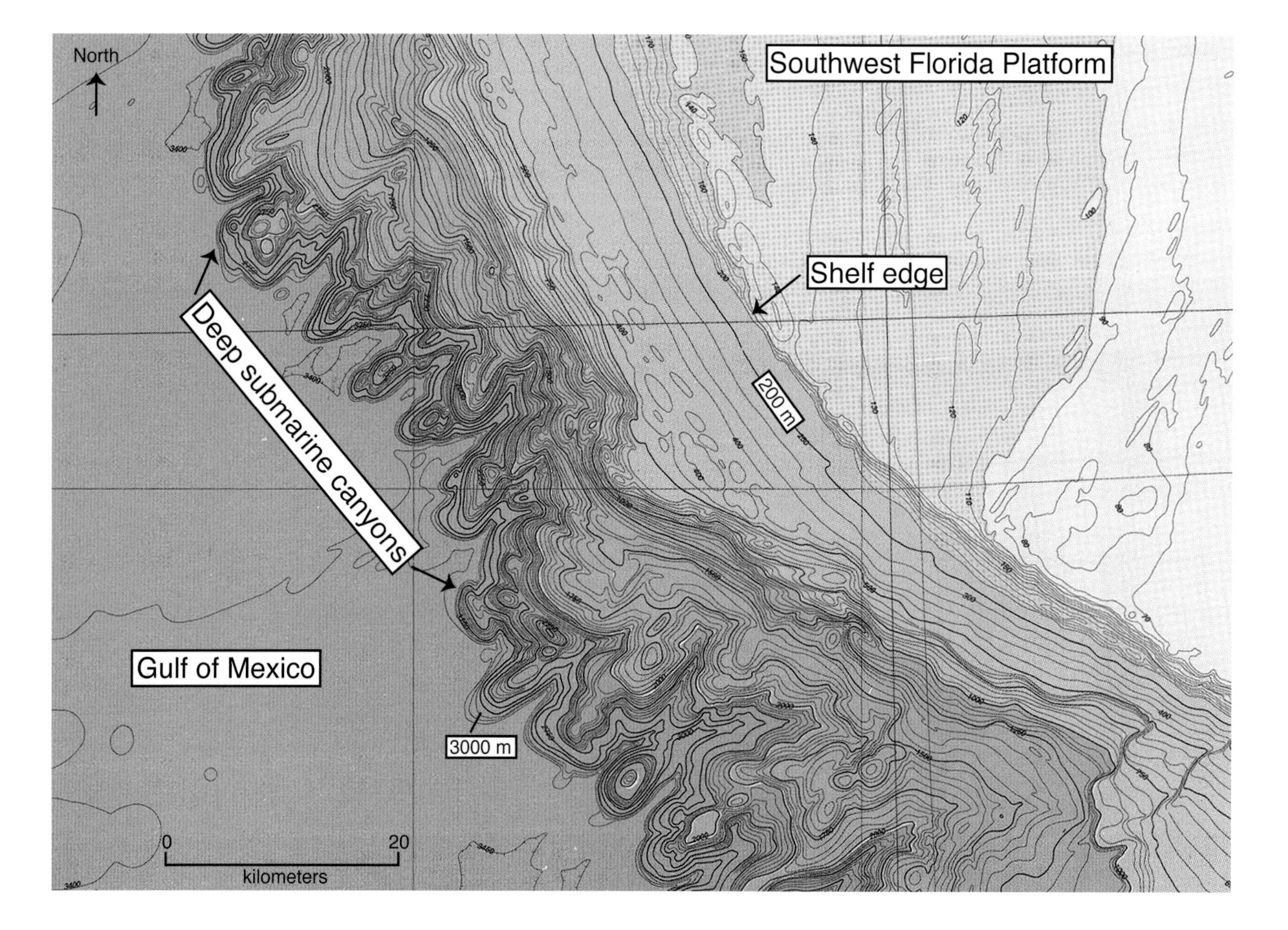

Figure 5.8. Bathymetric map of SW Florida margin illustrating large submarine canyon complex along lower section of West Florida Escarpment. Canyons were formed by limestone being dissolved by acidic fluids emanating from the base of the escarpment. (Source: NOAA National Ocean Survey map.)

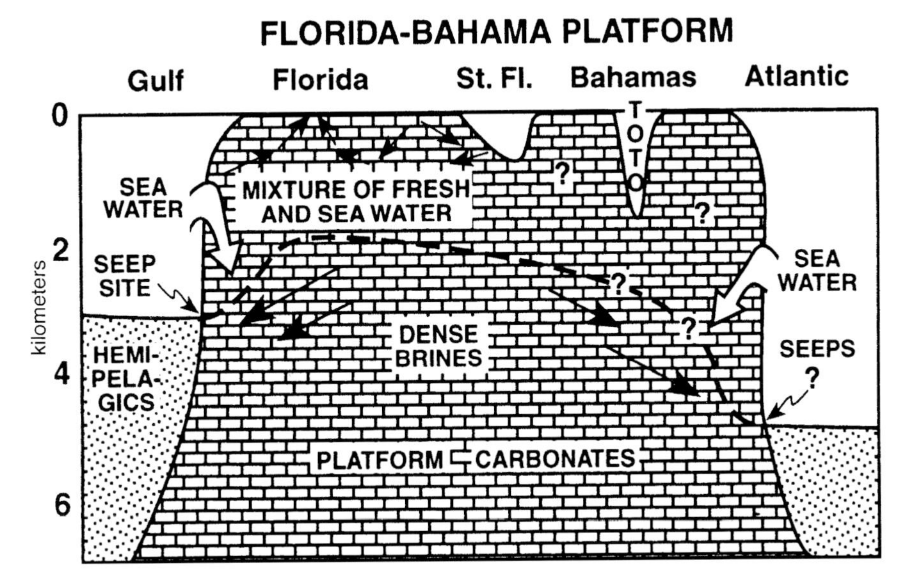

Figure 5.9. Cross section showing brine seepage that led to canyon development along the base of the SW portion of the West Florida Escarpment. (Source: Paull and Neumann 1987; used by permission from the Geological Society of America.)

Finding these vent-type benthic communities in ~3 km water depth at the base of the West Florida Escarpment has been one of the major oceanographic discoveries in Florida in recent decades.

The K/Pg Boundary Mass-Extinction Crisis Nearby Florida

One of the most spectacular geologic events was a ~10 km diameter meteor (or comet) traveling at ~30 km/sec, impacting Earth and releasing 2 million times more energy than the largest nuclear bomb ever detonated (fig. 5.10*A*). The impact formed a crater 180–200 km wide and 40 km deep in the northern Yucatan Peninsula. Remote sensing geophysical data clearly revealed the buried impact crater (fig. 5.10*B*). Additionally, cores (rocks) extracted from a hole drilled years ago into the then undiscovered crater contained shock features formed by meteorite impact.

The ejecta from the Chicxulub Crater, as it is known, created environmental conditions that most scientists now believe caused the mass extinction of plants and animals, including the dinosaurs, at 65.5 Ma. A thin layer of clay found in a number of terrestrial sites around the world and also found buried in ocean sediments is enriched in the element iridium—up to nearly 130 times normal Earth background concentration. This iridium-rich, thin clay layer defines the K/Pg boundary and marks this impact event. The *K* is geologic shorthand for Cretaceous, and *Pg* is shorthand for Paleogene (see Geologic Time Scale on inside cover). The timing of this event at 65.5 Ma is widely accepted. This point in geologic time used to be widely referred to as the K/T boundary. The crater is now completely buried, and only an arcuate line of karst lakes, called *cenotes,* on the northern Yucatan Peninsula indicates its subsurface presence.

Figure 5.12. Outcrop of brecciated, broken rocks of the Cacarajícara Formation in Cuba at a road cut from Soroa to Bahía Honda that probably resulted from submarine landslides destabilized by the impact of the meteorite forming the Chicxulub Crater. Note coin for scale. (Photo by A. C. Hine.)

catastrophic event exists within the Florida Platform, it lies deep in the subsurface unsampled by drilling or by remote sensing techniques. There have been sedimentary deposits located along the Texas coastal plain identified as tsunami deposits associated with this impact.

But the event, as imagined by Dr. Walter Alvarez (1997, 12) (first proponent of the mass extinction of many species by the impact, including the dinosaurs and the well-known *Tyrannosaurus rex*) must have been impressive:

> The impact occurred in the shallow water and coastal plains that flanked the Gulf, but it produced a huge disturbance in the waters of the Gulf through seismic shaking, submarine landslides triggered by the seismic waves, and the splashdown of the ejecta blanket. The result was a gigantic tsunami—a massive wave perhaps a kilometer high, which spread outward across the Gulf of Mexico at terrific speed. As the tsunami front reached the shallow water of Florida and the Gulf Coast, it was pushed higher and higher into a wall of water that towered above the shoreline. As this deluge crashed onto the coast, it not only ripped apart whole forests, but it shook the continental margin so violently that huge volumes of sediment were

mobilized into submarine landslides that flowed down into the deep Gulf, burying the impact debris that had just fallen.

However, as huge and devastating as this event must have been, the nearly complete disappearance by erosion and burial of this huge crater in the past 65.5 Ma clearly illustrates the potent ability of Earth surface processes to remove or heavily modify even the largest features including entire mountain ranges. With no atmosphere or water, ~4.0 Ga old craters are still easily seen on the Moon, Mercury, and other extraterrestrial bodies. On Earth, the Chicxulub Crater has disappeared in about 1.6 percent of the time those ancient lunar features were created.

Essential Points to Know

1. As the Florida Platform was being constructed by accumulating carbonate sediments, it was also being altered by other processes: the deepening of the west margin of the Florida Platform facing the Gulf of Mexico, and erosion of the west Florida margin, creating the steep West Florida Escarpment.

2. The west Florida shelf/slope is best defined as a ramp that terminates in ~1.8 km water depth at the top of the West Florida Escarpment.

3. Within the Florida Platform is a widespread unconformity or sequence boundary formed during the mid-Cretaceous (MCU; MCSB) that separates lower Cretaceous shallow-water carbonates from overlying upper Cretaceous deep-water carbonates. This surface defines a distinct change in sedimentation indicating a drowning event.

4. This drowning event resulted from global warming due to increased volcanic activity leading to elevated CO_2 levels, a greenhouse Earth, and elevated sea level. This greenhouse Earth caused sluggish ocean circulation leading to periods of intense anoxia—oceanic anoxic events (OAEs). With high sea level, these anoxic and nutrient-rich waters were introduced on top of shallow carbonate platforms, reducing their sediment production at many sites worldwide, including the west Florida Platform margin.

5. With reduced carbonate sediment production and accumulation, the west Florida margin fell below the photic zone into darkness and "drowned." Much more slowly accumulating deep-water carbonate sedimentation dominated. Sediment accumulation could not keep pace with persistent, tectonic subsidence. Thus drowning occurred over a relatively short period of geologic time, ~7.5 Myr. The signature "ramp" morphology resulted.

6. The edge of the west Florida margin was once in very shallow water. Now it lies in ~1.8 km of water. It is possible that the drowning event led to the formation of the interior seaways within the Florida-Bahamas Platform such as the Tongue of the Ocean, Exuma Sound, the NE/NW Providence Channels, and

the Straits of Florida. The northern Straits are important in that they separate Florida from the Bahamas, leaving two carbonate platforms that have quite different geologic histories afterwards. The southern Straits probably formed as a result of the collision with Cuba (see chapter 6).

7. As the west Florida margin gained relief, it became over-steepened and unstable, leading to slumping and erosion. With drowning, there was less shallow-water sediment production and transport to offset erosion, and the margin migrated back. The middle portion of the West Florida Escarpment features broad embayments resulting from large slabs of rock and sediment sliding off into the deep Gulf.

8. Along the SW Florida margin, large submarine canyons have incised as much as 25 km into the deeper portion of the escarpment. Instead of physical transport of large volumes of material causing erosion as is seen in the central-west Florida margin, here acidic brines seeping out from the base have dissolved large volumes of rock, causing undercutting leading to erosion. Where these seeps have been concentrated, these canyons have dissolved their way well back into the lower portion of the carbonate platform. The bathymetry of laterally merged canyons is highly complex. The sulfide-rich brines have also provided energy through chemosynthesis to unusual, deep-water biological communities that are found near mid-ocean ridge vents emanating from hot, mineralized waters.

9. The famous K/T (now called K/Pg) boundary event defined by a huge meteorite impact in the Yucatan Peninsula seemed to have had little effect on Florida's geology, even though the Chicxulub Crater, formed by the impact, was a relatively short distance away at 65.5 Ma.

Essential Terms to Know

anoxia: Total absence of oxygen; an extreme form of hypoxia or low oxygen.

cenote: A sinkhole with exposed rocky edges containing groundwater. It is typically found in the Yucatan Peninsula and some nearby Caribbean islands.

debris flow: Fast moving mass of unconsolidated debris that can carry particles ranging in size from clay to boulders. They move downslope under the force of gravity and are one type of sediment gravity flow.

distally steepened ramp: A gently sloping seaward surface that lacks the classic shelf/slope break. The ramp's gradient increases with depth until it reaches a very steep escarpment.

Ferrel cell: A circulation pattern that dominated the mid-latitude atmosphere of the Earth and is responsible for the prevailing westerly winds.

foraminifera: A large group of amoeboid protists. Also called forams. They typically produce a shell with multiple chambers, some becoming quite elaborate in structure. These shells are made of secreted calcium carbonate ($CaCO_3$) or agglutinating sediment particles. They may be planktonic or benthic. They are an important carbonate sediment producer.

greenhouse Earth: An extended period of geologic time (Cretaceous) when the Earth was much warmer than present due to the greenhouse effect of elevated atmospheric CO_2 and other gases. Ice sheets and glaciers were greatly reduced or absent.

Hadley cell: A circulation pattern that dominates the tropical atmosphere, with rising air near the equator, poleward flow 10–15 km above the surface, descending motion in the subtropics, and equator-ward flow near the surface but affected by Coriolis, thus forming the easterly Trade Winds.

mantle plume: An upwelling of abnormally hot rock within the Earth's mantle. As the heads of mantle plumes can partly melt when they reach shallow depths, they are thought to be the cause of volcanic centers known as hot spots and to have caused flood basalts on land and oceanic plateaus in the ocean.

nutrient: A chemical that an organism needs to live and grow or a substance used in an organism's metabolism that must be taken in from its environment. Nutrients generally contain the elements carbon, hydrogen, nitrogen, oxygen, phosphorous, and sulfur.

oceanic plateau: A large, relatively flat submarine region that rises well above the level of the ambient seabed and contains undersea remnants of large igneous provinces. These were formed by the equivalent of continental flood basalts.

olistostrome: A sedimentary deposit composed of a chaotic mass of heterogeneous material, such as blocks of mud, that moves as a semi-fluid body by submarine gravity sliding or slumping of the unconsolidated sediments.

paleoceanography: The study of the history of the oceans in the geologic past with regard to circulation, chemistry, biology, geology, and patterns of sedimentation. It is closely tied to paleoclimatology, the study of past climate on Earth.

photic zone: The depth of the water in the ocean that is exposed to sufficient sunlight for photosynthesis to occur. The depth of the photic zone can be greatly affected by water clarity. It is generally less than 100 m deep.

photosynthesis: A process that converts carbon dioxide and water into organic compounds using the energy from sunlight. Photosynthesis occurs in plants, algae, and many species of bacteria. Photosynthetic organisms create their own food. In plants, algae, and cyanobacteria, photosynthesis uses carbon dioxide and water, releasing oxygen as a waste product. Photosynthesis maintains the normal level of oxygen in the atmosphere, and nearly all life either depends on it directly as a source of energy or indirectly as the ultimate source of the energy in their food.

phytoplankton: The autotrophic (uses photosynthesis) component of the plankton community. The name comes from the Greek word for plant "wanderer" or "drifter." Most phytoplankton are too small to be individually seen with the unaided eye. These plants and algae live in the well-lit surface layer (photic zone) of the ocean. Phytoplankton account for half of all photosynthetic activity. Thus they are responsible for much of the oxygen present in the atmosphere—half of the total amount produced by all plant life. Their cumulative organic matter production (energy fixation in carbon compounds), called primary production, is the basis for the vast majority of oceanic food webs.

primary productivity: Production of organic compounds from carbon dioxide, principally through the process of photosynthesis by algae and plants near or on the ocean's surface.

pteropods: Small (mm to cm size) planktonic marine gastropod mollusks that produce a calcium carbonate skeleton. These skeletons settle and accumulate on the seafloor and may become the dominant sedimentary particle in some deep water environments.

secondary productivity: Amount of the biomass of zooplankton in a marine system.

sediment gravity flow: A broad class of at least four distinct sediment/water slurries driven by gravity. These include grain flows, debris flows, turbidity currents, and fluidized mud flows. The type of fluid/sediment interaction and sediment characteristics (grain size, etc.) define which type dominates. They each leave distinctive sedimentary deposits (e.g., turbidity currents produce turbidites).

sequence boundary: A boundary or buried surface that defines genetically related rock units. Sequence boundaries are identified as significant erosional unconformities and their correlative conformities.

thermocline: A thin but distinct layer in a large body of fluid (e.g., water, such as an ocean or lake, or air, such as an atmosphere), in which temperature changes more rapidly with depth than it does in the layers above or below. In the ocean, the thermocline may be thought of as an invisible blanket that separates the upper mixed layer from the less turbulent deep water below.

turbidites: Deposits from an underwater downslope movement driven by gravity that are responsible for distributing vast amounts of sedimentary material into the deep ocean. Sediments are carried downslope in a water/sediment slurry called a turbidity current. They commonly have a coarser basal unit that becomes finer-grained moving upward.

upwelling: A physical process in the ocean that involves wind-driven motion of dense, cooler, and usually nutrient-rich water toward the ocean surface, replacing the warmer, usually nutrient-depleted surface water.

unconformity: A buried erosion surface separating two rock masses or strata of different ages, indicating that sediment deposition was not continuous. In general, the older layer was exposed to erosion for an interval of time before deposition of the younger, but the term is used to describe any significant break in the sedimentary geologic record. They can form subaerially or underwater.

vent-type community: Consists of animals and plants that are found deep in the ocean near hot water or geothermal vents. Vents are found where the tectonic plates diverge. Molten lava lies deep within the cracks and heats up the water. Such ecosystems are teeming with life and rich in biodiversity. Many species found near the vent communities are not found anywhere else at the bottom of the ocean. Scientists discovered that these vents have hydrogen sulfide coming out of them. The hydrogen sulfide provides nutrients for the plants and animals. Vent animals have strange characteristics. They are giants. An example is the tube worm, which rapidly grows to be nearly 3 m long. Vent clams grow five times faster than regular clams.

western boundary current: Warm, narrow, and fast-flowing currents that form on the west side of ocean basins. They carry warm water from the tropics toward the poles. The Gulf Stream is an excellent example.

zooplankton: Small animals such as jellyfish and krill living in seawater whose complete life cycle is planktonic. They form a critically important food source for larger animals in the ocean. Zooplankton growth is called secondary productivity.

Keywords

Platform drowning, ramp, mid-Cretaceous unconformity/sequence boundary, greenhouse Earth, oceanic anoxic events, photic zone, platform segmentation, K/Pg boundary

Essential References to Know

Alvarez, W. *T. Rex and the Crater of Doom*. Princeton, NJ: Princeton University Press, 1997.

Arthur, M. A., and S. O. Schlanger. "Cretaceous 'Oceanic Anoxic Events' as Causal Factors in Development of Reef-Reservoired Giant Oil Fields." *AAPG Bulletin* 63, no. 6 (1979): 870–85.

Corso, W., R. T. Buffler, and J. A. Austin. "Erosion of the Southern Florida Escarpment." In *Atlas of Seismic Stratigraphy*, ed. A. W. Bally, 149–57. Tulsa: American Association of Petroleum Geologists, 1988.

Corso, W., J. A. Austin Jr., and R. T. Buffler. "The Early Cretaceous Platform Off Northwest Florida: Controls on Morphologic Development of Carbonate Margins." *Marine Geology* 86, no. 1 (1988): 1–14. doi: 10.1016/0025-3227(89)90014-5.

Dillon, W. T., and P. Popenoe. "The Blake Plateau Basin and Carolina Trough." In *The Atlantic Continental Margin, U.S.*, ed. R. E. Sheridan and J. A. Grow, 291–328. Boulder: Geological Society of America, 1988.

Mullins, H. T. "Part 4: Periplatform Carbonates." In *Carbonate Depositional Environments, Modern and Ancient*, ed. J. E. Warme and K. W. Shanley, 1–63. Golden: Colorado School of Mines Press, 1986.

Mullins, H. T., A. F. Gardulski, and A. C. Hine. "Catastrophic Collapse of the West Florida Carbonate Platform Margin." *Geology* 14, no. 2 (1986): 167–70. doi: 10.1130/0091-7613(1986)14<167:ccotwf>2.0.co;2.

Mullins, H. T., A. F. Gardulski, A. C. Hine, A. J. Melillo, S. W. Wise, and J. Applegate. "Three-Dimensional Sedimentary Framework of the Carbonate Ramp Slope of Central West Florida: A Sequential Seismic Stratigraphic Perspective." *Geological Society of America Bulletin* 100, no. 4 (1988): 514–33. doi: 10.1130/0016-7606(1988)100<0514:tdsfot>2.3.co;2.

Paull, C. K., B. Hecker, R. Commeau, R. P. Freeman-Lynde, C. Neumann, W. P. Corsco, S. Golubic, J. E. Hook, E. Sikes, and J. Curray. "Biological Communities at the Florida Escarpment Resemble Hydrothermal Vent Taxa." *Science* 226, no. 4677 (1984): 965–67.

Paull, C. K., and A. C. Neumann. "Continental Margin Brine Seeps: Their Geological Consequences." *Geology* 15, no. 6 (1987): 545–48. doi: 10.1130/0091-7613(1987)15<545:CMBSTG>2.0.CO;2.

Ruddiman, W. F. *Earth's Climate: Past and Future.* 2nd ed. New York: W. H. Freeman, 2008.

Seton, M., C. Gaina, R. D. Müller, and C. Heine. "Mid-Cretaceous Seafloor Spreading Pulse: Fact or Fiction?" *Geology* 37, no. 8 (2009): 687–90. doi: 10.1130/g25624a.1.

Tada, R., M. A. Iturralde-Vinent, T. Matsui, E. Tajika, T. Oji, K. Goto, Y. Nakano, H. Takayama, S. Yamamoto, S. Kiyokawa, K. Toyoda, D. Garcia-Delgado, C. Daz-Otero, and R. Rojas-Consuegra. "K/T Boundary Deposits in the Paleo-Western Caribbean Basin." In *The Circum-Gulf of Mexico and the Caribbean: Hydrocarbon Habitats, Basin Formation, and Plate Tectonics*, ed. C. Bartolini, R. T. Buffler, and J. F. Bilickwede, 582–604. Tulsa: American Association of Petroleum Geologists, 2003.

Takashima, R., H. Nishi, B. T. Huber, and R. M. Leckie. "Greenhouse World and the Mesozoic Ocean." *Oceanography* 19, no. 4 (2006): 82–92. doi: 10.5670/oceanog.2006.07.

6

Clash of Geologic Terrains

Colliding with Cuba (~56 Ma to ~40 Ma)

"One must observe to understand. . . . One must use a rock hammer to understand. . . . The more you understand, the more you see."

Stated to University of South Florida graduate students by Dr. Manuel A. Iturralde-Vinent, noted Cuban geologist, on wisdom required to conduct fieldwork on complex geologic problems in Cuba.

The Scenic Grandeur of Cuba

For several years during spring break, Dr. R. A. Davis Jr. at the University of South Florida and I had the good fortune to take our graduate classes to Cuba for a one-week field trip to study the western third of that 1,200 km long island member of the Greater Antilles chain. This contains the Guaniguanico Terrain and consists of Mesozoic carbonates that were once part of the Yucatan portion of the Yucatan-Florida-Bahama Platform complex (figs. 6.1 and 6.2).

The students who were lucky enough to participate said that it was a "trip of a lifetime." Yes, we drank mojitos, smoked cohíbas, sampled the food, and listened to the local music while walking around old Havana. We also learned the geologic history of Cuba presented to us by Dr. Iturralde-Vinent. But the scenery stole the show for everyone. The geology explains the scenery, and much of Cuban geology is intimately tied to Florida geology.

The topography of western Cuba is unlike anything seen in the United States as rows of ~300 m high limestone hills rise vertically from flat valley floors (fig. 6.2). The valleys contain red clay and sandy soils enriched in iron from African dust that has made its way across the Atlantic Ocean in the atmosphere. Weathered, impure limestone probably adds to some of the color (locally called terra rosa). Tobacco grows well in such soil, to be hand-rolled into the famous Cuban cigars. In great contrast, the mature karst topography forming *mogotes* or rounded, steep-sided hills is not seen in Florida and is found only in a handful of sites worldwide.

These carbonate formations have been uplifted from the Caribbean Sea and exposed to rainfall for ~50 Myr. The weak acid from rain over this time frame has dissolved these rocks to strange, rounded mounds flanked by vertical sides.

So the contrast from the flat Everglades in south Florida to these unusual carbonate formations only 150 km away could not be more striking.

The rocks that formed the mogotes in western Cuba once constituted the southern part of the Yucatan-Florida-Bahama carbonate gigaplatform mentioned in chapter 4. So Cuba and the Greater Antilles are important to the geologic history of Florida in that a component of the ancient Yucatan-Florida-Bahama Platform is now part of Cuba. Where did Cuba come from, and how did it become intimately involved with Florida geology?

To answer that question, we examine the merging two geologic terrains—a collision between a tectonically *active margin* and a *passive margin*—a collision that affected Florida starting at ~56 Ma. By crossing the 150 km from the Florida Keys to the northern coast of Cuba, one crosses into a drastically different geologic province. Whereas the Florida Platform consists of thick, flat-lying or gently dipping carbonate sedimentary rocks, Cuba features volcanic rocks exposed on the surface along with other types of rocks that have been folded and cut apart by thrust faults so complex that the details of the structural geology are still under debate and poorly understood. How can two vastly different landscapes, rocks, and geologic structures be so closely juxtaposed?

The Active Margin: Formation of the Greater Antilles

There are, perhaps, only a few geologic areas that have a history more complicated than that of the Caribbean Basin. Much is still to be debated, and the "devil will always be in the details." There are no eyewitnesses to interview—no photographic record to examine. All we have are the rocks and sediments that were deposited or altered during the formation of the modern Caribbean Sea and our ability to glean information from them. Much of that record is fragmentary and is missing altogether. Additionally, there are contrasting views about the origin and geologic history of the Caribbean Plate, which at one time supported the Greater Antilles along its northern margin. As we will see, Cuba is no longer part of the Caribbean Plate but now belongs to the North American Plate. How did that happen?

As mentioned in an earlier chapter, seafloor spreading forming basaltic ocean crust that separated North America from South America began ~175 Ma and stopped ~84 Ma, creating a ~3,000 km gap that formed the early or proto Caribbean Sea. This gap allowed free exchange of water from the western Atlantic Ocean to the eastern Pacific Ocean. An ancient current, the proto Gulf Stream, actually flowed to the west into the Pacific Ocean between North and South America.

As early as ~130 Ma, along the western entrance to this gap a *volcanic island arc* formed as a result of *subduction* of proto Caribbean Plate oceanic crust beneath the Pacific Plate. This curved line of active volcanoes would form part of the Greater Antilles. So that part of Cuba consisting of *volcanic rocks* was born in the eastern Pacific Ocean and eventually arrived at the doorstep of the southern margin of the Yucatan-Florida-Bahama Platform, according to a widely accepted theory.

A modern example of such a volcanic island arc would be Indonesia, a country that lies on a series of volcanic islands, some of which have produced the largest eruptions and explosions ever witnessed by humans (e.g., Krakatoa in 1883), huge earthquakes, and tsunamis associated with vertical displacement of the seafloor. The December 26, 2004, tsunami off Banda Aceh, Indonesia, resulting from a very large magnitude Sumatra-Andaman earthquake, killed ~230,000 people along the shores of the Indian Ocean due to tsunamis (Japanese word for "harbor wave") up to 30 m high. The Japanese Islands are also a volcanic arc associated with plate subduction, volcanic activity, huge earthquakes, and tsunamis. The enormous earthquake (one of the largest ever recorded—called Tohoku-oki) that occurred on March 11, 2011, produced huge tsunamis (maximum wave height ~40 m) killing some 20,000 people. Perhaps such events occurred during the early stages of Greater Antilles Island formation.

To provide a mental picture of one aspect of the Early to mid-Cretaceous age Greater Antilles, imagine many active volcanoes—some highly eruptive with lava flows, ash beds, huge landslides, and tropical vegetation forming around less active volcanic zones. In the surrounding marine realm, large blocks of rock and sediment moved downslope via huge submarine landslides into the surrounding deep basins—all very similar to what you would see today in Indonesia except that there were no people but probably an abundance of wildlife at that time. The frequent tectonic movement (vertical and horizontal movement of the Earth's crust), earthquakes, and eruptions once (but not now) defined this as an active margin.

From its original location in the eastern Pacific Ocean, the early Greater Antilles volcanic island arc migrated through the wide oceanic gap between the Americas, consuming the existing proto Caribbean/Atlantic ocean crust as a result of continued subduction and movement along transform faults, from ~84 Ma to 36 Ma (fig. 6.3).

With time, a new subduction zone formed to the southwest in the eastern Pacific Ocean that formed a second island arc system that would eventually create the Isthmus of Panama (fig. 6.3*C*). Now with subduction zones and volcanic island arcs along the leading and trailing edge and horizontal motion accommodated by two transform faults, a new, separate, distinct tectonic plate was formed—the Caribbean Plate (fig. 6.3*D*).

As a result of complex plate tectonic activity, the rocks that formed the ocean crust of the proto Caribbean Sea and the sediments that covered it sank into the mantle and were destroyed, only to be replaced by new ocean crust and

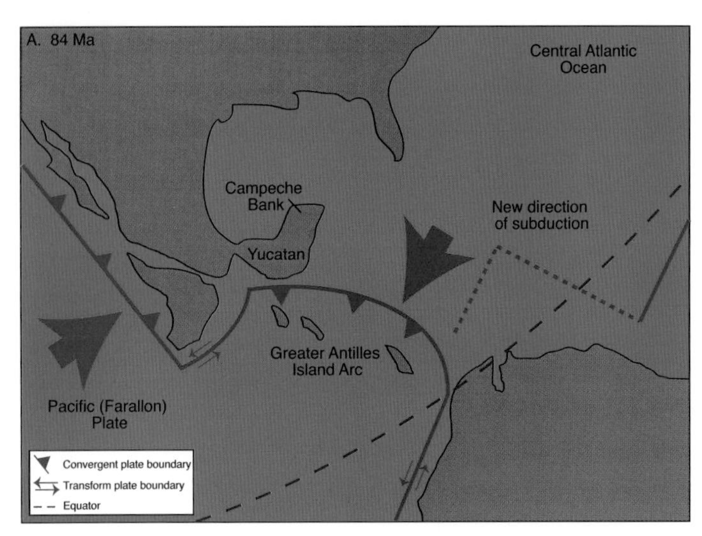

Figure 6.3*A*. Paleogeographic reconstruction of Caribbean Sea area ~84 Ma. (Modified from Redfern 2001.)

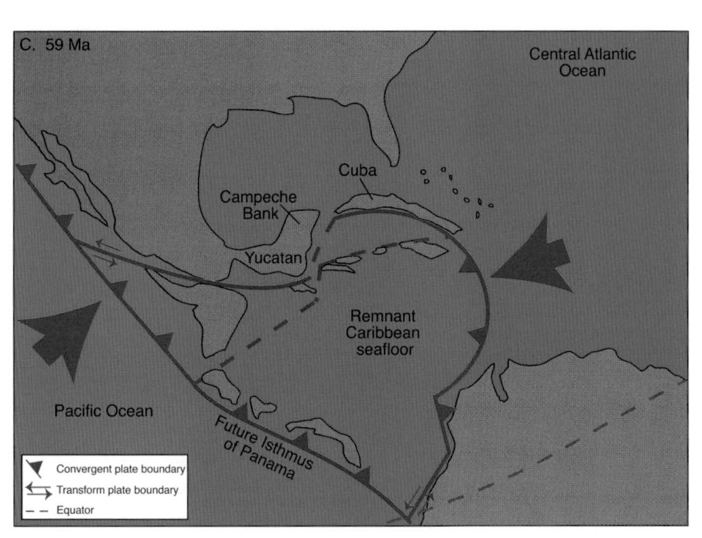

Figure 6.3*B*. Paleogeographic reconstruction of Caribbean Sea area ~72 Ma. (Modified from Redfern 2001.)

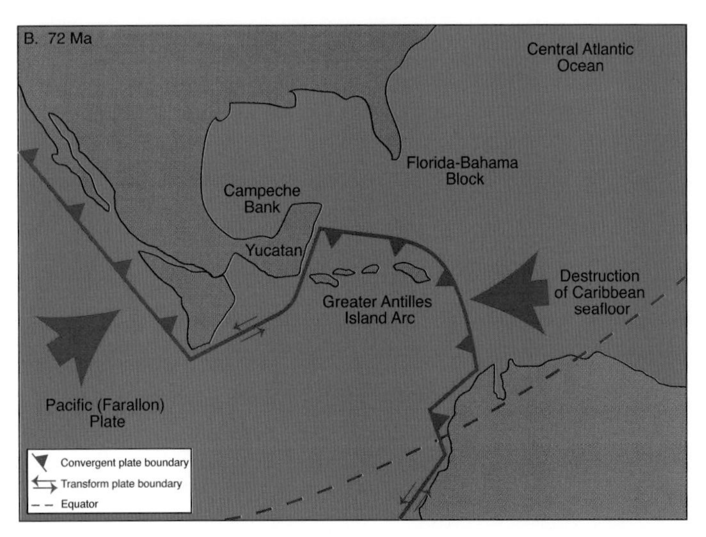

Figure 6.3*C*. Paleogeographic reconstruction of Caribbean Sea area ~59 Ma. (Modified from Redfern 2001.)

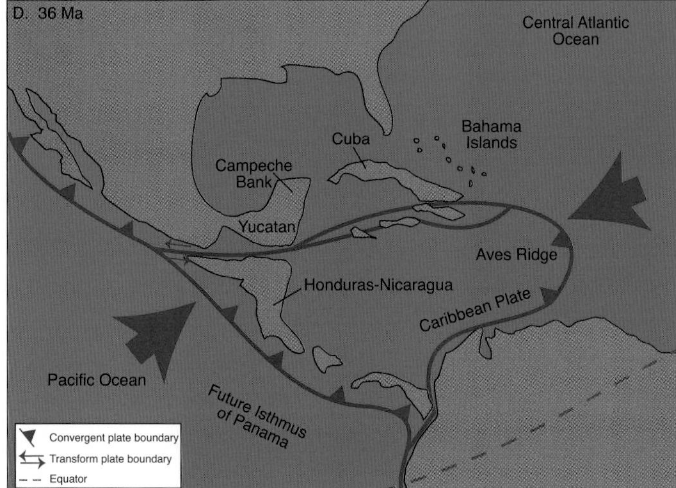

Figure 6.3*D*. Paleogeographic reconstruction of Caribbean Sea area ~36 Ma. (Modified from Redfern 2001.)

sediments that formed to the west. This new Caribbean Plate migrated some 2,000 km northeastward and eastward relative to North America.

It is still migrating eastward today, forming the active volcanic arc of the Lesser Antilles island chain (also called the Leeward and Windward Islands). Recent volcanic eruptions on Guadaloupe (early 1970s) and Montserrat islands (1995) as well as the deadly eruption on Martinique Island in 1902 that incinerated and/or buried 36,000 souls are the results of this continued relative eastward plate movement, subduction, and collision between the Caribbean and North/South Atlantic Plates.

The Passive Margin: Yucatan-Florida-Bahamas Platform

While the leading edge of the Caribbean Plate and its island arc were migrating through the gap between the Americas and consuming the ancient Caribbean/Atlantic seafloor, the Yucatan, Florida, and the Bahamas all formed a huge carbonate platform that shed carbonate-rich sediments into the deep basin along the southern margins. A large, contiguous slope extended down from the shallow platform margins to probably several thousand meters water depth.

As mentioned in previous chapters, this ~2,000 km long carbonate margin forming the edge of the Yucatan-Florida-Bahama Platform complex resulted from earlier rifting of North America from South America and therefore was a tectonically stable, passive margin undergoing only long-term subsidence. The conditions for a collision between a tectonically active volcanic arc and a tectonically passive continental margin covered by a carbonate platform were in place. The collision initially began about 84 Ma (grazing the Yucatan by 72 Ma) with most of the collision activity occurring from 56 to 45 Ma. Part of western Cuba was originally part of the Yucatan, whereas the rocks in north-central Cuba were part of the Florida-Bahama carbonate platform.

In stark contrast to the collisions that formed many of the major mountain ranges, this seemed to be a relatively "soft hit"—no Himalayan type peaks were formed, no great mountain glaciers ever existed, and no large rivers formed. Even though the collision was not Himalayan scale, there still was considerable uplift in the Greater Antilles along with significant accompanying submarine landslides and *debris flows*.

The very thick, relatively low-density carbonate rock forming the Yucatan-Florida-Bahama Platform complex passive margin was too buoyant to be subducted. The relative movement of the volcanic arc carrying the roots of the future Greater Antilles islands to the northeast essentially stalled, possibly as a result of this thick mass of less dense rock. The relative movement changed from northeast to the east over a period of several million years (probably as a path of least resistance) and still continues to the east as mentioned above (figs. 6.3*C*, 6.3*D*).

The resulting faulting and extension from this fundamental change in plate direction segmented the Greater Antilles into the separate and distinct islands that we see today. Cuba became part of the North American Plate, not the Caribbean Plate, as a new transform fault defined the northern margin of the Caribbean Plate, and Hispaniola, Puerto Rico, and the Virgin Islands which remained part of the Caribbean Plate.

Even though there was significant deformation of the rocks in Cuba such as uplift, vertically stacked, low-angle thrust sheets, high-angle strike-slip faults, folding, fracturing, and jointing (figs. 6.4, 6.5), there appears to have been no similar type or scale of deformation beneath Florida or the Bahamas.

Figure 6.4. Volcanic rocks in western Cuba folded as a result of the collision between the volcanic island arc and the carbonate platforms. (Photo by A. C. Hine.)

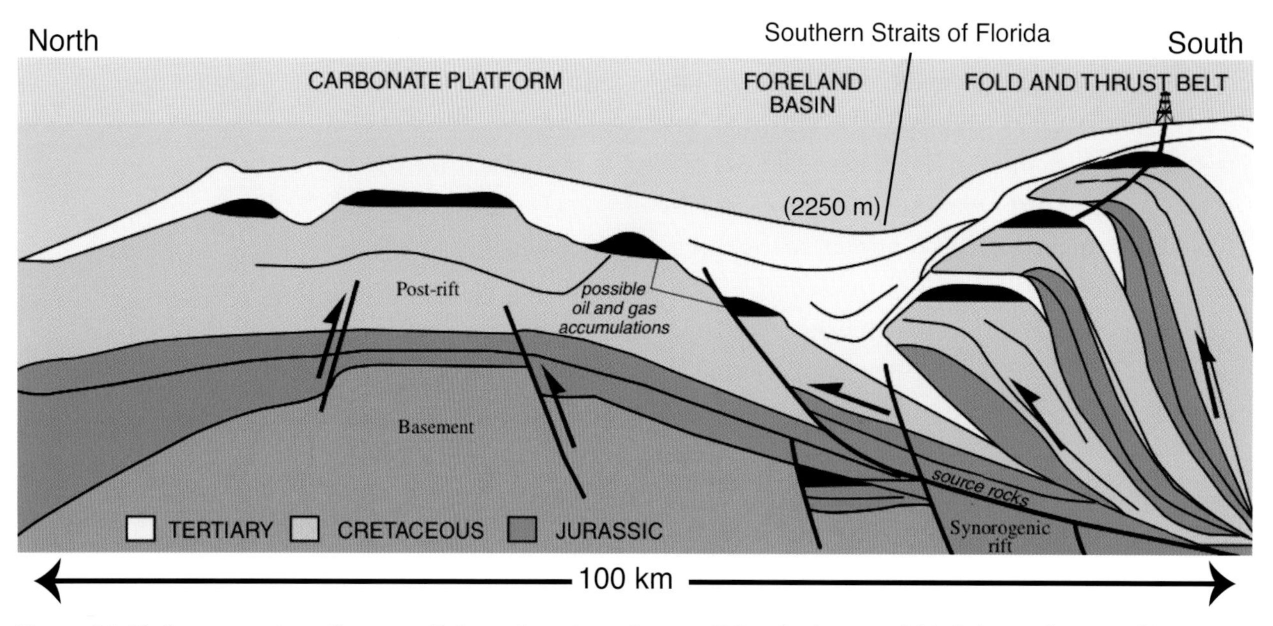

Figure 6.5. N–S cross section of western Cuba and southern Straits of Florida showing folded thrust sheets and location of potential hydrocarbon accumulations. The southern Straits of Florida is a foreland basin. (Source: USGS Fact Sheet 2005–2009.)

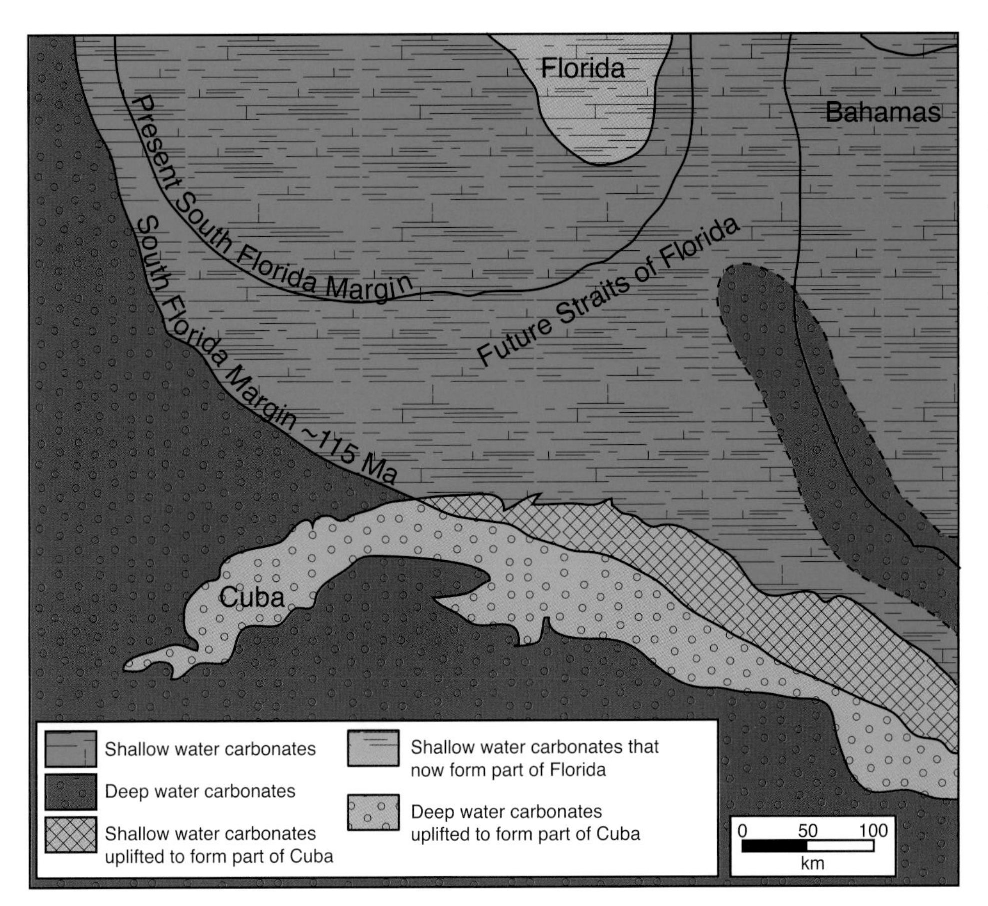

Figure 6.6. Paleogeographic map showing amount of the Florida-Bahama Platform that "overlapped" Cuba, thus becoming uplifted and expanding Cuba's land mass (or reducing Florida's land mass) to form the southern Straits of Florida. (Source: Denny et al. 1994; Hine 1997.)

During the collision, the southern margin of the Florida-Bahama carbonate platform was destroyed and became incorporated into the uplifted rocks in north-central Cuba (fig. 6.6). The Florida carbonate margin probably stepped back ~150 to 200 km with the deepwater slope and shallow-water margin rocks and sediments (e.g., reefs) being contorted/uplifted by the collision. As part of this collision, elements of the Yucatan platform margin were detached and became the present Cordillera de Guaniguanico of western Cuba, where the karstic "mogote" hills evolved. Additionally, the collision thrust rocks on top of the Florida-Bahama carbonate platform, depressing the area north of the collision. This is called *flexural loading* and probably contributed to creating the southern Straits of Florida and the Old Bahama Channel that separates the modern Bahama Banks from the Greater Antilles. Geologists would call this a *foreland basin.*

Evidence for this is seen as the former steep margin of the West Florida Escarpment extends north–south across the Straits of Florida and intersects the northern Cuban margin (fig. 6.7*A*). Additionally, the top of the mid-Cretaceous shallow-water platform is imaged in seismic data beneath the southern Straits of Florida (fig. 6.7*B*). If one drilled into the Straits of Florida (at ~2,000 m water depth) between Key West and Havana, for example, younger, deep-water, post-drowning (see chapter 5), and post-collision sediments would be encountered

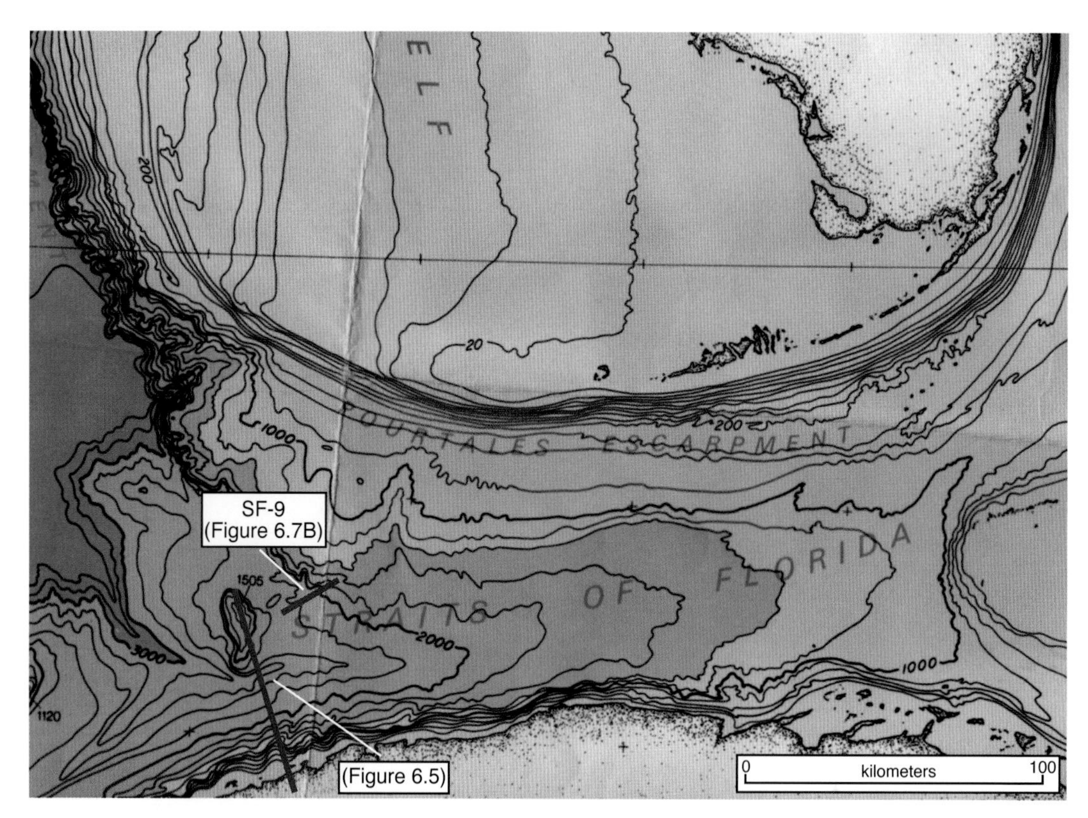

Figure 6.7*A*. Bathymetric map of the Straits of Florida. Note the N–S linear slope at the base of the Straits indicating the buried Cretaceous Florida carbonate platform. (Source: Bryant and Bryant 1991; reprinted by permission from the American Association of Petroleum Geologists, whose permission is required for further use.)

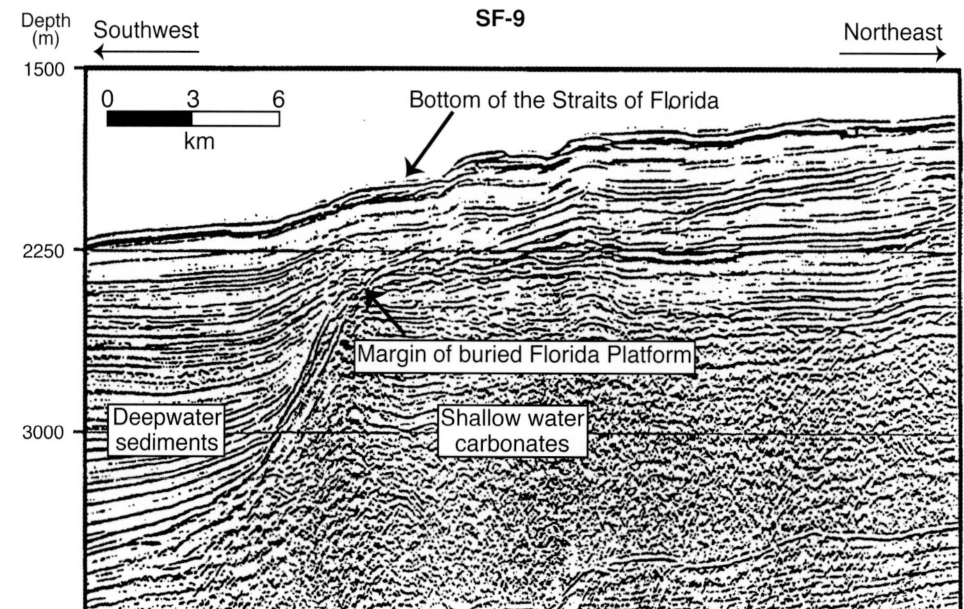

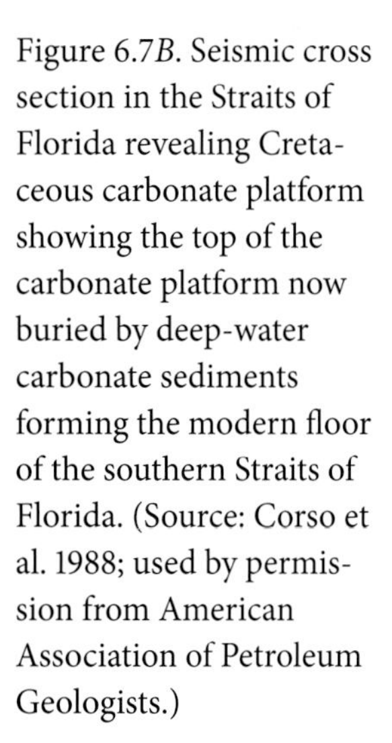

Figure 6.7*B*. Seismic cross section in the Straits of Florida revealing Cretaceous carbonate platform showing the top of the carbonate platform now buried by deep-water carbonate sediments forming the modern floor of the southern Straits of Florida. (Source: Corso et al. 1988; used by permission from American Association of Petroleum Geologists.)

before drilling deeper into older, lower to mid-Cretaceous shallow-water carbonate rocks that constituted the platform at that time.

Even though the geology and topography, and therefore the scenery, of Cuba and south Florida are vastly different, their geologic histories are intimately intertwined. Had this collision not occurred, Florida (as well as the Bahamas) would extend much further south, perhaps as much as 150–200 km and would face the open Caribbean Sea and not the relatively narrow southern Straits of Florida or the Old Bahama Channel. The city of Miami could have been located ~150 km further to the south in a truly tropical environment! The geography of the northern Caribbean would be vastly different than it is today if the Antillean Orogeny had not occurred.

Potential for Hydrocarbons

The Antillean Orogeny did occur, however, and these complex structures (folded thrust sheets, etc.) created traps for migrating hydrocarbons—oil and gas—both of which are being extracted along the northern coast of Cuba. Seismic reflection data collected by survey vessels show that these structures extend offshore into the Straits of Florida. There are probably hydrocarbon accumulations associated with these offshore structures, which were offered by the Cuban government to international companies in joint ventures to explore, drill, and recover oil and gas in the southern Straits of Florida. A 2004 USGS report indicates that there might be ~4.6 billion barrels of oil as well as natural gas off northern Cuba. This is about an order of magnitude less than what some experts think lies beneath the very deep Gulf of Mexico. As of early 2012, drilling platforms contracted by non-American oil companies and the Cuban government were conducting exploratory drilling offshore northern Cuba within the Cuban Exclusive Economic Zone in the southern Straits of Florida. The first exploratory hole completed in mid-2012 was dry, but exploratory drilling for hydrocarbon deposits continues as of this writing in deep water seaward of northern Cuba.

Such complex, hydrocarbon-bearing structures do not appear to extend northward beneath the modern south peninsular Florida. Due to dry (only a minor hint of hydrocarbon presence) exploratory bore holes drilled in the Florida Keys years ago as well as this abrupt lateral change in the geology beneath the Straits of Florida, one would not expect similar hydrocarbon potential in extreme southern Florida as along northern offshore Cuba.

Essential Points to Know

1. The geology and the scenery in western Cuba are vastly different than south Florida. Mature carbonate karst terrains featuring unique rounded, cavernous, steep hills up to 300 m high composed of carbonate rock resulted from ~50 Myr

of subaerial exposure. Additionally, deformed volcanic rocks are exposed. These rocks and topography do not exist in Florida, only ~150 km away to the north. Such close juxtaposition implies a significant geologic boundary between the two areas.

2. The Greater Antilles formed as a result of a collision between (a) a tectonically active margin—a volcanic arc formed by subduction of ocean crust belonging to one plate beneath ocean crust of another plate and (b) with a passive margin covered by the Yucatan-Florida-Bahama carbonate platform. The collision started ~56 Ma.

3. One major theory is that the Caribbean Plate formed in the eastern Pacific Ocean and migrated to the NE consuming the proto Caribbean ocean crust. The leading edge of this migrating plate was a subduction zone creating a volcanic island arc carrying components of the Greater Antilles. The collision between the Cuban part of the Greater Antilles affected the southern margin of the Yucatan-Florida-Bahama Platform by destroying it through multiple thrust sheets and uplift, exposing carbonate rocks. This uplift placed carbonates, subjected to karst dissolution processes, in close proximity to volcanic rocks of vastly different origin.

4. An additional result of the collision was to topographically depress the southern margin of the Florida-Bahama Platform forming the southern Straits of Florida and thus separating the two terrains by this deep-water seaway. Beneath the southern Straits of Florida lies the top of the Cretaceous shallow-water platform. As a result of the collision and the depression due to flexural loading, perhaps as much as 150–200 km of the original shallow-water Florida-Bahama Platform was destroyed (uplifted in Cuba or depressed and buried in the southern Straits of Florida). The southern Straits of Florida constitutes a foreland basin.

5. The collision stopped at ~45 Ma with the migration of the Caribbean Plate changing from its NE movement to an easterly movement that continues on today, forming active volcanoes along the Lesser Antilles located further to the southeast in the eastern Caribbean Sea.

Essential Terms to Know

active margin: Margins of plates that move toward each other through subduction forming a tectonic collision. Mountain ranges and volcanic island arcs are products of such plate margin collisions (also called collision margins). Plate margins defined by active transform faults such as the San Andreas fault are also tectonically active margins.

debris flows/debris flow deposit: Transport of mud, sand, gravel, and sometimes very large blocks of rock by a moving slurry of water and sediment creating a density-driven flow moving downslope. The flows and their deposits may occur on land or beneath the sea.

flexural loading: Bending of the Earth's crust resulting from a load being placed on it. Such a load could come from ice sheets, thick sediment accumulation, or thrust sheets being transported on top of the crust, thus creating a load.

foreland basin: A depression that develops adjacent to an orogenic belt (i.e., the Antilles) due to a mass of rock and sediments thrust upon the crust during the collision causing the lithosphere to bend by a process known as lithospheric flexure. The width and depth of the foreland basin is determined by the flexural rigidity of the underlying lithosphere. The foreland basin receives sediment that is eroded off the adjacent uplifted area.

island arc/volcanic island arc: A chain of volcanic islands or mountains formed by plate tectonics as an oceanic tectonic plate subducts under another tectonic plate and produces magma. There are two types of volcanic arcs: oceanic arcs (commonly called island arcs, a type of archipelago) and continental arcs. The arc or curved morphology results from the curvature of the Earth's spheroid.

mogotes: Geomorphologic structures encountered in the Caribbean, especially in Cuba. They are rounded, towerlike remnants of eroded limestone that result from tropical karst erosional processes.

obduction: Overthrusting of oceanic crust onto the continental crust.

passive margin: A continental margin that has been rifted and faces a seafloor spreading center; mature passive continental margins are low relief, have subsided slowly for long periods, and have accumulated a thick sedimentary cover.

subduction: The process that takes place at convergent boundaries by which one tectonic plate moves under another, sinking into the Earth's mantle, as the plates converge. A subduction zone is an area where two tectonic plates move toward one another and subduction occurs commonly resulting in volcanic activity.

tectonic movement: Movement of tectonic plates forming the surface of the Earth by forces occurring within the mantle.

thrust sheets/thrust faults: Sections of rock separated by low angle faults; sheets of rock are transported on top of each other by compressional tectonic forces.

volcanic rocks: Commonly formed from lavas (commonly basaltic in composition and texture), glassy shards (obsidian) from lava droplets quick-freezing in air, and ash associated with an eruption.

Keywords

Greater Antilles, mogotes, Cuban geology, karst topography, volcanic island arc, active and passive margins, subduction, Caribbean Plate, carbonate platform margin, flexural loading

Essential References to Know

Bryant, W. R., and J. R. Bryant. “Bathymetric Chart: Gulf of Mexico Region.” In *The Gulf of Mexico Basin*, ed. A. Salvador. The Geology of North America. Boulder: Geological Society of America, 1991.

Buffler, R. T. “Seismic Stratigraphy of the Deep Gulf of Mexico and Adjacent Margins.” In *The Gulf of Mexico Basin*, ed. A. Salvador, The Geology of North America, 353–87. Boulder: Geological Society of America, 1991.

Corso, W., R. T. Buffler, and J. A. Austin. “Erosion of the Southern Florida Escarpment.” In *Atlas of Seismic Stratigraphy*, ed. A. W. Bally, 149–57. Tulsa: American Association of Petroleum Geologists, 1988.

Denny, W. M., III, J. A. Austin Jr., and R. T. Buffler. “Seismic Stratigraphy and Geologic History of Middle-Cretaceous through Cenozoic Rocks, Southern Straits of Florida.” *Bulletin of the American Association of Petroleum Geologists* 78 (1994): 461–87. doi: 10.1306/BDFF90E6-1718-11D7-8645000102C1865D.

Iturralde-Vinent, M. A., and R.D.E. MacPhee. “Paleogeography of the Caribbean Region: Implications for Cenozoic Biogeography.” *Bulletin of the American Museum of Natural History* 238 (1999).

Iturralde-Vinent, M. A., ed. *Geología de Cuba para todos*. Habana: Editorial Científico-Técnica, Instituto del Libro, La Habana, 2009.

Kerr, A. C., M. A. Iturralde-Vinent, A. D. Saunders, T. L. Babbs, and J. Tarney. “A New Plate Tectonic Model of the Caribbean: Implications from a Geochemical Reconnaissance of Cuban Mesozoic Volcanic Rocks.” *Geological Society of America Bulletin* 111, no. 11 (1999): 1581–99. doi: 10.1130/0016-7606(1999)111<1581:anptmo>2.3.co;2.

Klitgord, K. D., P. Popenoe, and H. Schouten. “Florida: A Jurassic Transform Plate Boundary.” *Journal of Geophysical Research* 89, no. B9 (1984): 7753–72. doi: 10.1029/JB089iB09p07753.

Pardo, G. *Geology of Cuba*. Tulsa: American Association of Petroleum Geologists, 2009.

Pindell, J. L., and S. F. Barrett. “Geological Evolution of the Caribbean Region: A Plate Tectonic Perspective.” In *The Caribbean Region*, ed. G. Dengo and J. E. Case, The Geology of North America, 405–32. Boulder: Geological Society of America, 1990.

Pindell, J. L., W. Maresch, V. U. Martens, and K. Stanek. “The Greater Antillean Arc: Early Cretaceous Origin and Proposed Relationship to Central American Subduction Mélanges: Implications for Models of Caribbean Evolution.” *International Geology Review* 54, no. 2 (2012): 131–43. doi: 10.1080/00206814.2010.510008.

Poag, W. C. “Rise and Demise of the Bahama–Grand Banks Gigaplatform, Northern Margin of the Jurassic Proto-Atlantic Seaway.” *Marine Geology* 102, no. 1–4 (1991): 63–130. doi: 10.1016/0025-3227(91)90006-p.

Randazzo, A. F., and D. S. Jones, eds. *The Geology of Florida*. Gainesville: University Press of Florida, 1997.

Read, J. F. “Carbonate Platforms of Passive (Extensional) Continental Margins: Types, Characteristics, and Evolution.” *Tectonophysics* 81, no. 3–4 (1982): 195–212. doi: 10.1016/0040-1951(82)90129-9.

Sheridan, R. E., H. T. Mullins, J. A. Austin, M. M. Ball Jr., and J. W. Ladd. “Geology and Geophysics of the Bahamas.” In *The Atlantic Continental Margin: U.S.*, ed. R. E. Sheridan and J. A. Grow, Geology of North America, 329–64. Boulder: Geological Society of America, 1988.

7

Dissolution Tectonics

Sinkhole Development (~140 Ma to Present)

> This is the undiscovered Florida. A place where water and life begin from a source that bubbles up directly from the ground. Florida Springs flow the purest, clearest, and freshest water in the world with underwater views that are absolutely breathtaking. Ponce de Leon once thought Florida Springs were the "fountain of youth" when he first discovered them. There are over 600 freshwater springs throughout central and northern Florida.
>
> *Endangered Earth: The Caves of Florida; Florida's Underwater Caves and Springs; Existing Water Filled Sinkholes*

Caverns in Carbonate Platforms

During the 1950s, the oil industry thought that the Florida-Bahama carbonate platform complex might contain oil and gas. This notion was based, in part, on the successful extraction of hydrocarbons from west Texas in an area geologists refer to as the Permian Basin, a subsurface stratigraphic succession named after the age of the carbonate rocks that were producing the oil and gas.

Midland and Odessa are oil towns that sprang up after 1923 as a result of the discovery. About 250 km further west, the same type of Permian carbonate rocks from these oil-bearing strata crop out and form the Guadalupe Mountains featuring Guadalupe Peak, the highest point in Texas (2,651 m). These mountains display a massive, complete back reef/reef/reef slope/basin cross section exposed in canyons cut into the mountains (fig. 7.1). These outcrops constitute one of the most spectacular geologic sites to visit, perhaps second only to the Grand Canyon in the United States. It is a required "mecca-like" pilgrimage for earth scientists, retired experts, and undergraduates to view these rocks.

Experts in hydrocarbon exploration concluded that the Florida-Bahama system demonstrated strong geologic similarities to the Permian Basin and that possibly hydrocarbons could be discovered beneath these modern

Epigraph source: *Endangered Earth: The Caves of Florida; Florida's Underwater Caves and Springs; Existing Water Filled Sinkholes*, http://www.thelivingmoon.com/45jack_files/03files/Endangered_Earth_Underground_Water_Florida_02.html

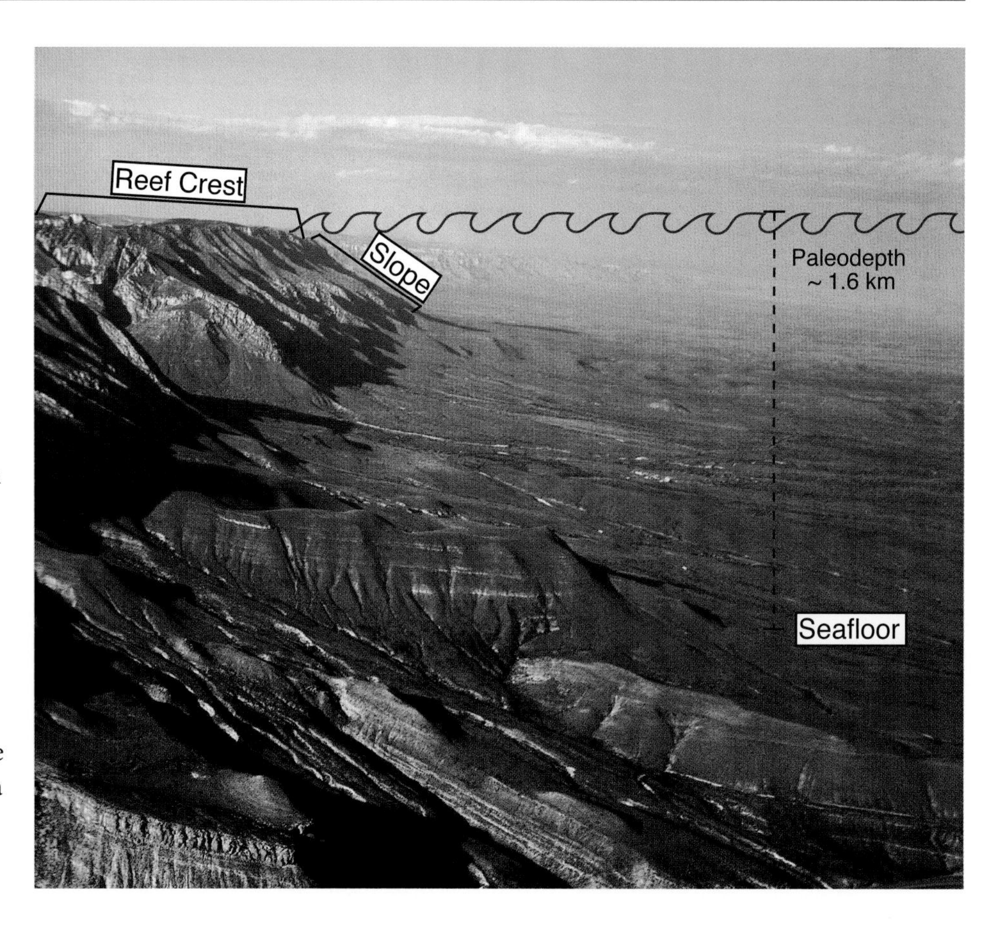

Figure 7.1. The Guadalupe Mountains of west Texas. The line of peaks is an ancient reef that once sloped off into a deep water basin to the right. Erosion has cut canyons into this Permian seascape, but for the most part the original depositional setting is intact. Fig. 7.1 illustrates approximate original sea level and depth of Permian age ocean. This was once a huge carbonate platform-basin system that evolved in warm tropical waters. Large caves, including Carlsbad Caverns, have formed within these limestone rocks. Such large caves exist with the Florida and Bahama Platforms as well. (Source: Redfern 2001; used by permission from University of Oklahoma Press.)

carbonate sedimentary systems as well. Additionally, the Florida and Bahama Platforms provided spectacular modern environments within which active sedimentary processes could be observed and measured directly. Such observations and measurements would logically lead to a better understanding of how ancient carbonate sediments were formed, ultimately aiding in the hydrocarbon exploration process.

Exploration drilling commenced, and a number of deep (like ~5.5 km deep!) exploration holes were drilled sporadically around the Bahama and Florida Platforms. Actually, the Sunniland Field (in lower Cretaceous carbonate rocks) in southwest peninsular Florida had already been discovered entirely by accident and was producing oil in the 1940s. But this was a very small operation, and it still is to this day, but enhanced drilling and extraction techniques could make the Sunniland more productive.

On Andros Island in the Bahamas, a drilling operation encountered a 55 m diameter water-filled hole at ~2,500 m depth. The drill string (length of pipe) fell into this hole and could not be retrieved. It was an expensive and unexpected loss. Ultimately, drilling recommenced with a new drill string, but the hole was abandoned after having penetrated nearly 5.5 km into the carbonate substrate and not having found any hydrocarbons.

Such was the story throughout the Bahamas and peninsular Florida—no oil and no gas—except for traces and for the relatively small Sunniland field. A total of 52 wells were drilled in the Florida Keys and the west Florida continental shelf. The oil companies sought more fertile ground elsewhere. One of the reasons little oil/gas has ever been found in Florida and the Bahamas is the fact that these carbonates are filled with holes, many of which lead to the surface (fig. 7.2). Thus if hydrocarbons had ever been formed, they are long gone, having reached the Earth's surface and/or dispersed to the ocean through seeps. However, the existence of that big subsurface cavern in Andros in the Bahamas was significant. Additionally, the spectacular springs in Florida linked to extensive cave systems explored by divers suggest that there is an enormous plumbing system that has been formed within the thick carbonate rocks (up to 6 km thick in south peninsular Florida) resting on top of Florida's basement rocks. Indeed, if one could part these limestone and *dolomite* formations with a huge knife and examine the subsurface as some immense outcrop, one would see a cross section of the Florida-Bahama Platform that might vaguely resemble poorly made Swiss cheese. The rocks contain millions of cavities, holes of many shapes, vertical pipes, enlarged fractures and joints, vugs, and caves, many interconnected.

Figure 7.2*A*. Divers explore Diepolder Cave, located on Sand Hill Boy Scout Reservation near Brooksville, Florida. Diving into Florida's gin-clear springs is a world-class adventure. (Photo by W. C. Skiles; used by permission from National Geographic Society.)

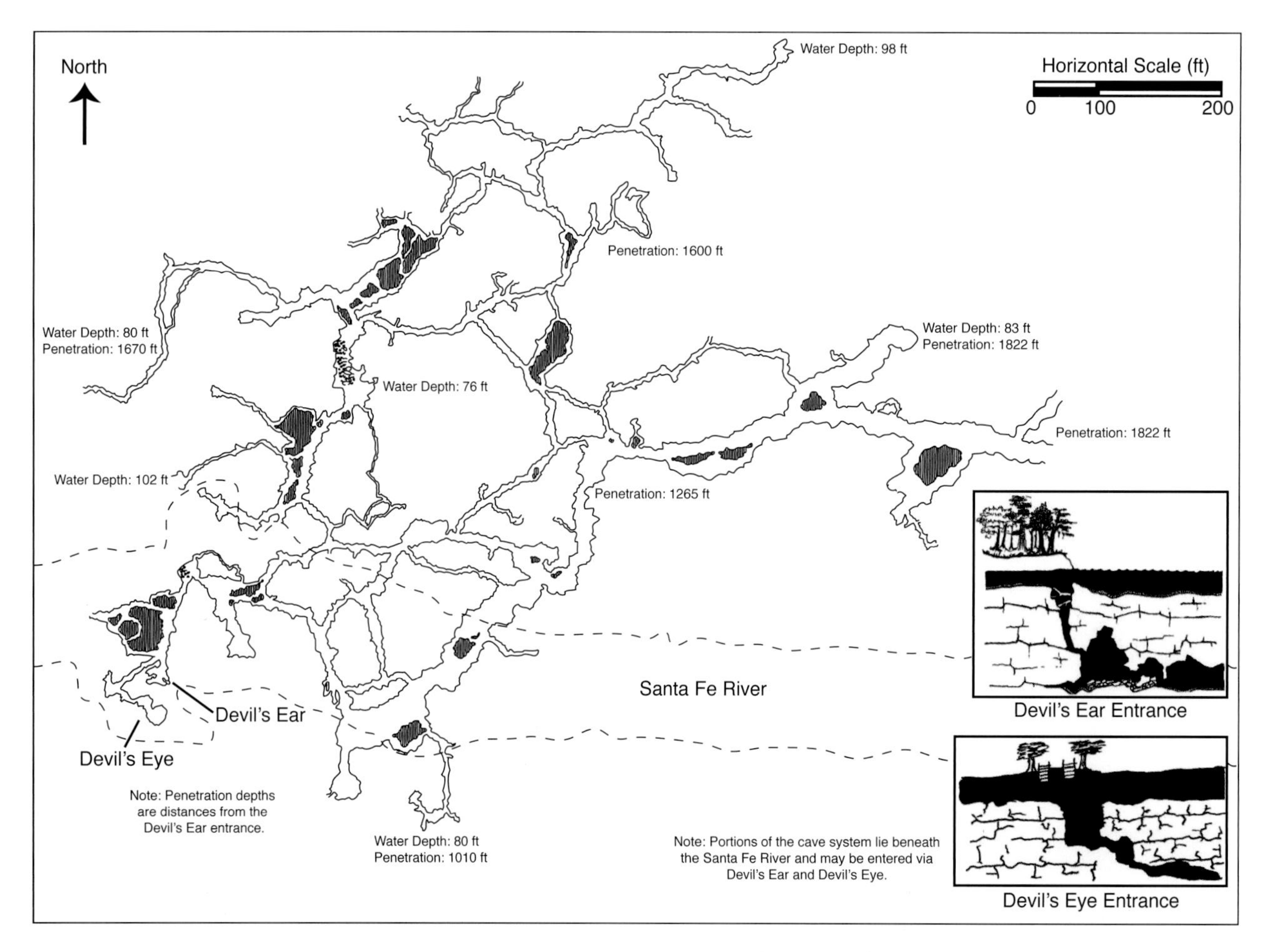

Figure 7.2*B*. Map of Devil's Eye and Ear cave system illustrating significant lateral and vertical extent, complexity, and intricacy of an interconnected cave system. (Cartography by Steve Berman; digitally reproduced by Gordon Roberts; modified by A. C. Hine.)

One of the world's largest cave systems, Carlsbad Caverns, is also in the Guadalupe Mountains—another carbonate platform, but older than the Florida-Bahama Platform (fig. 7.3). This series of 83 caves is located within the Permian reef itself. So we can expect large caves to be carved by dissolution deep within carbonate rocks.

To make matters even more interesting in Florida, these caverns are still forming and collapsing today near the ground surface as sinkholes that pose a real geologic hazard to homes and human infrastructure (fig. 7.4). Knowledge of this hazard was fully galvanized in May 1981 when a large sinkhole in Winter Park, Florida, rapidly appeared and consumed a municipal swimming pool, a house, a car dealership, and sections of two public streets. It grew to be 100 m wide and 30 m deep and eventually filled with water. Although there are much older sinkholes or sinkhole complexes that are larger and more numerous, this Winter Park event spurred significant research activity.

Sinkhole collapse again became public during the very cold Florida winter of 2010 when farmers pumped water furiously from wells to protect their crops from freezing through extensive spraying. This sudden lowering of the surface aquifer promoted a new round of sinkhole formation. As most

Figure 7.3*A*. One of the complex interior chambers of the Carlsbad Caverns formed within the Permian carbonate platform of west Texas and New Mexico. Note people for scale. (Source: www.archives.gov/research/ansel-adams/.)

Figure 7.3*B*. Another Carlsbad Cavern chamber. Undoubtedly, such caverns, now filled with water, occur throughout the Florida Platform. (Source: www.archives.gov/research/ansel-adams/.)

Figure 7.5*B*. Cross section down peninsular Florida showing aquifer system, including the Boulder Zone in southern Florida. (Sources: Miller 1997 and Florida U.S. Geological Survey, http://fl.water.usgs.gov/FASWAM/; modified by A. C. Hine.)

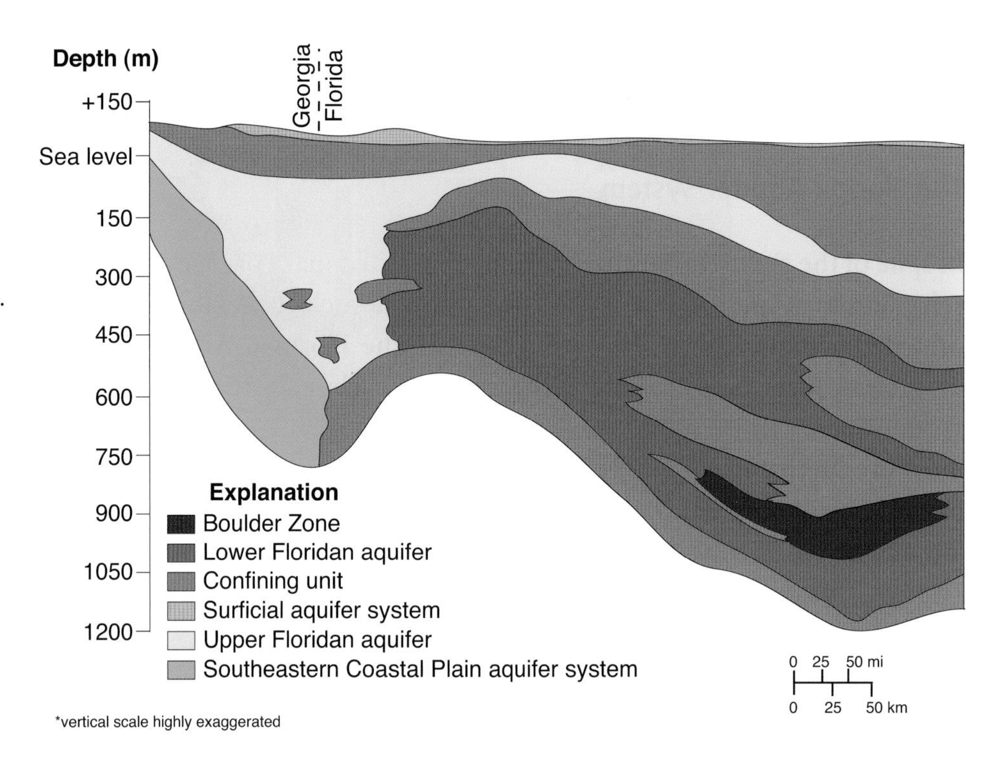

etc.) of tens of millions of people. Without this subterranean resource, life in part of the SE United States and certainly Florida would be unsustainable.

The Boulder Zone

The Boulder Zone, developed in fractured *dolomite*, is a deeply buried cavernous zone filled with saltwater extending down from 600 m to 1,000 m beneath 13 counties of southern Florida (fig. 7.6). Boulder Zone boulders are produced by chunks of extremely porous dolomite falling to the bottom of the drill hole where they are rolled around by the drill bit, making it difficult to continue to drill. The Boulder Zone represents caverns developed at several levels connected by vertical "pipes" or solution tubes similar to a modern cave system. A ~30 m diameter cavern reported in the subsurface in southern Florida probably is one of these vertical solution tubes.

The Boulder Zone has been used for years to store vast quantities of treated sewage injected into it by Miami, Fort Lauderdale, West Palm Beach, and Stuart. Because the salinity and temperature of the water in the Boulder Zone are similar to those of modern seawater, the zone is thought to be connected to the Atlantic Ocean. It is a good example of the many large solution features that have formed within the Florida Platform.

Surface and Shallow Subsurface Karst

Basically, two things are required to dissolve carbonate rocks: acidic water and a method to move this water through the rocks. The acidity comes from CO_2 in the atmosphere and soils, producing carbonic acid:

$$CO_2 + H_2O \leftrightarrow H_2CO_3 \leftrightarrow H^+ + HCO_3$$

The concentration of the H^+ defines the level of acidity known as pH. Water that has a pH of 7 is neutral; greater than 7, it is *alkaline*; less than 7, it is *acidic*. The average pH of Florida's rainwater is 4.77 and acidic. Rainwater becomes more acidic by dissolving pollutants in the air to produce sulfuric acid (H_2SO_4) and nitric acid (HNO_3) as well as carbonic acid. These pollutants are largely from our industrialization. This acid rain has severe environmental consequences. For example, there are large sections of dead and dying trees near the tops of the highest peaks in the Appalachian Mountains. These peaks commonly are above cloud level, and the trees are constantly bathed in acidic mist, greatly increasing their exposure and enhancing their

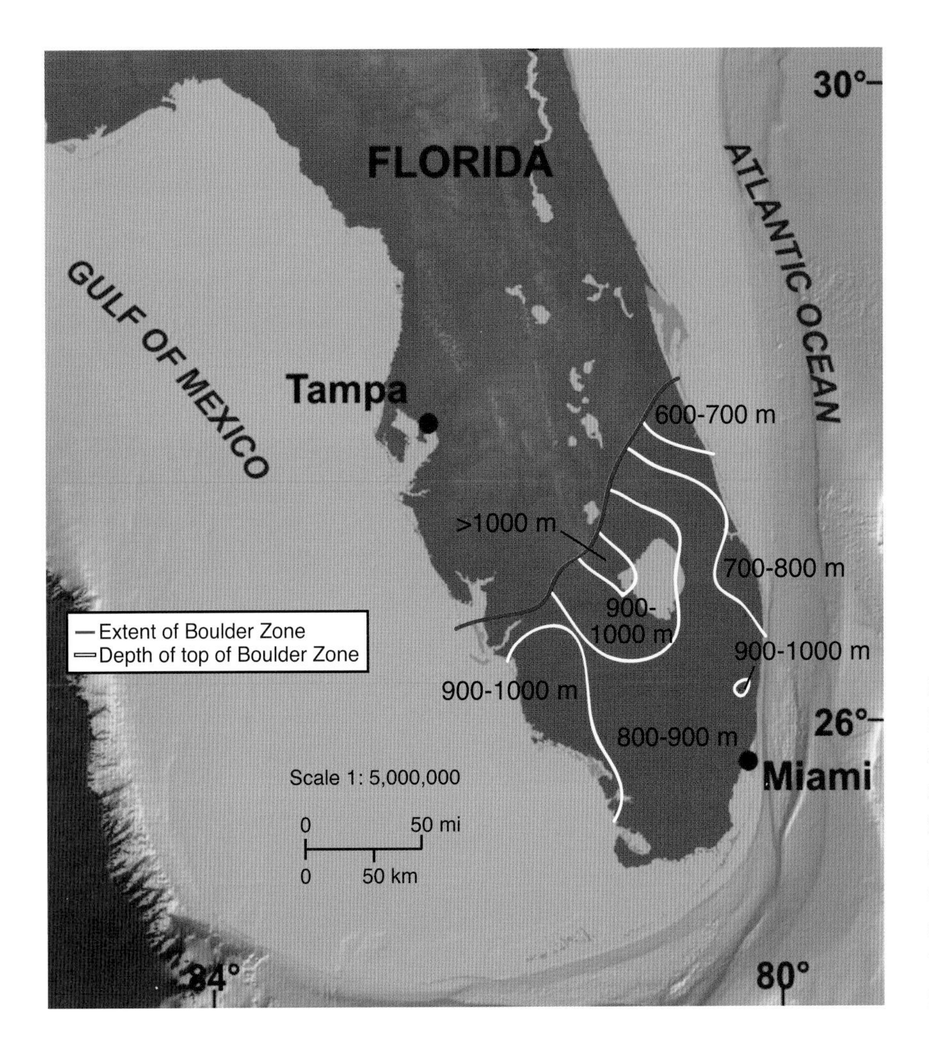

Figure 7.6. The top of buried Boulder Zone measured down from sea level. This cavernous zone is limited to SE Florida and is used for waste-water injection, as it contains saltwater not usable for human consumption. (Sources: Miller 1986 and Florida U.S. Geological Survey, http://fl.water.usgs.gov/FASWAM/.)

Left: Figure 7.7*A*. Surficial karst in Eocene age limestone along the Big Bend marsh coast near Ozello, Florida. (Photo by A. C. Hine.)

Above: Figure 7.7*B*. Close-up of surficial karst illustrating small, cavernous, fluted nature of limestone bedrock surface. (Photo by A. C. Hine.)

demise. However, natural rainwater has always been slightly acidic because of the omnipresent CO_2 in the atmosphere forming carbonic acid within the raindrops. Additionally, soils are acidic from natural *microbial activity* breaking down organic matter and producing CO_2.

By the time surface waters reach the underlying carbonates, they are capable of dissolving them. This shallow groundwater system is called the *surface aquifer*. In Florida, it has been calculated that surface dissolution can remove rock at the rate of 4 cm/kyr. The result is *karst topography* on the surface and cavities in the subsurface (fig. 7.7).

Sinkholes are closed depressions in the land surface formed by dissolution of near-surface rocks or by the collapse of underground channels and caverns (fig. 7.8). Sinkholes are a common geologic feature in places underlain by soluble rocks such as limestone and dolomite, which form the Floridan Aquifer system. Under natural conditions, sinkholes form slowly and expand gradually.

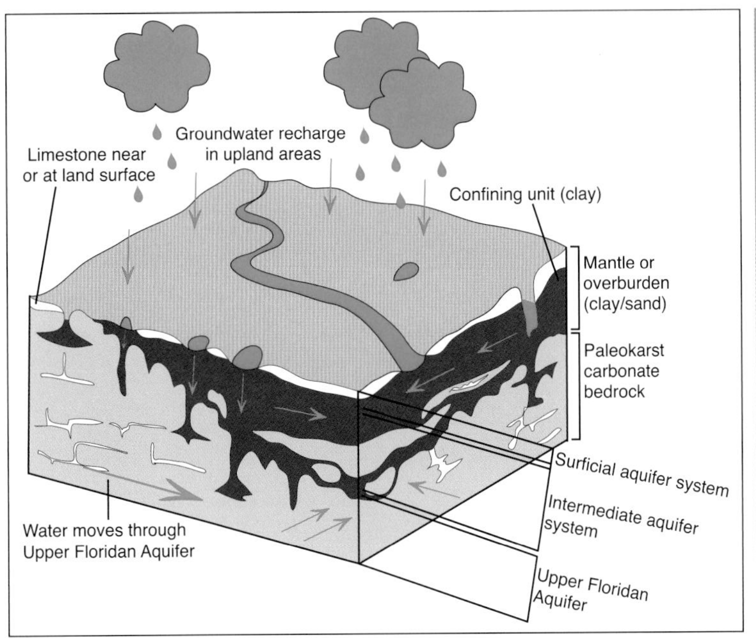

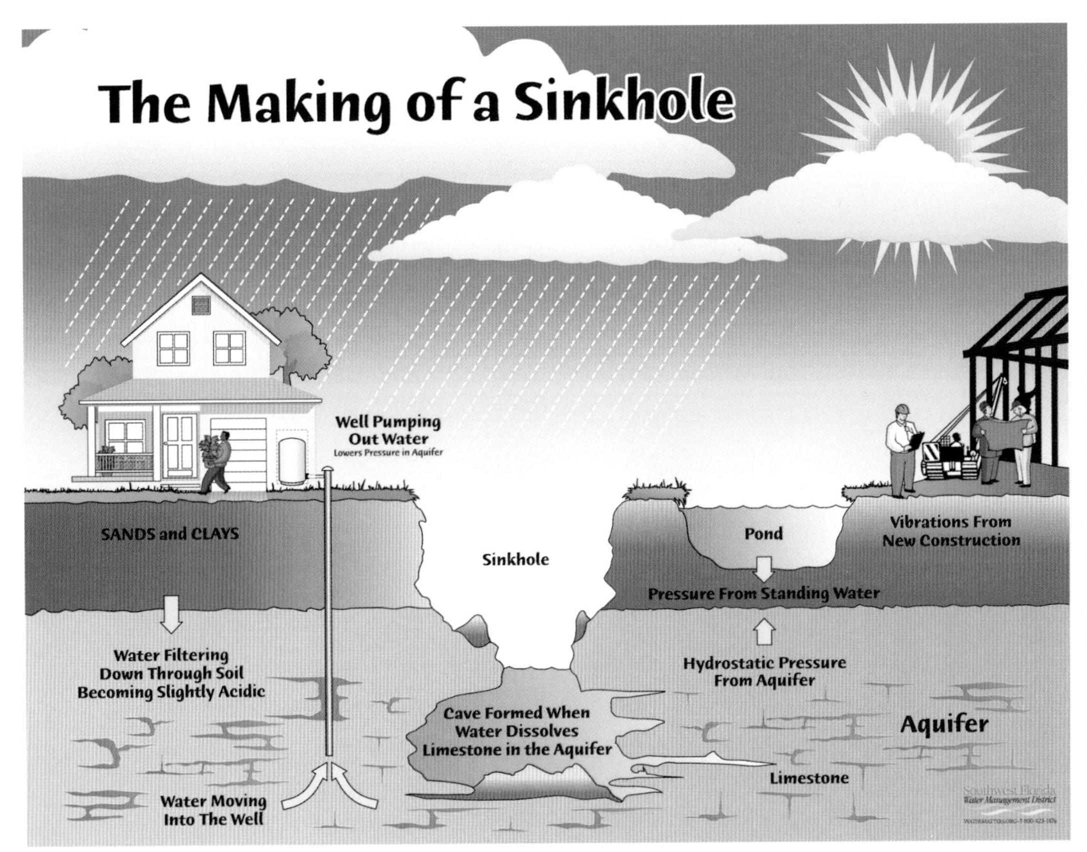

Top left: Figure 7.8*A*. Cartoon block diagram depicting water recharging surface aquifer and deeper aquifer systems including the upper Floridan Aquifer. Slightly acidic waters flowing through fractures, cracks, etc., dissolve the carbonate rock. Where cavities form near the surface and collapse, sinkholes are formed, creating circular depressions in the landscape. (Source: Southwest Florida Water Management District.)

Left: Figure 7.8*B*. The making of a sinkhole. (Source: Southwest Florida Water Management District.)

Top right: Figure 7.8*C*. Sinkholes in Florida since 1954. Note that most of the sinkholes were formed in north-central peninsular Florida where ancient limestone lies close to the surface covered with a relatively thin veneer of younger quartz-rich sediments. (Source: Florida Geological Survey.)

Collapse sinkholes form suddenly by the failure of the roof of a large solution cavity such as a cave. As the cavity expands laterally, its roof gradually flakes off under the effect of gravity. Continued dissolution and spalling of the cavity roof proceed until the roof suddenly collapses and a steep-sided, typically circular sinkhole forms. When these cavities in the shallow subsurface collapse forming sinkholes, they contribute to karst geomorphology on the surface. The term *karst* comes from the Karst Plateau of Yugoslavia, which is characterized by caves, sinkholes, and other types of openings caused by dissolution.

Groundwater flows through openings in rocks from rainfall being introduced to topographically higher recharge areas that are connected to aquifers, sediments at the surface (water table), or rocks in the subsurface that bear water because of their permeability. The elevation difference between the recharge area and the aquifer allows the water to seep downward by force of gravity, following the zone of highest permeability defined by the aquifer.

Florida is famous for its springs, which in their natural condition have nutrient-free, crystal-clear water of constant temperature, generally around 24°C (75°F; figs. I.2*D*, 7.2*A*). During the winter, when estuarine and coastal waters are cooler, Florida's manatees (commonly called sea cows) congregate around the larger, warmer spring openings. Many of the rivers in Florida are spring fed—the Suwannee River, for example, has at least 70 springs along its path all contributing to the river's discharge. Many of the springs create rivers that flow only short distances over land before reaching the ocean (e.g., Crystal River, Waccasassa River, Chassahowitzka River, Pithlachascotee River). Additionally, some springs discharge offshore onto the seafloor. But as sea level rises, these springs shut down and new ones open up on land. The increased hydrostatic pressure of the deepening seawater on top of the spring shuts down its vertically upward freshwater flow. Similarly, when sea level falls and more of the shelf is exposed or becomes shallower, old springs on land are reduced and newer ones start to form further offshore.

A fundamental aspect of freshwater is that it is less dense than seawater (1.0 gm/cc vs. 1.03 gm/cc). As a result, along the coast, the fresh groundwater lens will be supported or "float" on an underlying mass of seawater, which is slightly denser. The contact between these two subterranean water masses is actually a transition called the *mixing zone* (fig. 7.9). Both water masses may be *saturated with respect to* $CaCO_3$—that is, they cannot dissolve any more $CaCO_3$. But when mixed together, they become slightly *undersaturated* and can start to dissolve the carbonate rock within which they are in contact producing karstic porosity—making more holes in the rock.

As sea level fluctuates and the coastline moves back and forth across a wide, low-gradient shelf, the mixing zone migrates laterally as well, providing

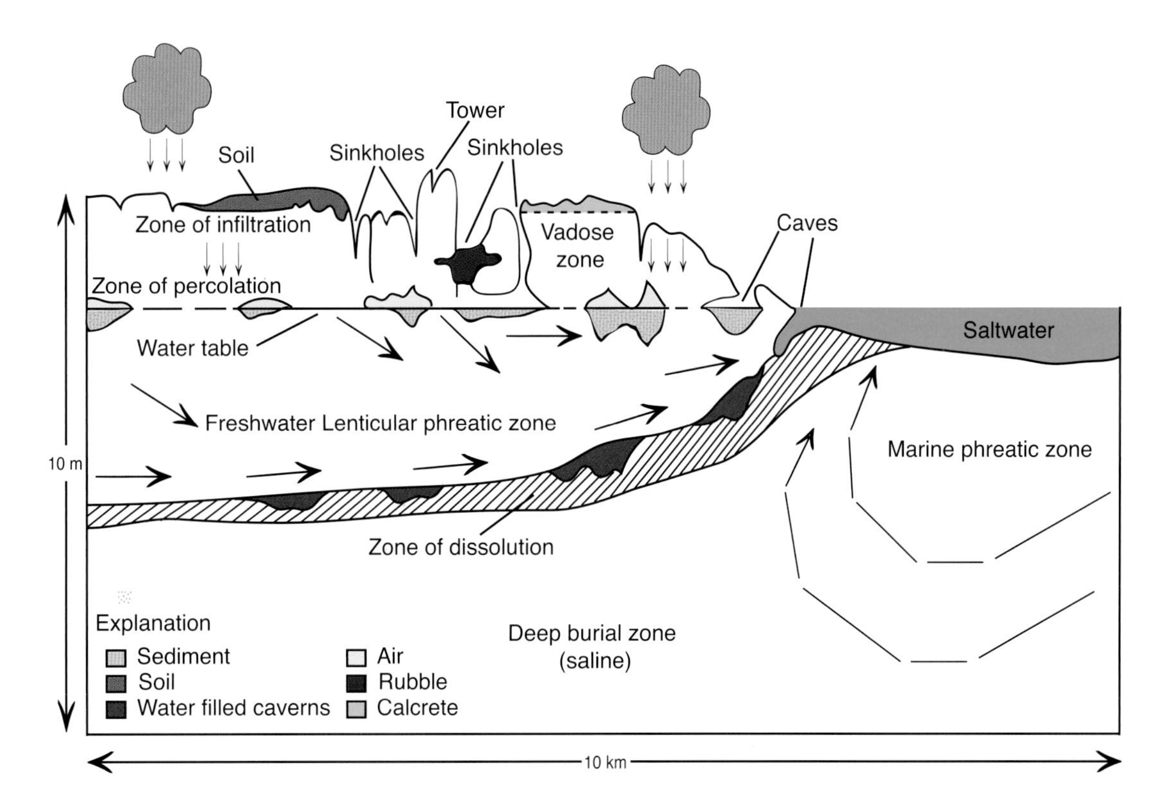

Figure 7.9. A mixing zone of brackish water between fresh groundwater and adjacent marine waters. Dissolution of carbonate occurs in this zone. As sea level fluctuates, this zone migrates laterally in response, thus dissolving a greater mass of rock. (Modified from James and Choquette 1988; used by permission from Springer-Verlag.)

a mechanism for rocks to dissolve in the shallow subsurface (down ~40 m below the rock surface). This mechanism may have contributed to the rugged but relatively low relief topography seen in the Ozello/Chassahowitzka area (fig. 7.10).

The 300 m high Cuban mogotes described in the previous chapter are an extreme example of a karst geomorphology on the Earth's surface. In Florida, the karst surface is more subtle, having as much as 10 m of relief, much of it buried by a thin sedimentary cover. A good example is shown in the Crystal River area where Eocene rocks (54.8 to 33.7 Ma) are directly exposed at the surface (fig. 7.7; see "Generalized Geologic Map of Florida" on page iii).

Many years of surface and shallow subsurface karst dissolution have produced a seemingly chaotic distribution of rocky nubs and basins. Many are sinkholes. At the coast, these rocky high areas form marsh islands or hammocks and are covered with vegetation—the Ozello/Chassahowitzka area (Florida's Big Bend coastline or Nature Coast) has been called a marsh archipelago due to the hundreds of rock-cored marsh islands. Additionally, many tidal creeks follow straight lines instead of normal meandering patterns. The water flow through marsh is controlled by fractures in the rocks that have been partially dissolved, creating a rectilinear pattern (series of 90° angles). If we could magically remove the water and thin (mostly 1–2 m, but nearly always <10 m) marsh sedimentary veneer, we would observe a rocky topography that probably could not be traversed by a human walking over it because of its rugged character (fig. 7.11).

Figure 7.10*A*. The Big Bend coastline near Ozello, Florida, showing rectilinear marsh creek pattern following jointing/fracturing in underlying limestone bedrock. Small vegetated islands (vegetation shown in red) are cored by rocky limestone nubs or pinnacles formed by surficial karst etching processes. (Source: USGS EROS Data Center.)

Figure 7.10*B*. A small sinkhole in the exposed Eocene limestone in the Ozello area. (Photo by A. C. Hine.)

Figure 7.10*C*. Rocky nubs or flat pinnacles of Eocene limestone that once supported vegetation. During sea level rise, these rocky high areas become flooded and form nucleation sites for oyster reefs in the Waccasassa area. (Photo by A. C. Hine.)

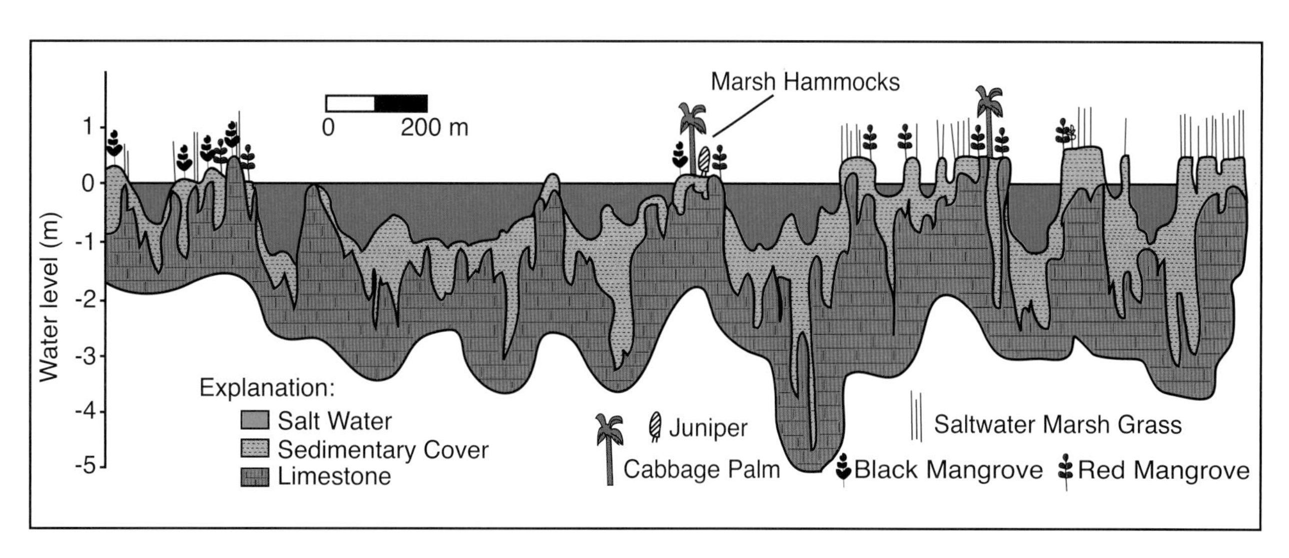

Figure 7.11. Probe-rod profile across section of marsh in Big Bend coastline. Stainless steel rods were pushed into the marsh to measure the thickness or depth to bedrock surface. When plotted, the rugged, irregular nature of the buried karst topography is revealed. In some places the relief exceeds 8 m. The rocky high areas form vegetated marsh islands supported by trees/shrubs. Storm waves and rising sea level eventually wash vegetation off rocky areas, forming exposed pinnacles seen in fig. 7.10*C*. When submerged, these rock pinnacles form nucleation sites for oysters. (Source: A. C. Hine.)

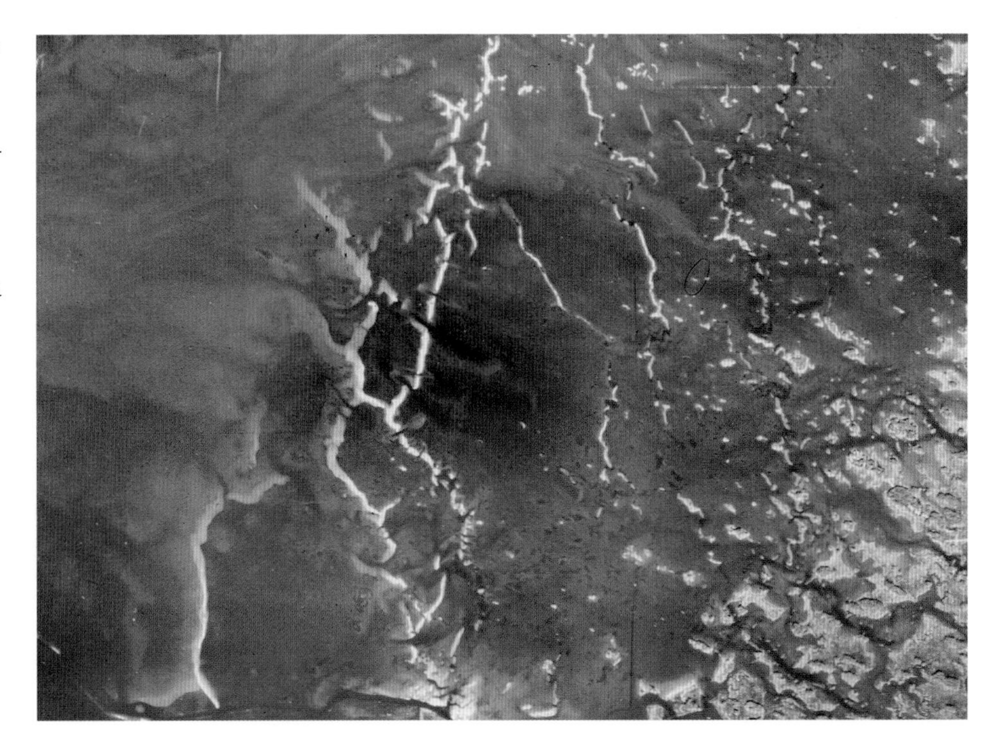

Figure 7.12. Oyster reefs in the Big Bend area. Large, linear oyster bioherms align normal to tidal flushing and occur in areas of low salinity due to spring-fed freshwater input. (Source: USGS EROS Data Center.)

When sea level rises, the vegetation and soil are washed away, and these rocky high zones become encrusted with oysters forming impressive shelly reefs that may extend for tens of kilometers (fig. 7.12). So in this manner surficial karst processes have strongly controlled the geomorphology and sedimentary processes along this important part of the Big Bend coastline. In this part of Florida, due to the many sea level fluctuations in the geologic past, the surficial karst is effectively biologically eroded by boring marine organisms, so no significant relief is ever attained. Additionally, these Eocene rocks were probably not exposed to acidic rainfall and groundwater as long as those carbonates in western Cuba. The Florida carbonate rocks lie in a temperate climate—not a tropical climate such as Cuba, so chemical weathering, in general, has not been as great in Florida. Finally, the carbonates composing the Cuban mogotes were uplifted hundreds of meters above sea level, exposing them to enhanced surficial karst dissolution. The Florida limestone has not been similarly uplifted.

The Deep Subsurface: How Florida Obtained Its Internal Plumbing System

Now that we know how the carbonate rocks exposed on the surface of Florida have been carved, etched, and shaped by karst processes, what about deeper down into the substrate? It is obvious that cavities near the surface collapse and form sinkholes on the surface, but how does circulation far deeper into this subterranean world work?

As we know from tectonic plate motions producing earthquakes and movements ranging from building mountains to slow subsidence, the Earth's crust is constantly in motion, both vertically and horizontally—more active in certain areas than in others to be sure. All sections of our planet's crust move or have moved. This is one of the key factors that separate Earth from the other planets in our solar system—a highly mobile crust due to the heat engine at the Earth's core! There are even Earth tides that cause movement in the crust roughly akin to how ocean tides move water. Since the rocks for the upper crust are brittle, movement will develop joints, fractures, and faults (movement along fractures). Indeed, it is rare, if not unheard of, for rocks in the upper crust, including the sedimentary rock cover, not to have been fractured to some extent. So we can expect fractures and minor faulting to be nearly ubiquitous in brittle rocks.

Fracturing can be driven by events such as the collision (most vigorous from ~56 Ma to ~50 Ma) of the Greater Antilles Cretaceous volcanic arc with the Bahamas, Florida, and Yucatan passive margin (see chapter 6). Such compressional forces translated into the Florida Platform could have stimulated deep-seated fracturing and faulting in the older carbonate rocks. Or just the normal differential subsidence commonly associated with passive margins could produce fracturing. Either way, the carbonate rock cover in Florida has been thoroughly broken through time, and these fracture traces can be seen where the rocks are exposed on the surface and can be mapped from the air (fig. 7.13).

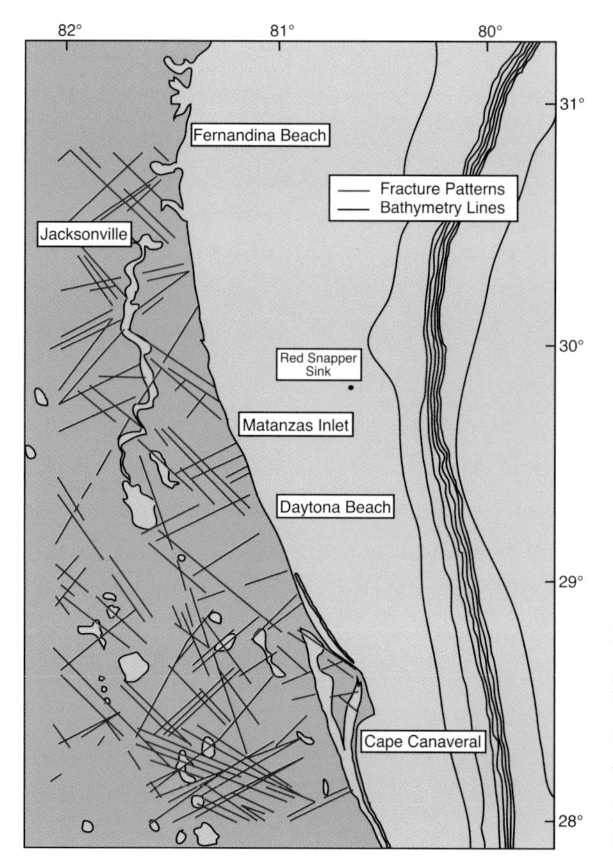

Figure 7.13. Fracture patterns in rocks as seen from aerial photos along the east-central portion of the state. The carbonate rocks constituting the Florida Platform are probably filled with fractures, cracks, and joints that all allow groundwater to easily flow through the platform and ultimately produce cavities, caverns, and caves. (Source: Hine 1997; modified originally from Popenoe et al. 1984.)

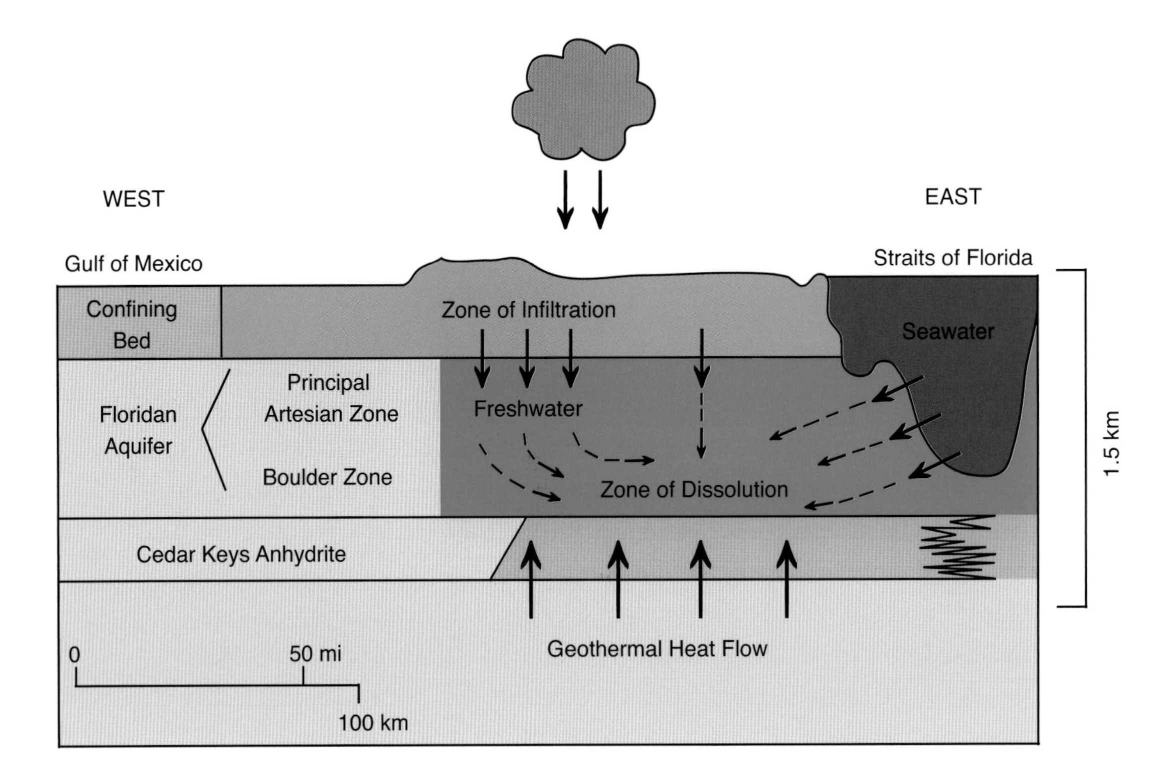

Figure 7.14. The Kohout convection concept. Geothermal heat from below drives the mixing of seawater ingested from the flanks of the carbonate platform with freshwater introduced from above. The mixing of these two water types stimulates dissolution of carbonates. (Modified from Kohout et al. 1988.)

Additionally, there are much older faults associated with the basement rocks that were active during rifting between North America and Africa. It is possible that these much deeper faults became reactivated and caused fracturing in the carbonate rocks that have been deposited above them. Although Florida is not considered to be an earthquake prone area, there are *microseisms* occurring all the time. Every so often there is a ~6.0 magnitude earthquake like the one that was widely felt along the west coast of Florida in 2006, even though the *epicenter* was hundreds of kilometers offshore in the deep Gulf of Mexico. Such movement can continue to fracture the brittle carbonate rocks of the Florida Platform.

Rock fractures, particularly where they intersect, provide subterranean pathways through which fluids can flow at depth. This is called *secondary porosity*, leading to enhanced permeability. Dissolution of carbonate rocks is greatest where groundwater circulation is most vigorous, eventually creating caves, solution channels, and large-diameter pipes or channels that allow tremendous volumes of water to pass quickly through the aquifer with little resistance. *Transmissivity*, or the capacity of an aquifer to transmit water, is one way of measuring the relative ease with which groundwater moves. The greater the transmissivity, the more readily water is able to move through the aquifer. The groundwater enlarges preexisting openings ranging from pore spaces between limestone particles to fractures in the rock. The enlarged spaces eventually form a network of caves, pipes, and other types of conduits, all of which collect and channel even larger volumes of groundwater.

Groundwater is flushed through the Floridan Aquifer by gravity due to elevation differences between the recharge areas, which may be as far away as South Carolina, and the point of withdrawal. Another mechanism, called Kohout convection, was first described by a groundwater geologist, Francis Kohout. He postulated that heat-flow emanating from deeper in the Earth draws seawater into the margins of the Florida Platform, which is mixed with freshwater entering from above. This creates a circulation whereby the two water masses mix, stimulating dissolution at depth. This is a mixing zone, but different than the one that exists along coastal areas (fig. 7.14). The Kohout convection allows for mixing at a much greater depth and on a larger scale.

Geologists have further suggested that where faults and fractures in the rocks are concentrated, creating greater transmissivity, the dissolution is accelerated, allowing for bigger, deeper, and larger cavities. If a large number of these cavities are concentrated, the overlying rocks may collapse, filling the space formerly filled with water. This collapse may propagate upward until it reaches the surface, creating a broad, topographically lower area, which may be hardly distinguishable from the surrounding terrain.

We have called this process *dissolution tectonics*. But the deformation does not result from regional movement of the Earth's crust that defines plate tectonics. This deformation is due to vertical collapse of overlying, younger rocks and sediments collapsing into dissolution-hole complexes and causing fold and sag structures. Where there has been regional subsidence due to this vertical collapse, broad, shallow topographic depressions may form (figs. 7.15, 7.16).

This large-scale collapse from below may be a possible mechanism for the origin of Tampa Bay and Charlotte Harbor. *Seismic reflection profiling* in both of these estuaries reveals numerous sinkholes, sags, warps, and folds in the strata, indicating deformation from deep-seated collapse. These basins then begin to fill with quartz-rich sediments carried into them by local streams, thus filling them to about 90 percent capacity (see chapter 8). These basins may be as deep

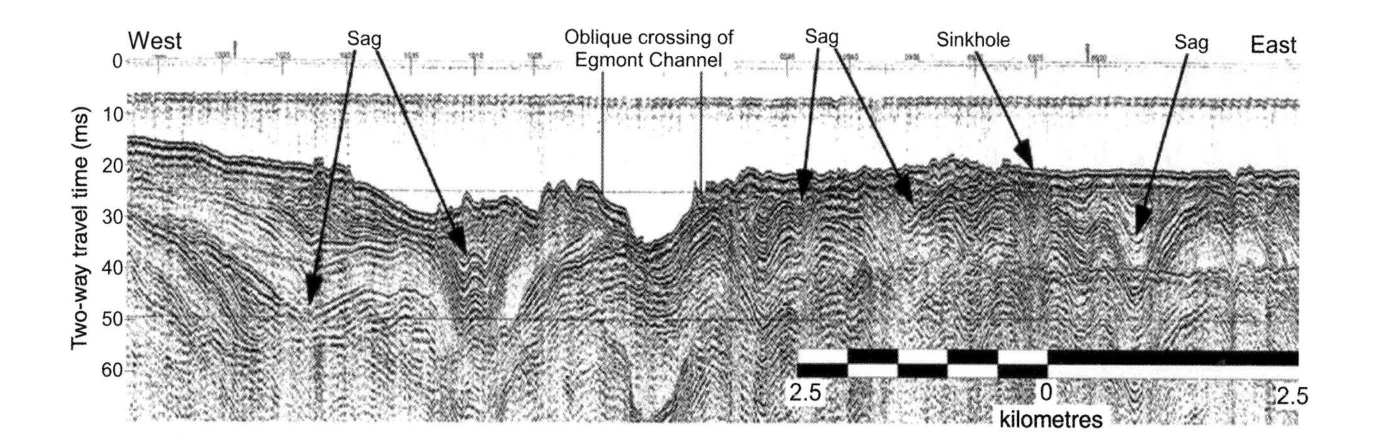

Figure 7.15. High-resolution seismic reflection profile across mouth of Tampa Bay showing folds, warps, sinkholes, and sags—deformation of strata due to deep-seated dissolution of limestone. (Source: Hine et al. 2009; used by permission from Elsevier.)

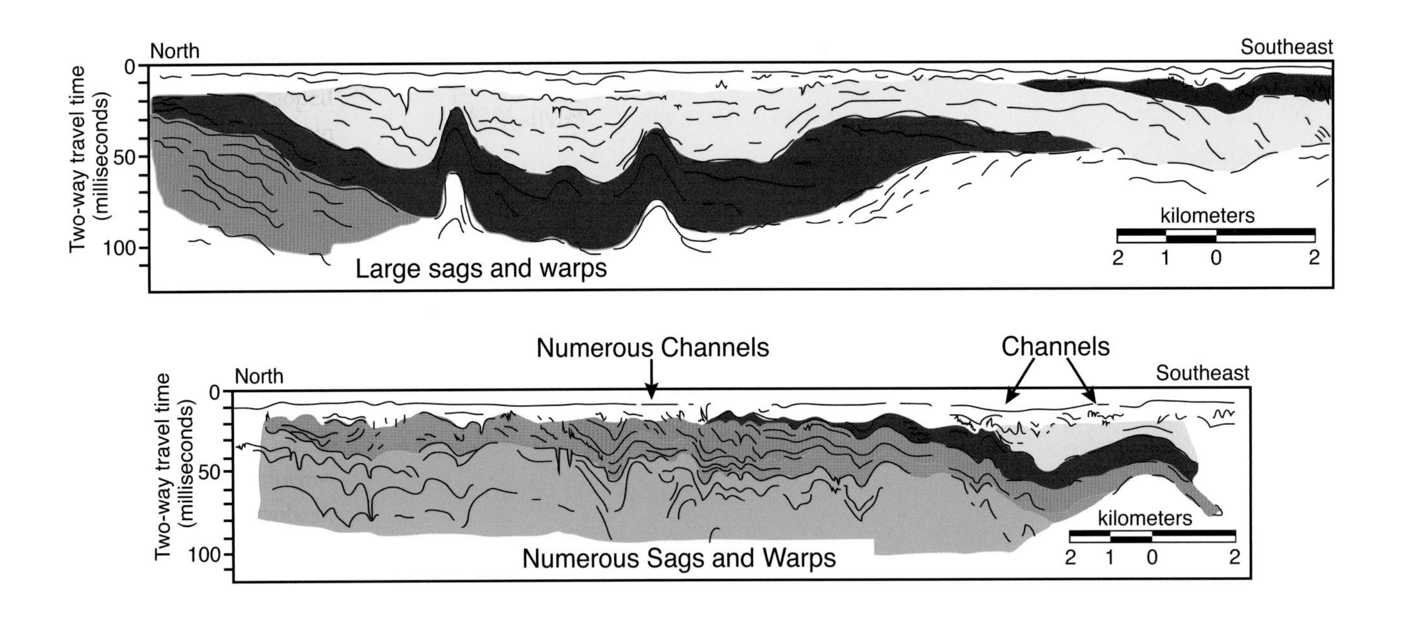

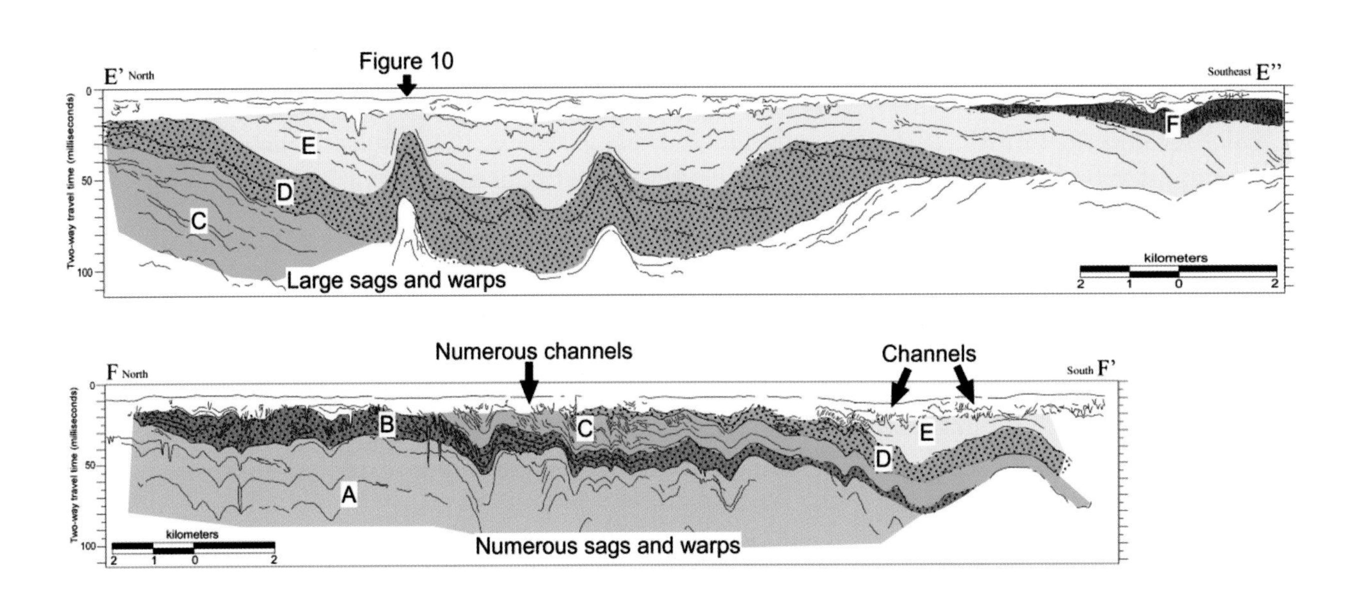

Figure 7.16. Series of interpreted high-resolution seismic reflection profiles from Charlotte Harbor illustrating similar deformation of same age and type of rocks seen in Tampa Bay. Warps, sags, and folds form as a result of deep-seated dissolution of carbonate rocks with the overlying stratigraphy collapsing down into these voids. We have termed this process "dissolution tectonics." (Modified from Evans 1989.)

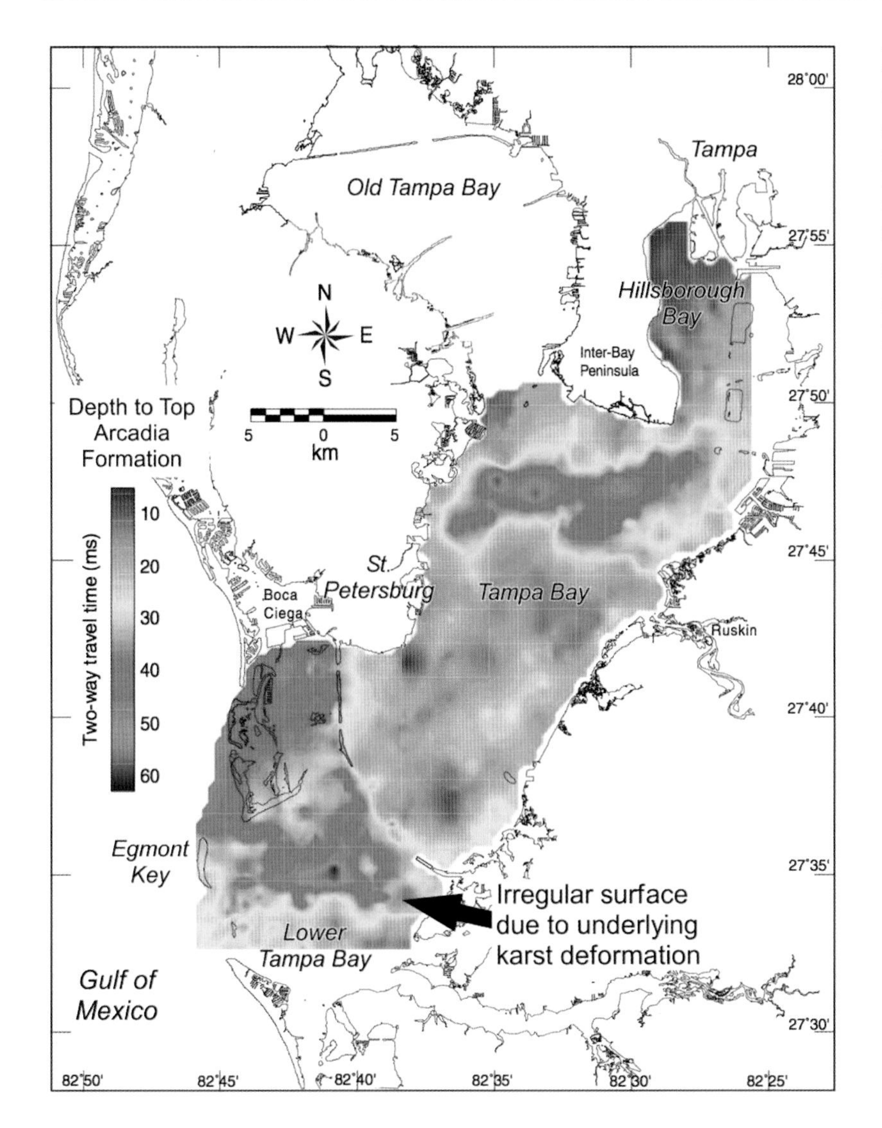

Figure 7.17. Map depicting depth to the top of the deformed limestone surface beneath Tampa Bay—Arcadia Formation—Early Miocene in age. Dimpled surface shows extent of deformation from below by dissolution of deeper carbonate. (Source: Hine et al. 2009; used by permission from Elsevier.)

as 100 m, and they have been filled in with as much as 90 m of sediment. In the case of Tampa Bay, the underlying basin complex has 40–50 m in relief, but the average water depth is only 4 m. If the basins had been completely filled in, there obviously would be no Tampa Bay or Charlotte Harbor—they would all be dry land or wetlands. During sea level lowstand, these depressions would support lakes and bogs, and streams would have flowed into them. During sea level highstand, these broad surficial depressions became flooded with seawater, forming the estuaries we see today (fig. 7.17).

Such collapse may be stimulated during periods of extensive sea level lowstands (subterranean caves filled with air, not water, thus becoming structurally weaker), which have occurred at least three times in the Cenozoic (since 65.5 Ma)—Oligocene, Early Miocene, and Late Miocene. However, deformation at depth is still ongoing as deep, unfilled sinkholes tied to dissolution tectonics 200 m in the subsurface can be seen on the seafloor out on the shelf. One such sinkhole, Red Snapper Sink (off the east coast of Florida, named because of the large numbers of red snapper fish that find refuge in it) was measured to be 133.2 m deep (fig. 7.13). Data reveal sinkholes on the seafloor that have been completely filled. This suggests that if collapse ceases, the holes fill in rapidly with marine sediment. So the process of collapse must be ongoing. It is unclear if the depth of these sinkholes is an indicator of the magnitude of sea level fall.

Comparison between the Florida Platform and the Bahama Platform

Even though we pointed out at the beginning of this chapter that there are large cavities deep in the Bahama Platform, high-resolution seismic data do not reveal the fold, sags, warps, and flexures in the subsurface that we see in the Florida Platform. One explanation might be that the large-scale groundwater system beneath the Bahamas must have been quite different than Florida. A large part of the Florida Platform has been *subaerially* exposed since the Middle Miocene (at ~15 Ma) and is underlain by the Floridan Aquifer, which extends well to the north beneath large portions of the southeast United States. In contrast, there is no huge area of freshwater recharge that is connected to the offshore Bahama Banks. As a result, groundwater movement and mixing-zone dissolution, even during periods of sea level lowstand, was not nearly as great as in the Florida Platform. Hence it appears that the subsurface dissolution activity was not nearly as active within the Bahama Platform as within the Florida Platform. Additionally, freshwater recharge is limited to the relatively small Bahama islands (as compared to the entire banks) during periods of sea level highstand.

With the dissolution of carbonate rock beneath the Florida Platform, a loss of mass has resulted, allowing for some isostatic rebound (uplift). This would allow much older rocks to be exposed and subjected to surficial karst dissolution for significant lengths of time in Florida. For example, Eocene shallow-water limestone (at ~50 Ma) is exposed at the surface in west-central Florida. Rock of the same age, deposited in the same depositional setting (shallow water) on the Bahama Platform, now lies ~1 km buried below sea level, indicating that these two carbonate platforms have had quite different subsidence histories. This differential was possibly due to significant differences in their internal plumbing, water movement, subsequent rock loss due to dissolution, and physical location.

Essential Points to Know

1. Carbonate platforms typically have many large caves, caverns, solution pipes, vugs, cavities, and innumerable holes creating significant porosity and permeability through which water flows.

2. The Floridan Aquifer is one of the most productive in the world, supplying

water for millions of people and supporting their activities in Florida and much of the Southeast.

3. Florida's world-famous freshwater springs result from water emanating upward from the Floridan Aquifer.

4. The dissolution of Florida's carbonate rocks comes from slightly acidic freshwater and the ability to move large quantities of this water through the porosity/permeability of the rocks. The acidity comes from dissolved CO_2 in the water as well as acids from decaying organic matter and pollutants.

5. The shallow groundwater, or surface aquifer, dissolves carbonate rock forming distinctive topographic irregularities and a surface morphology known as karst.

6. The Florida Platform also has an elaborate internal plumbing system deep within the carbonate rock whereby large caverns have formed. The deep internal dissolution is formed by mixing of fluids of different chemistries along specific zones. Some of this deeply buried karst was probably inherited when the rocks were exposed at the surface.

7. Carbonate rocks in Florida originally deposited at the same water depth as rocks of similar age in the Bahama Platform now occupy significantly different elevations—the rocks in the Bahamas are much deeper. This suggests that the Bahama Platform has subsided much more rapidly than the Florida Platform over geologic time scales, possibly because Florida lost much more rock material due to enhanced surface and subsurface rock dissolution. The Florida Platform, being lighter (more caves, vugs—less dense), would not subside as much as the denser Bahama Platform. Indeed, the Florida rocks may have isostatically been uplifted due to its relatively greater loss of rock mass.

Essential Terms to Know

acid/acidic: Any chemical compound that, when dissolved in water, gives a solution with a hydrogen ion (actually the hydronium ion H_30^+) activity greater than in pure water; a pH less than 7.0.

alkaline: Having a pH greater than 7.0; a substance that has a low concentration of hydronium ions and conversely has a high concentration of hydroxyl ions (OH^-).

aquifer: A water-bearing layer of rock or unconsolidated sediments that will yield water in a usable quantity to a well or spring.

dissolution tectonics: Collapse of overlying strata into large cavities created by dissolution of deeper carbonate rocks.

dolomite: Carbonate rock and a mineral, both composed of calcium magnesium carbonate $CaMg\ (CO_3)_2$ found in crystals. Limestone is converted in place to form dolomite by migrating fluids containing Mg. The trans-

8 Sands from the North

The Quartz Sand Invasion (~30 Ma to Present)

> Florida's beaches rank among the best beaches in the country for beauty, accessibility, and facilities. Families, couples, and active singles enjoy year-round sunshine, glittering white sand, and clear Gulf of Mexico or Atlantic Ocean waters.
> Visit Florida

Florida's Sandy Coastline: The Obvious Presence of Abundant Sand

Ever since there have been tourists, people have come to Florida to enjoy its beautiful sandy beaches, vegetated dunes, barrier islands, crystal-clear aquamarine water, and coastal breezes (fig. 8.1).

The interior of the state appeared, at first, to offer little to the casual visitor other than mosquitoes, alligators, snakes, and dense undergrowth (thus prompting the old gag about selling swampland in Florida to gullible Yankees). As a result, Florida's economic development over the early years became heavily dependent upon its unspoiled, alluring coastline. Indeed, many towns and cities in Florida have adopted the word *beach* into their names, thus signaling the main attraction. Miami Beach heads the list, which also includes West Palm Beach, Cocoa Beach, Daytona Beach, Clearwater Beach, and Vero Beach.

However, few tourists probably ever wondered where all this sand came from to create such a coastal paradise. After all, we grow up just naturally assuming that where waves crash on land, there must be sand to dissipate this energy. There are rocky coastlines defined by outcrops, cliffs, and boulders and other coastlines dominated by mud, marsh plants, and mangroves around the world. So it is not necessarily a "given" that all coastlines must be sandy. But much of Florida's coastline has been blessed with sandy barrier islands fronted by gorgeous beaches. The remainder of Florida's coastline consists of the Florida Keys string of rocky islands (see chapter 10) and two plant-dominated coastlines—the mangroves of the Ten Thousand Island

---

Epigraph Source: http://www.visitflorida.com/beaches

Figure 8.1*A*. The foredune ridge and beach at Caladesi Island. Most of the sediments that built this scenery are quartz-rich. (Photo by A. C. Hine.)

Figure 8.1*B*. Sanibel Island, one of the widest barrier islands along Florida's Gulf of Mexico coastline because it lies at the end of a longshore transport system. Most of the sediments composing this barrier island are quartz-rich. But there are a few outstanding shell collecting sites along the northern end of this island. The shells are made of carbonate minerals. (Photo by A. C. Hine.)

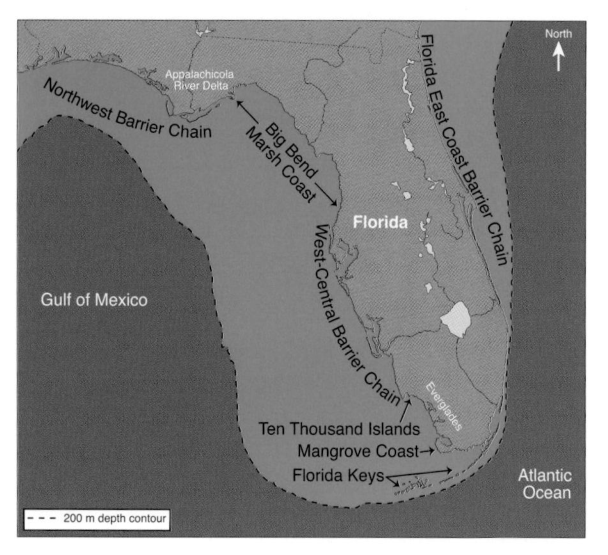

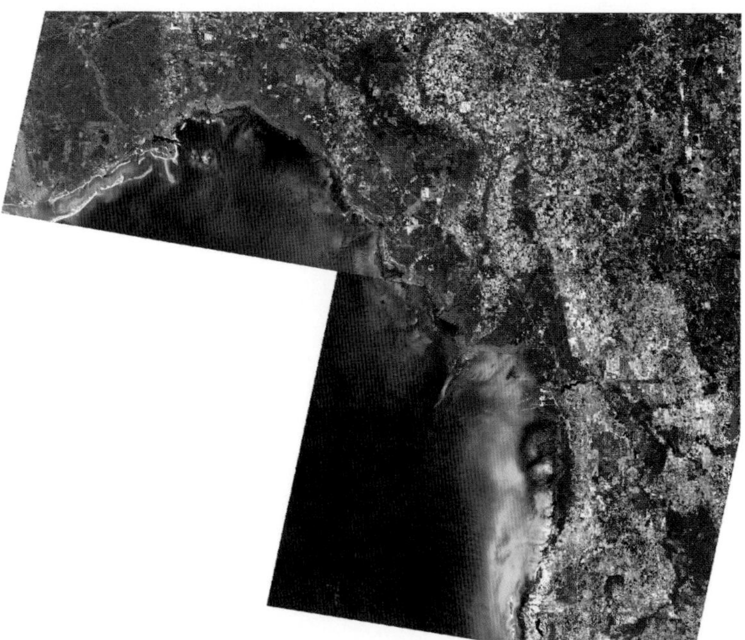

Figure 8.2*A*. Classification system of Florida coastline illustrating both sandy coastlines and large sections dominated by plants. (Source: Davis 1997; used by permission from author.)

Figure 8.2*B*. The Big Bend coastline dominated by marsh plants due to low wave energy and lack of sand-sized sediments to form beaches. (Source: Digital orthophoto quarter quadrangle imagery; http://gisdata.usgs.gov/metadata/doqq.htm.)

Figure 8.2*C*. Part of the northwest barrier chain (St. Vincent Island) shown in fig. 8.2*A*. These are quartz-rich sand barrier islands that derived from the Apalachicola River delta. (Source: USGS EROS Data Center.)

Figure 8.2*D*. The Ten Thousand Island coast is dominated by mangrove colonies. Low wave energy and a lack of sand preventing barrier islands from forming allow this plant-dominated coast to flourish. (Source: Digital orthophoto quarter quadrangle imagery; http://gisdata.usgs.gov/metadata/doqq.htm.)

coast in SW Florida and the marsh grasses of the Big Bend coastline, both facing the open Gulf of Mexico (fig. 8.2). It may surprise some, but ~40 percent of Florida's coastline has no sand.

Quartz Sand Cover

As we have seen in previous chapters, Florida consists primarily of limestone, a type of sedimentary rock, resting upon a much older *basement* of igneous, metamorphic, and some sedimentary rocks such as sandstone and shale (see chapter 3).

We never see these basement rocks on the surface as they have been covered by carbonate (thicker limestone and dolomite) and thinner evaporite strata—in some places up to 6 km thick. Since Florida is primarily a large limestone block, so to speak, formed from the accumulation and *cementation* of countless animal skeletons and plants secreting calcium carbonate ($CaCO_3$) occurring over millions of years (see chapter 4), would it not be reasonable to assume that the modern sandy beaches might consist of similar skeletal material? Yes, of course. And if we go to any beach anywhere, we can see many shells (exoskeletons of mollusks). Additionally, in the Florida

Figure 8.3. Microscope view of mostly carbonate sand from a beach consisting of broken shells. Carbonate sand is generally the coarsest material on Florida's beaches, sometimes forming thick carbonate gravel-sized (>2 mm diameter) deposits. (Source: Dr. S. D. Locker.)

Keys, the small beaches consist almost entirely of broken plant and animal skeletal debris (fig. 8.3). These are paradoxically rare in Florida and are confined to that string of rocky islands terminating at Key West. Similar beaches are also found in the Dry Tortugas area located further to the west.

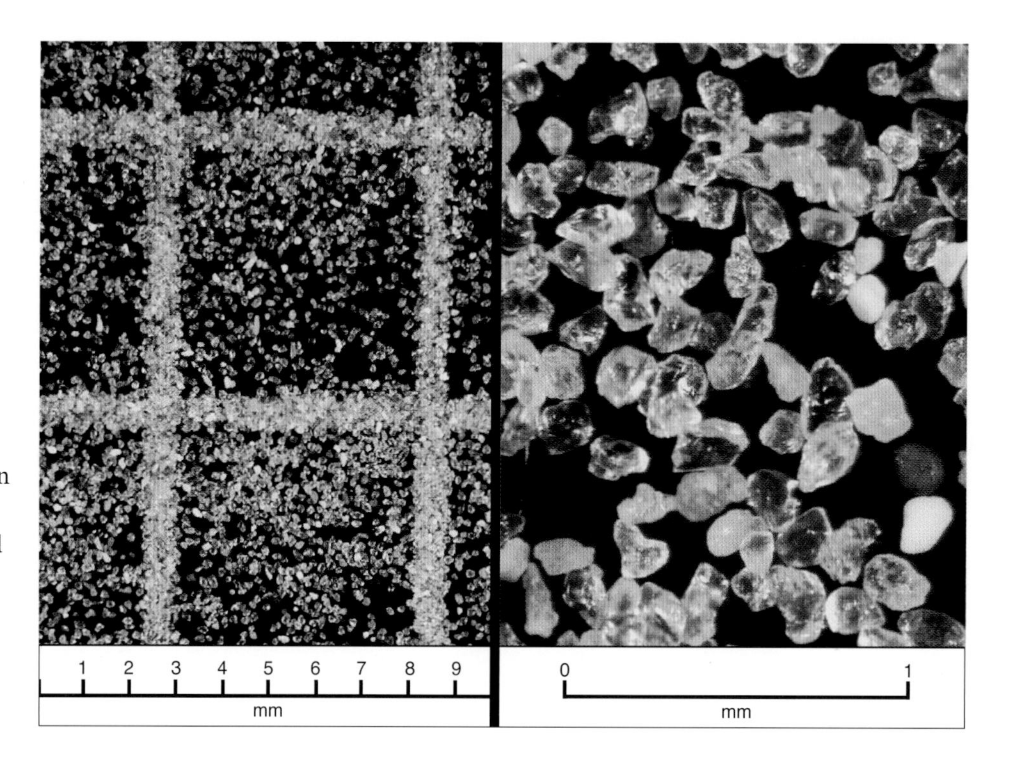

Figure 8.4. Two views of the same quartz sand through a microscope at different magnifications. In Florida, the quartz sands are generally finer-grained than the carbonate fraction and well-sorted. All are ~.25 mm. They have a clear, glassy appearance. (Source: Dr. S. D. Locker.)

From Pensacola to the Apalachicola River delta along the panhandle, from Anclote Key to Cape Romano along the west-central peninsular Florida Gulf of Mexico coast, and from Jacksonville to just south of Miami on Florida's east coast, most of the sand constituting the beach is not the limestone-producing, calcium carbonate skeletal material, but grains of the mineral quartz (SiO_2)—one of the most durable and most abundant of all minerals on Earth. These are *siliciclastic* beaches—*silici* meaning silicon-based and *clastic* meaning fragments of preexisting rocks (fig. 8.4).

Most of the Florida coastline is dominated by quartz-rich sand. Indeed, most of the state, except for the Everglades and the Keys area, is covered with quartz-rich sediment—generally a relatively thin cover of a few meters on top of the limestone. The quartz sand even extends offshore out onto the continental shelf both off the east coast and as far as 40 km offshore onto the west Florida shelf. Here, the quartz sand cover is defined by numerous

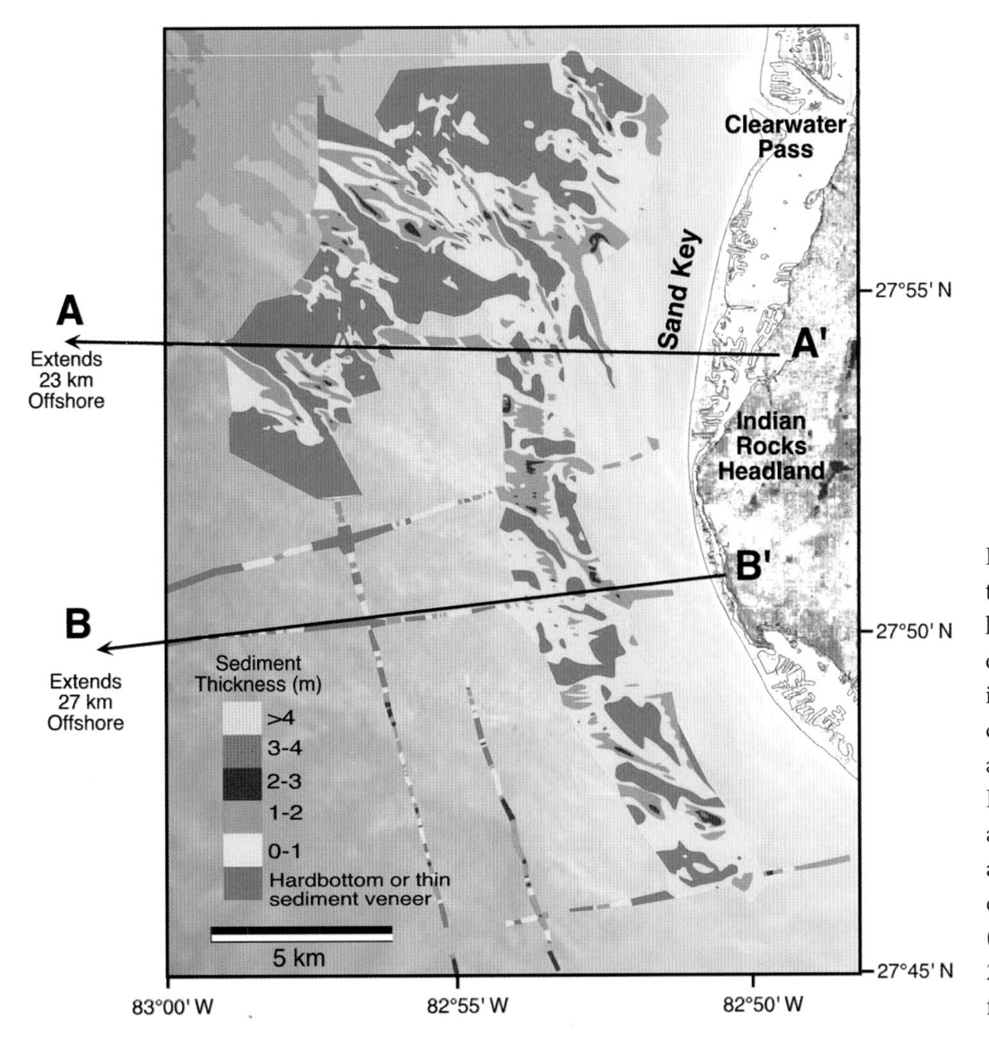

Figure 8.5*A*. Sediment thickness map off Pinellas County in the Gulf of Mexico. The sediment in these ridges is mostly quartz sand. The ridges are generally <3 m thick. In between the ridges are areas of coarse carbonate sand and gravel or exposed Miocene bedrock (Source: Edwards et al. 2003; used by permission from Elsevier.)

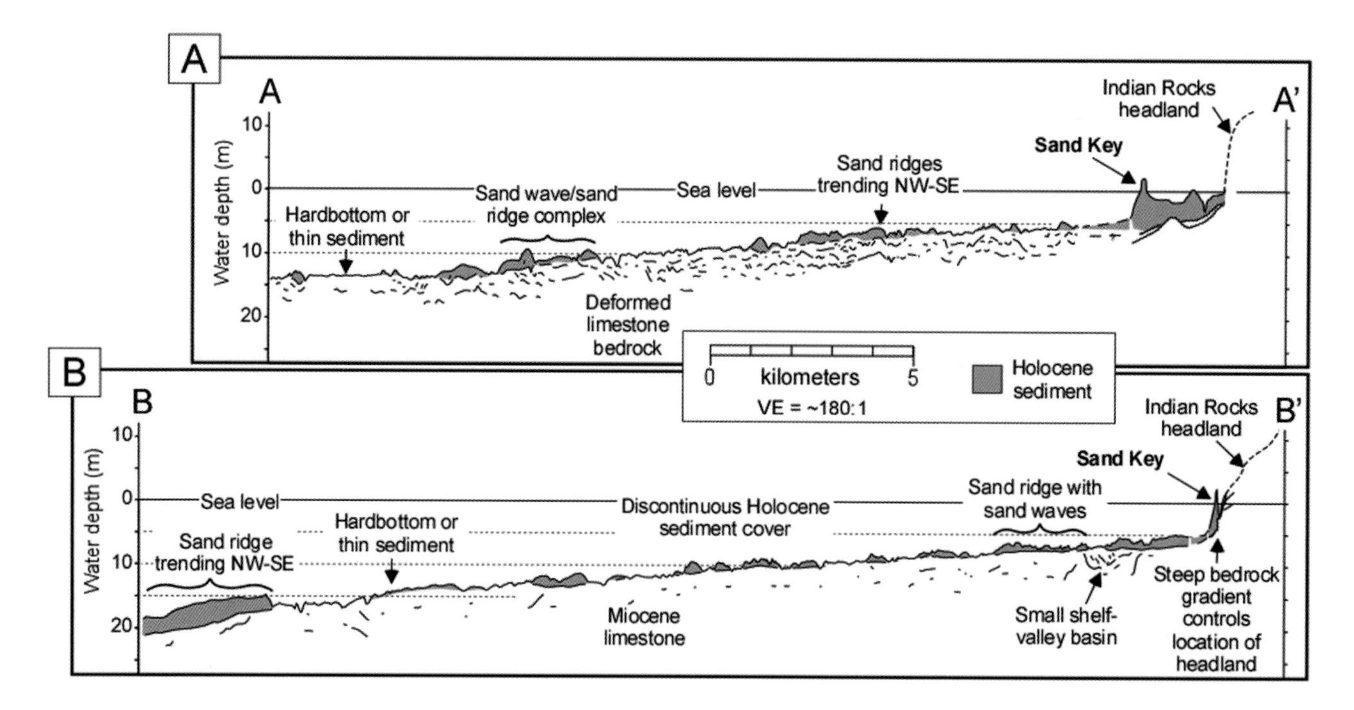

Figure 8.5*B*. Two cross sections traversing the inner shelf off Pinellas County. Note the overall thin sedimentary veneer lying on top of deformed ancient limestone. These sediments are mostly quartz sand with some coarse shell fragments. Note that the thickest portion along these cross sections constitutes the barrier islands (Sand Key) themselves. (Source: Edwards et al. 2003; used by permission from Elsevier.)

long linear sand bodies up to 4 m thick separated by exposed Cenozoic age limestone (fig. 8.5). But, amazingly, in a few areas, quartz-rich *stratigraphic units* are up to ~150 m thick.

Source of the Quartz Sand

Quartz sand grains come from the weathering of igneous rocks such as granite, metamorphic rocks such as quartz-rich *schists*, *quartzites*, and *gneisses*, and well-cemented, sedimentary rocks called *quartz-rich sandstones*. These early rocks upon which later sedimentary rocks have been deposited are commonly called basement rocks or bedrock. However, if Florida's basement is inaccessible to weathering by being located thousands of meters below the surface, we have to look elsewhere for the source.

The closest exposed bedrock to yield quartz-rich sediments constitutes the southern Appalachian Mountains and the adjacent piedmont in northern Georgia, South Carolina, and North Carolina. The Appalachian Mountains were formed in the mid- to late Paleozoic era during the assembling of then supercontinent Pangea, and they probably rivaled the modern Himalaya Mountains in elevation and grandeur (fig. 8.6).

Over the past ~250 Myr, the Appalachian Mountains have been reduced from peaks possibly exceeding 8,000 m (Mount Everest is 8,852 m) to peaks that are less than 2,000 m high. The approximate elevation difference between the Appalachian Mountains at their zenith and today's worn-down peaks is about 6,800 m. So at least this much material has been eroded. However, due to isostatic rebound effects during the long-term erosion process, more than

Figure 8.6*A*. Photograph of Mt. Everest, the highest point on Earth at 8,852 m. This relatively young geologically mountain scenery exists because of the large volume of rock that has been removed, mostly by physical erosion (glaciers) and continued uplift due to tectonic forces. These huge peaks are the only remnants of this vigorous erosion process. Most of the removed rock materials are sediments that now lie in northern India, Bangladesh, or out on the seafloor of the northern Indian Ocean. (Photo by Ron Blakey; used by permission from Panoramic Images.)

Figure 8.6*B*. Mt. Mitchell, the highest peak east of the Mississippi River valley in the southern Appalachian Mountains of North Carolina at 2,038 m elevation. This much older mountain chain has had many kilometers (e.g., 6.8 km) of rock removed, which has formed much of the continental margin of the Carolinas and Georgia and provided quartz sand (and some gravel) to cover the Florida Platform. The Appalachian Mountains may have been as high as the Himalaya Mountains. (Photo by A. C. Hine.)

6,800 m of rock have been weathered away. As rock mass is removed, the Earth's crust rebounds vertically due to the reduction in the lithostatic load, thus exposing even more rock mass for removal (think of cutting off the top of an iceberg at the waterline—the iceberg will still rise up as a result of its buoyancy, but not as high as before). So how does this erosion process work? How can that much rock be stripped off, and then where does it all go?

Removing Rocks: The Weathering Processes

The processes acting on the surface of the earth are amazingly active and can literally move mountains, piece by piece, given enough time. Through chemical and physical weathering upon exposure to the atmosphere and water, solid, massive *crystalline bedrock* can be broken down into smaller sedimentary components. Some minerals, such as *mica*, *feldspar*, and other *silicates,* in the igneous and metamorphic rocks constituting the mountains break down and produce silt and clay-sized sedimentary particles. Because Georgia, South Carolina, and North Carolina lie in a warm temperate climate zone, *chemical weathering,* required to produce these fine-grained particles from bedrock, is enhanced. In both the mountains and the *piedmont,* the bedrock is heavily weathered chemically so that it is soft and friable, commonly retaining the structure of the parent rock—this altered profile is called a *saprolite*. Even at the highest elevations in the North Carolina mountains, for example, thick, reddish soil profiles are evident (fig. 8.7).

As a result, the streams and rivers that now flow from the piedmont through the coastal plain and eventually to the ocean are heavily laden with suspended, mud-sized particles, particularly during high rainfall events. The streams along the coastal plain of the southeastern United States sometimes flow brick-red in color after floods due to this high-suspended load of chemically weathered bedrock. Just compare the color of these southern rivers to the rivers in New England where cooler climate dominates, enhancing *physical weathering* and producing less mud. In relative terms, these freshwater conduits up north are much clearer and more transparent than their southern counterparts.

Importance of Eroded Continent Bedrock to Coastlines

There is a remarkable benefit to this high *suspended sediment load* and input to the coastal zone of the Carolinas and Georgia in particular. The Georgia coastline is dominated by offshore barrier islands (fig. 8.8*A*). Between the barrier islands and the mainland are lagoons that have been filled by these fine-grained sediments (*mud-sized*), supporting one of largest marsh systems in the world. If this coastline had been starved of these fine-grained

Figure 8.7*A*. Trench cut into mountain slope at ~1700 m elevation in the Blue Ridge Mountains of North Carolina, revealing highly weathered, broken bedrock due to physical processes and reddish fine-grained sediment, which is clay derived from chemical processes. (Photo by A. C. Hine.)

Figure 8.7*B*. Outcrop of heavily weathered bedrock showing original folded strata in the Blue Ridge Mountains. These were hard rocks that have now been converted to easily removed clay particles by chemical weathering. (Photo by A. C. Hine.)

Figure 8.8*A*. Vertical infrared image of South Carolina–Georgia coastline with the Savannah River defining the boundary between the two states. Note the extensive marsh system shown in green behind and in between the barrier islands (modern and Pleistocene) shown in red. Without the large suspended load of muddy sediments originally derived from the piedmont and the mountains, these marshes could not have formed. These muddy sediments ultimately came from the sources shown in fig. 8.7. (Source: Pandion Books, from Hayes et al. 2008.)

Figure 8.8*B*. Aerial image of Caladesi Island off the west-central coastline of Florida illustrating an open lagoon with no marsh system, largely due to the lack of muds being introduced from the land. Much of Florida is devoid of these continental weathered muds. Hence there is no material to fill in the back barrier island lagoons like those along the coast of the SE United States as shown in fig. 8.8*A*. (Source: Digital orthophoto quarter quadrangle imagery; http://gisdata.usgs.gov/metadata/doqq.htm.)

sediments, most likely there would be open water between the mainland and these barriers, preventing one of the most important coastal ecosystems from developing.

In contrast, the lagoons and bays behind the barrier islands in Florida are open bodies of water lacking marshes because there are no suspended sediment-laden streams of mud to provide the needed substrate (fig. 8.8*B*). Many of the streams entering Florida's coastal systems are coffee-colored and light-impenetrable—not due to suspended sediments from faraway mountains, but from locally derived, dissolved organic matter. Of course, much of the freshwater entering the ocean along Florida's coastal system is crystal clear emanating from nearby springs. Either way, most of Florida's coastal system does not receive the fine-grained, weather-derived sediments as do the Carolina and Georgia coastlines.

The *bedload* of these southeastern rivers and streams carry the quartz-dominated *sand-sized* particles to the ocean that now form the offshore barriers. So what were once crystals locked together forming mountain bedrock have been neatly partitioned into mud (mostly clay minerals) and quartz sand, each forming a different but critical portion of our coastline.

Through millions of years and numerous climate changes, weathering processes have released sediments bound by the bedrock. Streams have carried these sediments to rivers, and rivers have carried these sediments to the sea, creating one of the longest and widest coastal plain/continental shelf/continental slope systems in the world. In places, the coastal plain/continental shelf/continental slope consisting of sediments eroded from the mountains off to the west forms a wedge that is at least several kilometers thick.

So the rock mass that disappeared to reduce the elevation of the southern Appalachian Mountains so dramatically over time was transported *east* to build up the Georgia, South Carolina, North Carolina coastal plain/continental shelf/continental slope. And by doing so, it extended the continental margin into the Atlantic Ocean many tens of kilometers.

And these sediments went *south* to cover the Florida carbonate platform—about 1,000 km to the south (fig. 8.9)!

Introducing Quartz Sediments to Florida

A limestone or carbonate platform, like the modern-day Bahama Banks, requires a stable environment consisting of clear, well-illuminated, shallow, warm water of normal salinity (~35 psu—practical salinity units, traditionally called parts/1,000) for the carbonate-secreting biota to produce carbonate sediments. Allowing for long-term slow (1 cm/kyr or 10 m/Myr) subsidence, which is normal in tectonically inactive areas, and the interstitial cementation of the skeletal sedimentary particles together, thick limestone

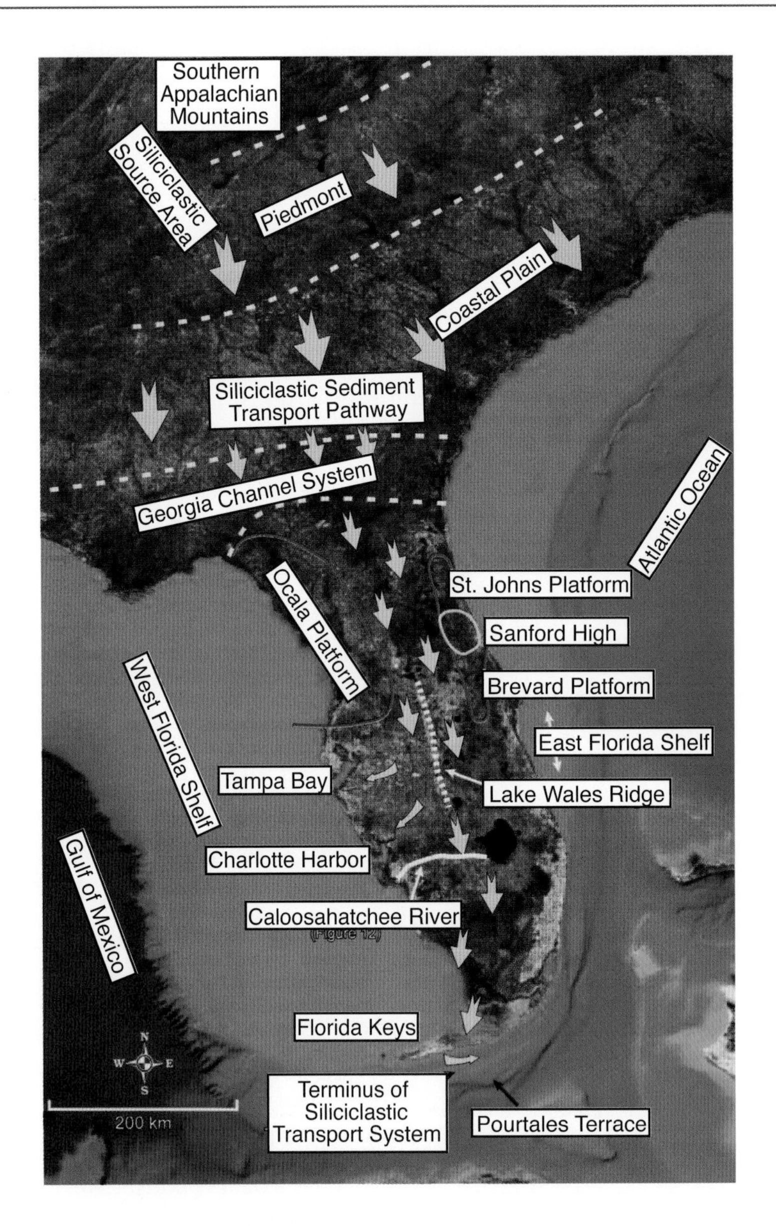

Figure 8.9. The siliciclastic sediment transport pathway from the weathered source rocks in the Appalachian Mountains to beyond the south tip of peninsular Florida. These sediments built up the broad coastal plain that defines eastern North Carolina, South Carolina, and Georgia. (Source: Hine et al. 2009; used by permission from International Association of Sedimentologists and Wiley-Blackwell.)

strata can be produced (see chapter 4). This shallow-water carbonate factory will be productive as long as it is not stressed by large changes in water depth, water temperature, nutrients, and overall water clarity (light is needed for photosynthesis). The carbonate factory can be buried by the influx of other sedimentary material. The carbonate factory cannot survive close to environments such as river deltas discharging *turbid*, nutrient-rich freshwater that is associated with a large land mass.

The Florida carbonate platform was isolated from the rivers draining the southern Appalachian Mountains by the Georgia Seaway Channel from the Early Cretaceous (~100 Ma) to the Oligocene (~30 Ma). This seaway, through which the early Loop Current flowed from the Gulf of Mexico northeast to the Atlantic Ocean, formed a dynamic barrier, which prevented turbid, sediment-laden, and nutrient-rich waters from reaching the Florida platform (fig. 4.2). As a result, the broad, shallow seas covering Florida during extended periods of sea-level highstand remained crystal clear, allowing the carbonate factory to flourish.

However, as a result of multiple prolonged sea level lowstands starting about 30 Ma (Early Oligocene—see sea level curve on page ii), the Georgia Seaway began to fill up with sediments being shed off the mountains to the north. Eventually, the seaway was completely buried, allowing river deltas carrying muddy water to reach the Florida platform. As a result, the carbonate factory began to shut down in northern peninsular Florida. The Florida Platform became buried by quartz-rich sediments. The filling of the Georgia Seaway also provided a land bridge, allowing a variety of terrestrial animals to occupy Florida during the mid-Cenozoic.

Geologists have debated whether erosion of the southern Appalachian Mountains and subsequent sediment production were accelerated due to regional *tectonic uplift* temporarily increasing mountain elevation and enhancing weathering. Alternatively, warmer climates earlier in the Eocene (40 Ma) could have stimulated chemical weathering, erosion, and sediment release. By the time the extended sea level lowstands occurred some 10–15 Myr later, there was abundant siliciclastic sediment poised to fill in the Georgia Seaway and enter the Florida Platform. There is ample evidence for a global Eocene warming, but little geophysical evidence for regional tectonic uplift and a period of rejuvenated mountain building in the southern Appalachian Mountains during this period. Undoubtedly, steady isostatic uplift occurred continuously as erosion stripped the mountains of their rock mass.

Regardless, the advent of significant siliciclastic sedimentation onto the Florida Platform during the Early Oligocene marked a fundamental and permanent change in sedimentation and depositional patterns in the state's (and entire SE United States) geologic history, which was ultimately related to global events. The extended sea level lowstands were probably related to a general cooling of the Earth's climate and the accumulation of ice on Antarctica. Additionally, this was a period of global ocean circulation change as the equatorial tropical circulation became restricted due to the closing of the *Tethys Ocean*. This led to major extinctions of Gulf of Mexico faunas. So the Late Eocene to the mid-Oligocene (~40 Ma to 28 Ma) was a period of significant transition.

Eventually, quartz-rich sediments covered all of the present state including the continental shelf to the east and about 40 km offshore to the west. However, explaining how the quartz-rich sand and gravel moved from the former location of the Georgia Seaway to south-central peninsular Florida has been controversial. Some geologists have said that the rivers draining the southern Appalachian Mountains continued to the south carrying these sediments great distances down the peninsula. Others have pointed out that peninsular Florida is a topographically high area and that it would have been impossible for rivers to flow south essentially along a ridge. Instead, the regional drainage must have been diverted to the west into the northern Gulf of Mexico or to the east into the northern Straits of Florida. The presence of the locally elevated Ocala Platform and the Sanford High (fig. 8.9) in central Florida probably would have prevented any long-distance north-to-south river drainage down peninsular Florida.

Sediment Transport Processes

If one examines the topography of the state, the presence of numerous former beaches, scarps, and shorelines is evident.

This strongly suggests that sediment movement from north to south down this portion of peninsular Florida must have occurred by breaking waves moving sand through *longshore transport*—exactly the same way sand is moved along modern beaches today. This happened when sea level was elevated. During periods when sea level was lower, local streams and small rivers probably eroded into these former shorelines and moved a small amount of sediment to the west and to the east.

Overall, the breaking-wave-driven longshore sediment transport system probably extended all the way down to the southern end of the Lake Wales Ridge (figs. 8.9, 8.10). This ridge is a paleo-shoreline complex complete with sand dunes that abruptly ends just north of Lake Okeechobee. As a result, there could not have been any significant longshore sand or gravel transport south of this area as Florida becomes much flatter and topographically lower. These elevated paleo-shorelines (linear sand ridges) disappear.

This fundamental topographic change has allowed the vast wetlands constituting the modern Everglades to form. This huge area is so low and so flat that only a modest rise in sea level (~2 m) would turn the Everglades into a shallow marine lagoon, greatly expanding Florida Bay. As a result, a different sediment transport mechanism must have been available to continue the movement of sand and gravel to the south—eventually to the southern Straits of Florida. So how were quartz sediments transported further to the south?

From ~6 Ma to ~3 Ma, an enormous deposit of siliciclastic sediment, some of it gravel-sized, was deposited from the Caloosahatchee River area to the top of the Pourtales Terrace now lying in 200–300 m of water (figs. 8.11, 8.12).

This siliciclastic-rich sedimentary unit lies beneath the modern Everglades and extends south seaward beneath the Florida Keys. It is ~100 km wide and ~150 m thick and is called by geologists the upper Peace River Formation and the Long Key Formation. It represents an important geologic event that is not well understood—the emplacement of all this quartz sand. Additionally, this thick quartz-rich unit is capped by thinner limestone units that now cover most of south Florida and immediately underlie towns and cities in Broward, Dade,

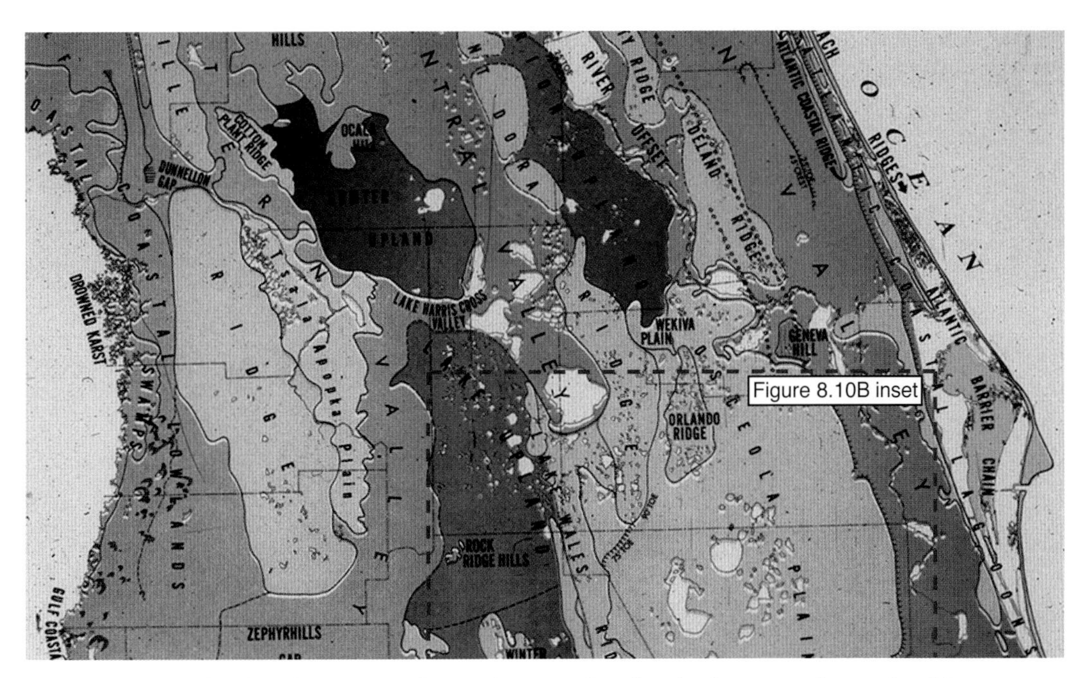

Figure 8.10*A*. Surficial geologic map of central peninsular Florida showing paleo sea level features such as shorelines and scarps. The different colors represent different coastal deposits formed by different elevations of sea level at various times in the geologic past. (Source: White 1970; Florida Geological Survey.)

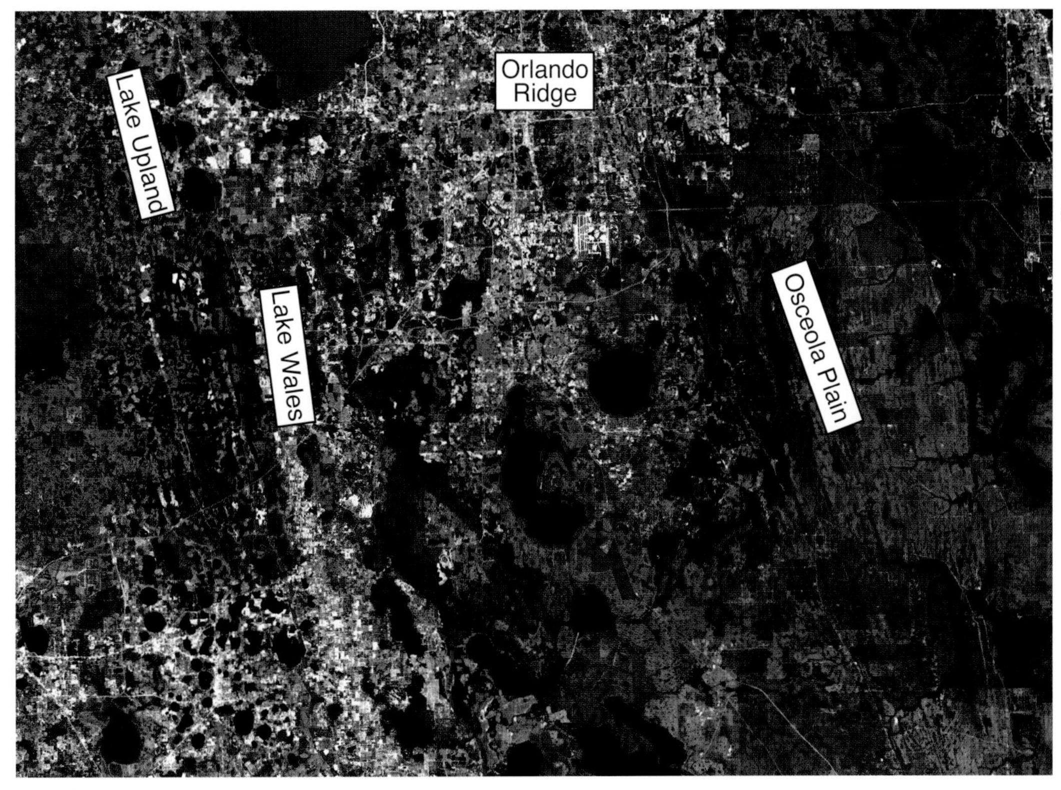

Figure 8.10*B*. Vertical image of eastern Florida illustrating section of fig. 8.10*A* showing linear features that are paleo-shorelines deposited during periods of high sea level. (Source: Digital orthophoto quarter quadrangle imagery; http://gisdata.usgs.gov/metadata/doqq.htm.)

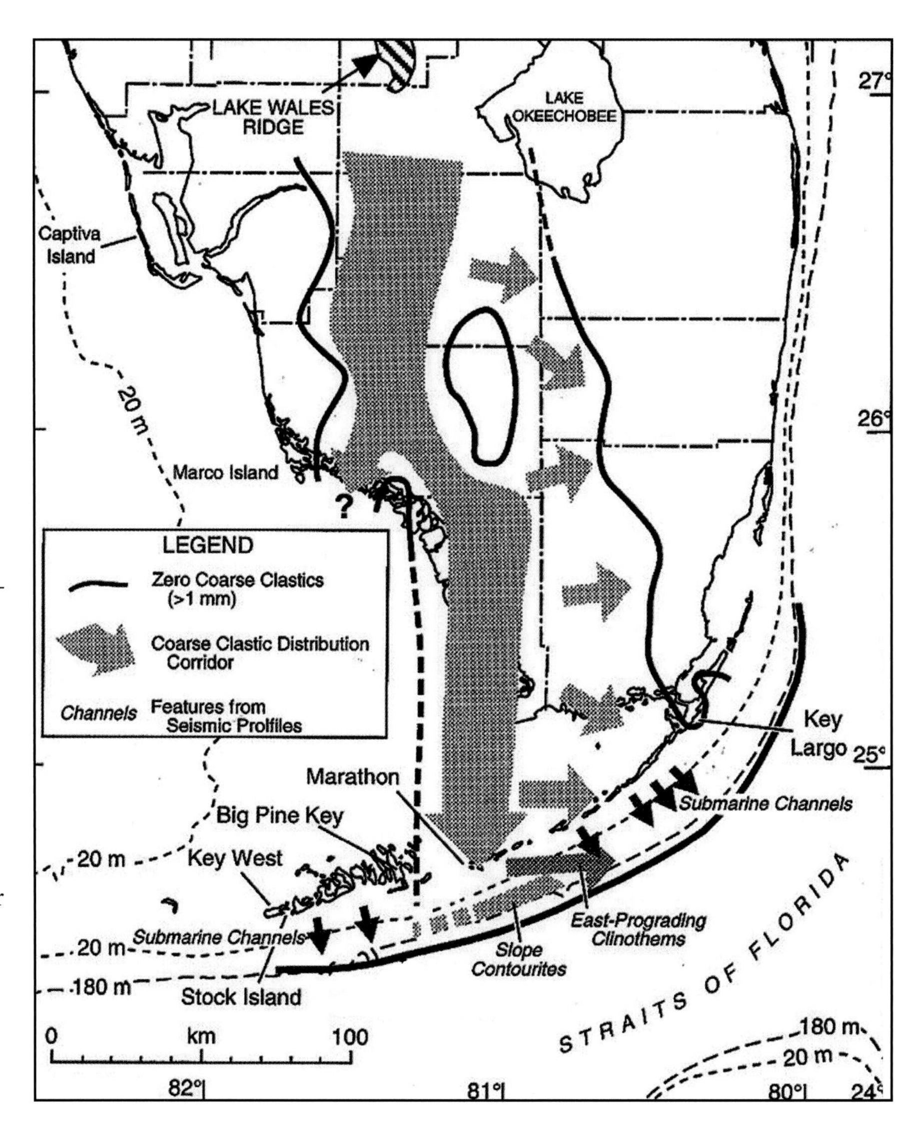

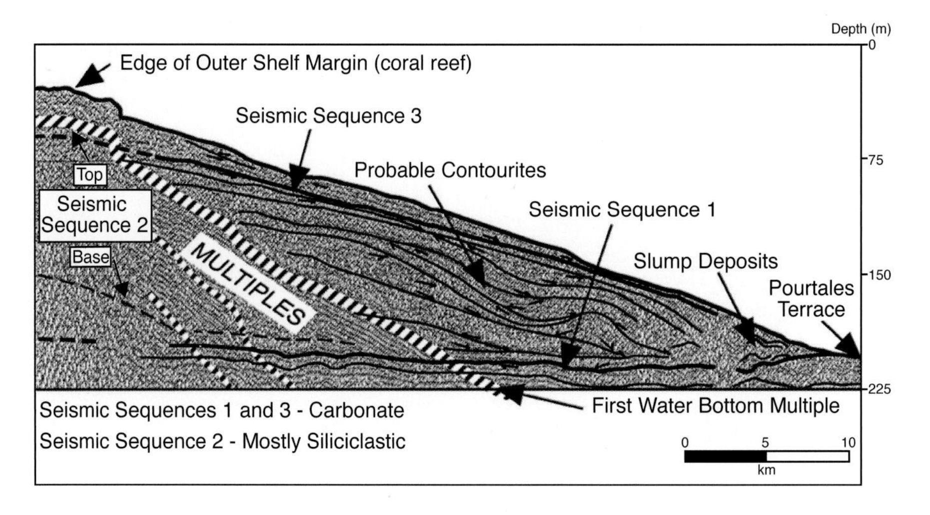

Top: Figure 8.11. Map of quartz-rich sediment transport pathway via prograding river deltas south of the Lake Wales Ridge. These sediments were transported all the way to the Straits of Florida from 5–3 Ma. This is not the modern quartz sediment transport system that now lies within the longshore transport system along Florida's beaches. (Source: Warzeski et al. 1996; used by permission from the Society for Sedimentary Geology.)

Bottom: Figure 8.12. Interpreted seismic profile illustrating quartz-rich sediments downlapping (Sequence 2) and burying a portion of the Pourtales Terrace—an older marine erosional surface—here shown to be 200 m water depth in the Straits of Florida. This was probably the southernmost extent of siliciclastic sediment transport on peninsular Florida. Pleistocene and modern carbonate rocks underlying the Everglades and Florida Bay and forming the Florida Keys and its reefs and sediments overlie these quartz-rich sediments. The overlying Sequence 3 is composed of carbonates shed from the Pleistocene/modern south Florida margin. (Source: Warzeski et al. 1996; used by permission from the Society for Sedimentary Geology.)

and Monroe Counties as well as the Everglades and the Florida Keys. This return to carbonate deposition indicated a fundamental change in geologic conditions from siliciclastic sediment transport to *in situ* carbonate sediment formation and probably began at ~3 Ma (see chapter 10).

Using sound energy to penetrate geologic strata (*seismic-reflection profiling*—same principle using acoustic energy that oil companies employ to find hydrocarbons), geologists have discovered buried river deltas consisting of mud, quartz sand, and quartz gravel beneath the northern Everglades (fig. 8.13). These

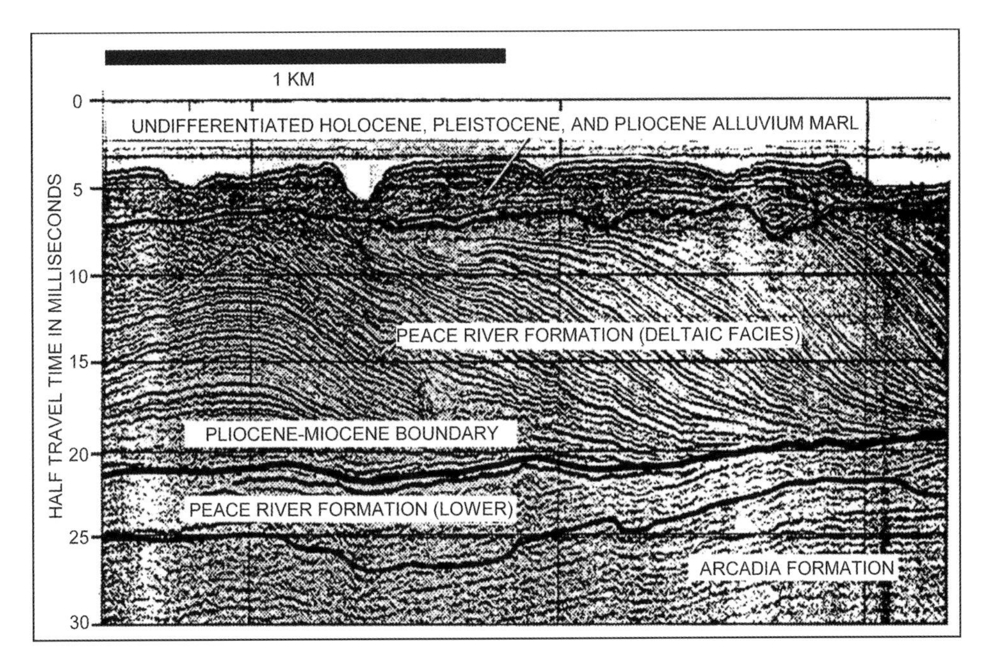

Figure 8.13*A*. Seismic profile illustrating an ancient buried river delta consisting of quartz sand and probably quartz pebbles. This delta lies beneath modern Florida just east of Sanibel Island in the western Caloosahatchee River. Features like this one reveal that enormous volumes of this sediment type were transported by rivers southward down the lower portion of peninsular Florida. (Source: Missimer and Gardner 1976; USGS; used by permission from T. M. Missimer.)

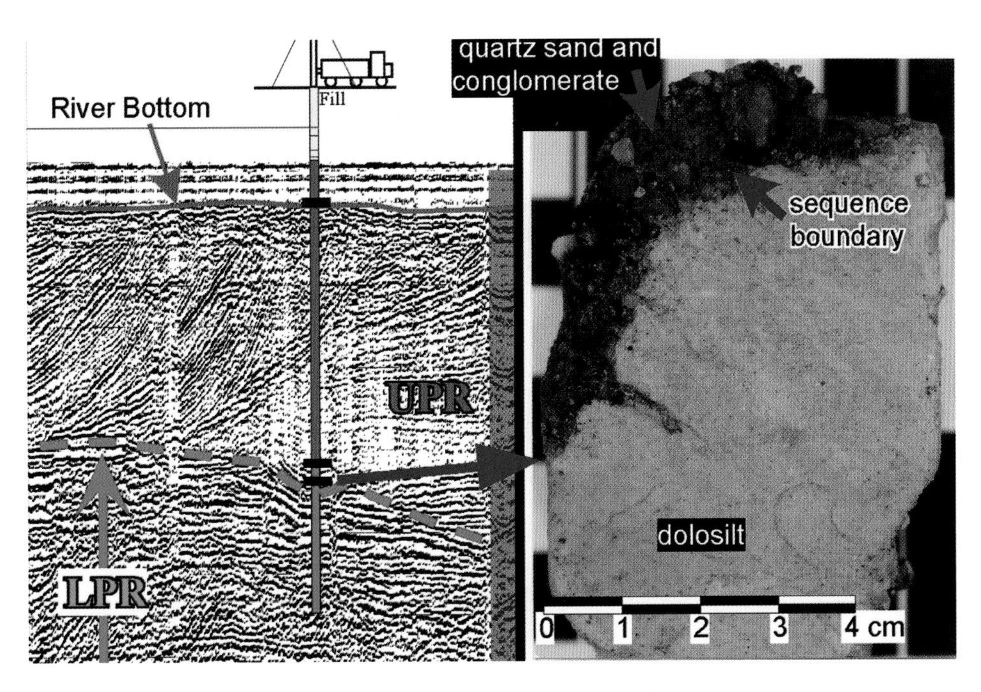

Figure 8.13*B*. Seismic line in the central Caloosahatchee River showing dipping reflectors typical of a prograding river delta similar to that shown in fig. 8.13*A*. Drilling and core retrieval adjacent to seismic line verifies the siliciclastic, deltaic interpretation. Note that the core section contains the sequence boundary between the upper Peace River and the lower Peace Formations. Note the gravel-sized quartz pebbles lying on top of this boundary. (Source: S. D. Locker and K. Cunningham.)

ancient deltas *prograded* southward in waters that must have been nearly 100 m deep. The river deltas migrated nearly 200 km between 5 Ma and 3 Ma all the way to the southern Straits of Florida. Geologists tell us that sea level was as much as ~35 m higher during portions of this time frame. But, more important, sea level fluctuated widely, causing southward delta migration to decelerate during periods of higher sea level and accelerate during periods of lower sea level. Perhaps maximum deltaic progradation occurred during brief intervals or pulses of enhanced sediment discharge during the falling stages of higher-frequency sea-level events (see sea-level curve on page ii). Additionally, while this was going on, smaller streams flowed to the west, filling in karst-generated basins with quartz-rich sediments. These basins did not completely fill with sediments and now form the modern Tampa Bay and Charlotte Harbor estuaries during periods of sea level highstand.

A Different Climate

As we have pointed out, there probably was no long river carrying sands down the length of the north-central peninsula. The longshore transport process was primarily responsible for this movement. So in south-central peninsular Florida, the river deltas must have come from short, high-sediment discharge rivers that eroded the Lake Wales Ridge and earlier quartz-rich sedimentary units. The high-sediment discharge in these local rivers may have come from abnormally high rainfall resulting from more intense thunderstorms and a longer thunderstorm season than seen today.

This epoch of time, called the Pliocene, had a warmer climate than today, and the surrounding marine waters in the Gulf of Mexico and the Straits of Florida were warmer as a result (+2 to +4 °C warmer)—thus possibly significantly stimulating thunderstorm activity and local rainfall. If true, the rainfall runoff provided much more sediment to the local rivers. So it is possible that this unusually warm climate drove the river delta system south from south-central peninsular Florida to the south margin of the platform. This idea, of course, is highly speculative.

Finally, since these land-derived sediments eventually found their way into deep water resting on top of the Pourtales Terrace, which was eroded by the Loop Current–Florida Current (fig. 8.12), we can only surmise that there must have been another fundamental shift in the sediment transport mode. From river deltas, the quartz-rich sediment must have been transported across a shallow-marine continental shelf by waves and currents and then transported downslope by gravity processes—eventually coming to rest in 200–300 m of water or even deeper.

This period of warmth ceased as a result of the Earth approaching the ice ages of the past ~2 Myr—thus turning off the enhanced thunderstorm activity and reducing river transport and river delta migration. Additionally, the added thickness of the quartz-rich sedimentary deposits caused water depths in south Florida to become much shallower, providing the shallow platform for the Everglades, Florida Bay, and the Florida Keys. This decreasing depth allowed the carbonate factory to turn on, covering the siliciclastics with a thin veneer (up to 50 m) of limestone (see chapter 10).

The Entire Siliciclastic Transport System: Beginning to End

This return to carbonate sediment production ended a ~1,000 km long transport system that began with sediments eroding from mountains in the southern Appalachians and finally terminating in a deep seaway within ~150 km of Cuba. This was a sediment transport system that began with small streams feeding into large rivers from the mountains to the coastal plains. These rivers forced deltas to migrate south, filling in the Georgia Seaway and lapping onto the Florida Platform. From there, sediment movement largely occurred in the longshore transport system driven by breaking waves on beaches, extending down peninsular Florida to about the Caloosahatchee/Okeechobee area. There, the transport returned to migrating river deltas, which carried quartz-rich sediments another 200 km, eventually ending in an open-marine shelf and slope depositional environment. This did not occur as one continuous, even process. Rather, it was discontinuous both in time and space with the initial invasion of quartz-rich sediments occurring in northern Florida much earlier than the final transport downslope covering a portion of the Pourtales Terrace at the south end of this system. Most certainly, there were many remobilizations or re-sedimentation events.

This pathway consisted of multiple sedimentary compartments (coastal plains, river deltas, coastlines, karst entrapment basins, and ultimately the open marine shelf/slope) and multiple sedimentary transport processes within and between these compartments. This transport pathway was modulated by multiple sea level fluctuations and climate changes as well, making this source-to-sink process complex and still poorly understood due to the lack of data.

Today, the net southward movement of quartz sand exists in the form of longshore movement on the modern beaches of peninsular Florida. The direction and magnitude of longshore transport is complex due to numerous inlets, headlands, and cuspate forelands and is highly variable on short time scales of weeks to decades and over the past centuries and several millennia. However, the overall net sand movement has been to the south as shown by the southward migrating spits such as Cape Florida along the east coast and Cape Romano and the multiple beach ridges of Sanibel Island along the west coast. This is due to the prevailing northeasterly and northwesterly winds creating a net wave-energy flux to the south.

However, at this moment in geologic time, there is virtually no new sand being introduced from the north either by rivers or by longshore transport across Florida's coastal borders. Along the Gulf of Mexico, the Apalachicola and Suwannee Rivers now discharge very little sand. The small rivers discharging into Tampa Bay and Charlotte Harbor have no *bay-head deltas* that would indicate significant bedload sand movement. For sure, there is no longshore sand transport from the Panhandle beaches to the west-central Florida barrier islands. Such a transport system would have to span the huge sediment-starved, marsh plant–dominated Big Bend coastline (fig. 8.2*B*). Along Florida's east coast, the only source of new beach sand would be that relatively small amount that makes it around the St. Mary's River delta defining the Florida/Georgia border. The lack of new sediment is a first-order reason why more beaches are eroding than not, thus posing chronic problems for many coastal communities.

Without the siliciclastic invasion from the north over the past 30 Myr, perhaps the first tourists would have found little to recommend to their friends, and human history in Florida would have turned out quite differently.

Essential Points to Know

1. The emergent portion of the Florida Platform is mostly covered with a relatively thin veneer of quartz-rich sand, which constitutes most of the beaches around the state.

2. Beneath this veneer is a much thicker limestone section that accumulated for millions of years.

3. These quartz-rich or siliciclastic sediments were formed by weathering of ancient basement rocks such as granitic igneous or metamorphosed rocks constituting the southern Appalachian Mountains.

4. These sediments were transported by streams and rivers to the south and east where they built the coastal plain.

5. Eventually, migrating river deltas filled the Georgia Seaway that separated the Florida Platform from the rest of the southeast portion of North America. This happened at ~30 Ma during a pronounced sea level lowstand with many remobilization and re-sedimentation events. It was not a continuous process.

6. Siliciclastic sediments were then able to enter and cover the Florida Platform during periods of sea level highstand in the coastal longshore transport system. This longshore transport system extended to south-central peninsular Florida at the southern end of the Lake Wales Ridge.

7. At this point, a significant river delta migrated another 200 km carrying quartz-rich sediments to the southern margin of the Florida Platform. Marine sediment transport processes carried these sediments across the shelf and downslope to end on top of the Pourtales Terrace, thus ending a ~1,000 km long siliciclastic sediment transport system.

8. The river deltas were fed by local streams that had probably been augmented by enhanced rainfall during a 2–3 Myr time frame starting at ~5 Ma.

9. Quartz sand still makes its way down peninsular Florida in the longshore transport system associated with the beaches, but very little new sediment is being transported from the southeastern United States onto the Florida Platform.

Essential Terms to Know

basement rocks: The thick foundation of ancient metamorphic and igneous rock that forms the crust of continents, often in the form of granite. Basement rock is contrasted to overlying sedimentary rocks, which were laid down on top of the basement rocks after the continent was formed.

bay-head deltas: River deltas prograding into the head of estuaries or bays as opposed to river deltas that are prograding across a continental shelf.

bedload: Describes particles in a flowing water (usually a river) that are transported along the bottom as opposed to a suspended load where sedimentary particles, generally much finer grained, are carried entirely in suspension.

cementation: Process in which chemically enriched water (fresh groundwater or seawater) flows within spaces between sedimentary grains, precipitating minerals such as calcite to bind or cement the grains together, forming a sedimentary rock.

chemical weathering: Breakdown of basement or crystalline bedrock (or any rock for that matter) by chemical reactions to form sedimentary particles as well as dissolving some of the rock.

crystalline bedrock: Basement rocks that are igneous or metamorphic in origin; essentially the same as basement.

downlapping: Physical relationship between geologic strata or beds whereby a dipping bed terminates downslope onto a flat surface—dipping beds downlap onto and on top of another.

feldspars: A group of rock-forming silicate minerals that make up as much as 60 percent of the Earth's crust. They crystallize from magma in both intrusive and extrusive igneous rocks, as veins, and are also present in many types of metamorphic rock and certain types of sedimentary rock.

gneiss: A common and widely distributed type of metamorphic rock formed by deeply buried igneous and sedimentary rocks exposed to heat and pressure; resembles schist, except that the minerals are arranged into bands and contain fewer mica-like minerals.

isostatic rebound: Earth's crust moving vertically as a result of mass (rocks, sediments, water, ice) being removed. Flow of rock within the mantle compensates for the volume change.

Evans, M. W., A. C. Hine, and D. F. Belknap. "Quaternary Stratigraphy of the Charlotte Harbor Estuarine-Lagoon System, Southwest Florida: Implications of the Carbonate-Siliciclastic Transition." *Marine Geology* 88, no. 3–4 (1989): 319–48. doi: 10.1016/0025-3227(89)90104-7.

Guertin, L. A., D. F. McNeill, B. H. Lidz, and K. J. Cunningham. "Chronologic Model and Transgressive-Regressive Signatures in the Late Neogene Siliciclastic Foundation (Long Key Formation) of the Florida Keys." *Journal of Sedimentary Research* 69, no. 3 (1999): 653–66.

Hayes, M. O., J. Michel, and J. M. Holmes. *A Coast for All Seasons: A Naturalist's Guide to the Coast of South Carolina.* Columbia: Pandion Books, 2008.

Hine, A. C. "Structural and Paleoceanographic Evolution of the Margins of the Florida Platform." In *The Geology of Florida*, ed. A. F. Randazzo and D. S. Jones, 169–94. Gainesville: University Press of Florida, 1997.

Hine, A. C., B. C. Suthard, S. D. Locker, K. J. Cunningham, D. S. Duncan, M. W. Evans, and R. A. Morton. "Karst Subbasins and Their Relation to the Transport of Tertiary Siliciclastic Sediments on the Florida Platform." In *Perspectives in Carbonate Geology: A Tribute to the Career of Robert Nathan Ginsburg*, ed. P. K. Swart, G. P. Eberli, and J. A. McKenzie, 179–97. Hoboken: Wiley-Blackwell, 2009.

McKinney, M. L. "Suwannee Channel of the Paleogene Coastal Plain: Support for the "Carbonate Suppression" Model of Basin Formation." *Geology* 12, no. 6 (1984): 343–45. doi: 10.1130/0091-7613(1984)12<343:scotpc>2.0.co;2.

Missimer, T. M. "Stratigraphic Relationships of Sediment Facies within the Tamiami Formation of Southwestern Florida: Proposed Intraformational Correlations." In *Plio-Pleistocene Stratigraphy and Paleontology of Southern Florida*, ed. T. M. Scott and W. D. Allmon, 63–92. Tallahassee: Florida Geological Survey, 1992.

———. "Sequence Stratigraphy of the Late Miocene: Early Pliocene Peace River Formation, Southwestern Florida." *Gulf Coast Association of Geological Societies Transactions* 49 (1999): 358–68. doi: 10.1306/2DC40C53-0E47-11D7-8643000102C1865D.

Missimer, T. M., and R. A. Gardner. "High-Resolution Seismic Reflection Profiling for Mapping Shallow Water Aquifers in Lee County, Fla." In *Water-Resources Investigations Report* 30. Tallahassee: U.S. Geological Survey, 1976.

Scott, T. M. "Miocene to Holocene History of Florida." In *The Geology of Florida*, ed. A. F. Randazzo and D. S. Jones, 57–68. Gainesville: University Press of Florida, 1997.

Warzeski, E. R., K. J. Cunningham, R. N. Ginsburg, J. B. Anderson, and Z.-D. Ding. "A Neogene Mixed Siliciclastic and Carbonate Foundation for the Quaternary Carbonate Shelf, Florida Keys." *Journal of Sedimentary Research* 66, no. 4 (1996): 788–800. doi: 10.1306/d426840a-2b26-11d7-8648000102c1865d.

White, W. A. *The Geomorphology of the Florida Peninsula.* Vol. 51. Tallahassee: Bureau of Geology, Division of Interior Resources, Florida Department of Natural Resources, 1970.

Willard, D. A., T. M. Cronin, S. E. Ishman, and R. J. Litwin. "Terrestrial and Marine Records of Climatic and Environmental Changes during the Pliocene in Subtropical Florida." *Geology* 21, no. 8 (1993): 679–82. doi: 10.1130/0091-7613(1993)021<0679:tamroc>2.3.co;2.

9

Erosion in the Ocean, Marine Fertility, and Huge Sharks

The Florida Phosphate Story (~22 Ma to ~5 Ma)

> In 1950, Florida's highest land point was neither Hernando County's Chinsegut Hill (280 ft) nor Walton County's Lakewood (345 ft) but rather a phosphate dump heap atop Sand Mountain in Polk County . . . rising to 350 ft.
>
> G. R. Mormino, *Land of Sunshine, State of Dreams*

> Perhaps at one time when this land lay under the sea, a place where sharks lurked, their teeth were buried in the muddy sea-bed, but now they are found in the middle of the island owing to a change in the position of the sea-bed.
>
> St. Nicolai Steno, from A. Cutler, *The Seashell on the Mountaintop*

Artificial Scenery

When driving east from Tampa on Route 60, we pass by some of the topographically highest areas in Florida (fig. 9.1). Do not expect to see some relict shoreline once formed by a *sea level highstand*. Likewise, do not expect to see some jagged rocky outcroppings left behind from some past tectonic event or some indurated, *erosional remnant*.

Rather, these elevated areas are man-made *phosphogypsum* stacks—huge mounds of $CaSO_4{\cdot}2H_20$ (hydrated calcium sulfate; plus unuseable phosphate) left over as a by-product from the chemical processing of phosphate mined in central peninsular Florida to make phosphoric acid (HPO_3), which ends up in solid form (P_2O_5) to make fertilizer. Phosphogypsum contains radionuclides left over from the chemical processing, hence the term *phosphogypsum* and not just gypsum.

Approximately 32 million metric tons of new gypsum are created each year, and the current total stockpile is nearly 1 billion metric tons. The average phosphogypsum stack occupies 135 acres (100 football fields), and many reach 60 m in elevation. So something important has been going on in central Florida to generate this huge amount of waste product. What could be so important that

Right: Figure 9.1. Space image with approximate area of phosphate-rich units and phosphate mining operation located in west-central peninsular Florida. (Source: USGS EROS Data Center; Florida Institute for Phosphate Research data.)

Below: Figure 9.2. A phosphogypsum stack near the port of Tampa. The gypsum is a waste product of the chemical processing of phosphate ore. There are a number of these stacks, some forming the highest elevations in west-central Florida. Due to contaminants, this source of gypsum cannot be used for other industrial purposes. (Photo by A. C. Hine.)

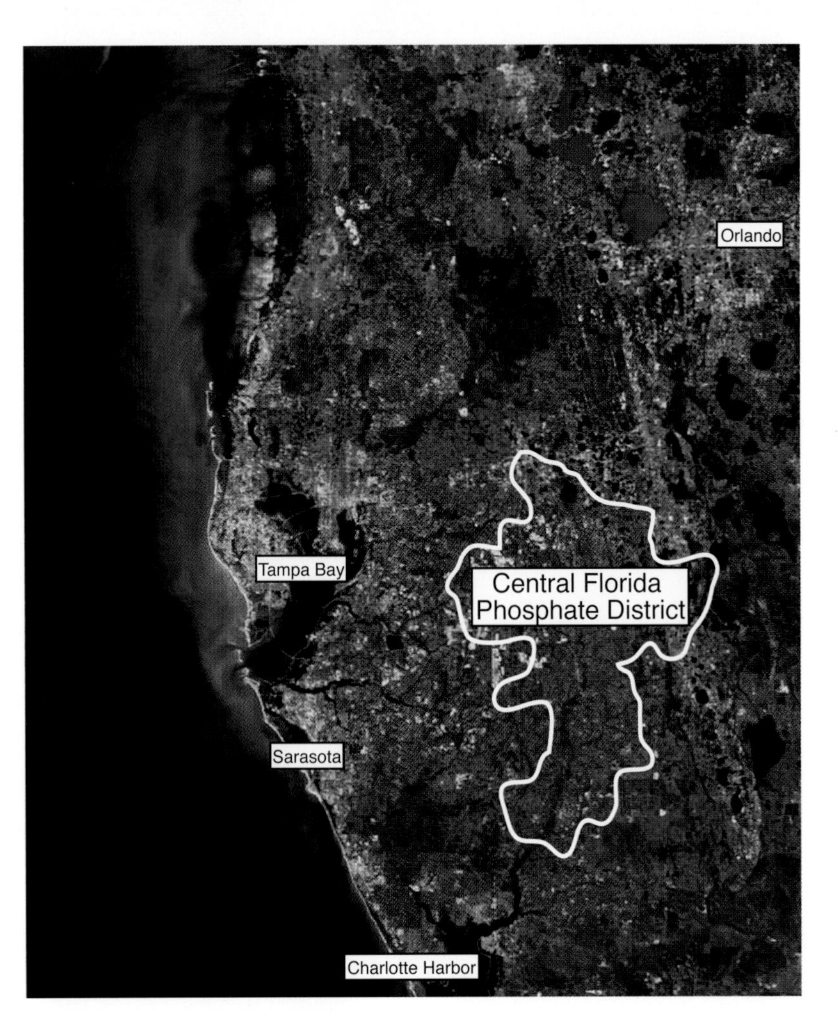

Floridians have tolerated such a massive accumulation of a useless man-made product?

Mechanical Beasts

Traveling through portions of central peninsular Florida we see enormous mechanical beasts looming off in the distance. These metal monsters are 600 times heavier than the largest Jurassic T-Rex (*Tyrannosaurus rex*; who tipped the scales at ~6 tons), swinging huge buckets that could hold two cars parked side by side (70 yds^3) at the ends of booms longer than a football field. It appears that these seemingly shy mechanical monsters never venture too close to the highway for easy viewing. They are off in their own habitat, working in eerie silence consuming the Earth, having appetites that never seem appeased. They are both noiseless and smokeless (all electrical).

They are some of the largest moving mechanical devices ever made—so big that they have to be dismantled and reassembled if they are to be moved any substantial distance. They can move short distances by picking up one foot, moving it forward a few meters, and then setting it down. Then the other foot moves. It is a slow process. Electrical cables ~7 cm in diameter snake across the ground feeding these monsters with energy—hence the noiseless and smokeless emissions—actually, no emissions. But they use more than 6 megawatts of electricity when in operation.

These are the draglines (fig. 9.3). Draglines are used to strip-mine to remove ore bodies that lie just beneath an unusable *overburden,* in this case Plio-Pleistocene age quartz-rich sands (fig. 9.4). Beneath about 10 m of sand lies a *stratigraphic unit* called the Bone Valley Member of Middle Miocene age (~15 Ma). It is part of the Peace River Formation, which, in turn, belongs to the Hawthorn Group. This geologic hierarchy (Member-Formation-Group) is the way geologists define distinctive but related stratigraphic units. Contained within the Bone Valley Member lies one of the richest phosphate deposits in the world. Once the phosphate-rich unit has been exposed by removing the overburden, the dragline dumps sediments from the ore body into large pits where high-pressure water hoses break down the material (fig. 9.5) and send it off for further processing.

Phosphate was discovered in the 1880s, and Florida now generates as much as 30 percent of the world's output and at one time employed 15,000 people (down to 6,000 in 2008), but it is still a $5.9 billion industry in Florida. In fact, there are about 10 billion tons of *economic grade* phosphate in Florida and the southeast United States. If you happen to produce 30 percent of anything globally, such as iron, oil, gold, chromium, or lead, you are a major economic player in that commodity. Florida is a major player in the phosphate business. Approximately 75 percent of the phosphate fertilizer used in the United States comes from Florida.

Left, top: Figure 9.3*A*. A dragline bucket used in the strip-mining process. Two school buses could fit inside the largest buckets used by the phosphate mining industry. (Photo by A. C. Hine.)

Right, top: Figure 9.3*B*. Typical electrically powered dragline machine that deploys the buckets from a large boom. These machines are so big that they have to be taken apart and reassembled if moved any significant distance. But they can move short distances slowly using two large mechanical "feet." (Photo by A. C. Hine.)

Left, bottom: Figure 9.3*C*. Geologists examining stratigraphy inside a large section scooped out by the dragline above. (Photo by A. C. Hine.)

Right, middle: Figure 9.3*D*. Same scooped-out section taken from the viewpoint of the dragline operator from far above. (Photo by A. C. Hine.)

Right, bottom: Figure 9.3*E*. First type of machinery used by the phosphate industry back in the 1930s to extract phosphate ore. (Courtesy of Mulberry Phosphate Museum.)

Figure 9.4. Geologist standing in trench dredged out by dragline revealing very thin section of Bone Valley phosphate-rich unit. (Photo by A. C. Hine.)

Left, middle: Figure 9.5*A*. High-powered hoses used to break up the sedimentary material containing the phosphate ore dumped into a pit by the dragline. The slurry is then piped to a separating facility. (Photo by A. C. Hine.)

Right: Figure 9.5*B*. Unused, fine-grained suspended sediment pumped into settling ponds. These ponds are contained by large above-ground earthen dikes. (Photo by A. C. Hine.)

Left, bottom: Figure 9.5*C*. Vertical image showing size of settling ponds. (Source: Digital orthophoto quarter quadrangle imagery; http://gisdata.usgs.gov/metadata/doqq.htm.)

Figure 9.6*A*. Large trench dug by dragline to remove overburden and reach phosphate-rich ore bed. (Photo by A. C. Hine.)

Figure 9.6*B*. Landscape after strip-mining. (Photo by A. C. Hine.)

Figure 9.6*C*. Landscape now restored to natural environment. (Photo by A. C. Hine.)

The mining industry removes 25 million metric tons of phosphate within 4,000 acres annually with the mining companies owning the mineral rights to 442,000 acres in central Florida. The Port of Tampa exports more phosphate fertilizer than any other port in the world. U.S. consumers used 8.5 million metric tons in 2007. For a state known for its pleasant climate, beaches, satellite launches, and theme parks, this fame and stature might be a surprise to many. But the deposition of these phosphate deposits constitutes a major event in Florida's geologic history.

Phosphate mining and processing activity have posed some of the sternest challenges to Florida's environment over the decades (fig. 9.6). The importance of the phosphate industry is the reason why Floridians have tolerated the emplacement of the giant phosphogypsum stacks. No one yet has come up with an effective, economically viable process to make useful and safe products of this waste.

Phosphate: What Is It?

What is phosphate, and why do we need it? Phosphate is a general term that defines sediment or sedimentary rock that contains the element phosphorous (P) in a family of minerals, but primarily the carbonate-fluorapatite having the chemical formula $Ca_5(PO_4)_{2.5}(CO_3)_{0.5}F$.

This looks complicated, but it merely says that the primary ingredients (actually, elements) are calcium (Ca), oxygen (O), carbon (C), phosphorous (P), and fluorine (F). Minerals rarely have phosphorous and fluorine in them, so carbonate-fluorapatite is a bit unusual and requires special circumstances to produce it. Actually, there is a whole family of sedimentary minerals that contain phosphorous. Their composition can be highly variable, and the origin is not particularly well understood. Most likely this chemical variability is dependent upon the specific environmental factors (e.g., lack of oxygen, pH) at work during the mineral formation and alteration after formation.

We will see below what circumstances are required. The fluorine is extracted as an important by-product used to fluoridate public water supplies. Fluoridation of city water supplies for health purposes has become controversial, but in west-central Florida, the fluorine used has come as a by-product of the nearby phosphate industry.

These elements (Ca, O, C, P, F) are tied together by *chemical (electrical) bonds* that form a predictable internal structure of its atoms and molecules, forming a mineral. Minerals are defined by how certain elements are all linked together into a predictable 3D pattern with recognizable physical properties such as hardness, color, cleavage, and density. These are all fundamental features that people use to identify their favorite minerals.

Most sedimentary deposits contain a small amount (~0.3 percent by weight) of phosphorous (expressed as P_2O_5). Anything >1 percent P_2O_5 is unusual and requires special conditions and environments to form. Where a phosphate deposit contains >15 percent P_2O_5, making it profitable to mine, the term *phosphorite* is used to describe the sedimentary deposit. But the term *phosphate* (technically <15 percent P_2O_5) seems to have universal appeal as it applies to all deposits that have some P_2O_5.

Why Is Phosphorous Important?

The element P is an important nutrient for all life, and it stimulates plant growth in particular. As such, it is one of the key ingredients in fertilizer—90 percent of phosphate is used to make fertilizer. Another 5 percent is used in animal feed, and the remaining 5 percent is used to make a variety of products including toothpaste, metal coatings, oil and gasoline additives, plastics, insecticides, dyes, steel alloys, vitamins, light bulbs, film, flame resistant fabric, and food preservatives. (Look at the list of ingredients on a can of Coca Cola!)

More specifically, phosphorous gives shape to the DNA molecule, the blueprint of genetic information for every living cell. Phosphorous plays a vital role

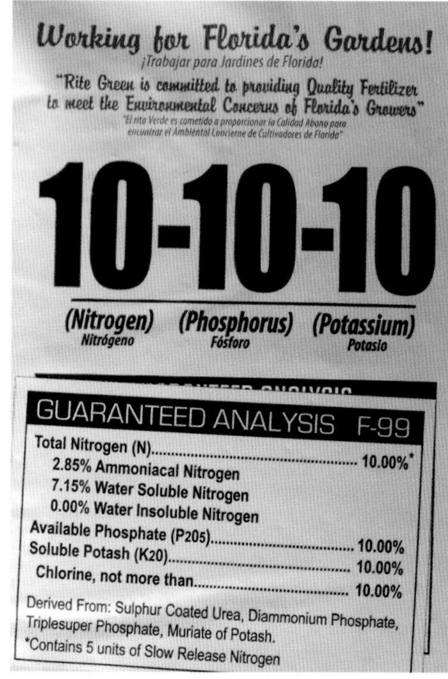

Left: Figure 9.7*A*. Typical bag of fertilizer sold at many outlets. Note the numbers 10-10-10. These numbers refer to amount of primary nutrients required by all plants: nitrogen, phosphorous, and potassium.

Right: Figure 9.7*B*. Back of fertilizer bag. Note other minor elements used in fertilizers. The sequence of numbers is always the same (nitrogen, phosphorous, and potassium), but the numbers may change depending upon the type of plants requiring different concentrations of these nutrients. (Photo by A. C. Hine.)

in the way living matter provides energy for biochemical reactions in cells and strengthens bones and teeth. Although phosphate does not have the global strategic and geopolitical importance of oil and gas (we do not fight wars over phosphate, for example), fertilizers are critical for feeding the ~7 billion of us now inhabiting Earth. So assuredly phosphate mining is important and globally relevant. And with another 3+ billion humans due to join us by 2100, using fertilizers to grow food will become even more important.

If you go to a garden center and buy a bag of fertilizer, you will note that the bag generally has three numbers printed on it, such as 10-10-10, or 29-3-4 (fig. 9.7). These numbers indicate the weight percent of nitrogen (N), phosphorous (P_2O_5), and potash (expressed as K_2O; K is potassium). All three of these nutrients form the basis of most fertilizers along with additional trace ingredients such as iron (Fe), copper (Cu), and sulphur (S). With a 10-10-10 fertilizer (30 percent), the remaining 70 percent is mostly inert material such as quartz sand. Different fertilizers have different nutrient concentrations and are based on the needs of specific plants.

Some Problems and Unintended Consequences

As you can imagine, removing the sedimentary material (overburden) lying on top of the phosphate-rich layers devastates the *terrestrial ecology*. So restoring the land to its natural state, or land reclamation, is required.

Radon gas is also a health hazard posed by the strip-mining process. The gas is produced by *radioactive decay* of naturally concentrated uranium associated with the phosphate deposits. Structures built on reclaimed land have to be carefully ventilated to prevent radioactive radon gas accumulation inside. Radon gas (Rn 222) is much heavier than normal air and is easily trapped in buildings, exposing humans to radiation.

An unused portion of the ore body is its mud-sized fraction, which is removed at a grain-size sorting floatation facility that separates the sand-size quartz from the sand-size phosphate grains. These fine-grained sediments are sent to huge settling ponds (used to be called slime ponds) held back by earthen dams, where the tiny sediments eventually settle out. These settling ponds can be seen from space and form important habitats for birds and other wildlife.

At a chemical processing plant, the phosphorous is removed from the sand-sized phosphate grains using sulfuric acid (H_2SO_4). It is this process that produces large quantities of phosphogypsum waste, which also contains uranium as an unwanted by-product. Hence our highly visible and unusable phosphogypsum stacks grow larger and larger.

The chemical plants in Florida and elsewhere finally make artificial grains of soluble phosphate to be mixed with the other nutrients and poured into fertilizer bags.

Growing plants and using fertilizer to do so are essential to our lives. But another unintended consequence of fertilizer use has recently appeared in the Gulf of Mexico. Over the years, the increased use of fertilizers increased nutrient runoff into the Mississippi River, which receives water discharge from 32 states. The nutrient-enriched water enters the northern Gulf of Mexico, where it promotes algal blooms. The organic matter from these blooms decomposes, consumes O_2 in the process, and ultimately creates O_2 "dead zones," threatening the Gulf's $2.6 billion/year fishing industry. The U.S. government stimulus to increase ethanol as a fuel caused farmers to increase corn production by 20 percent in 2007. So a cruel irony is that our phosphate wealth and good fortune on land in Florida is now creating an environmental problem in our adjacent ocean.

Florida's Fossil Treasure Trove

Central peninsular Florida is also one of the best fossil hunting areas in the world. The rivers that run through these mid-Cenozoic age deposits erode and concentrate the fossilized remains of aquatic animals such as whales, manatees, alligators, rays, and sharks. During periods of *sea level lowstand*, central Florida supported large numbers of terrestrial animals such as horses, rhinoceroses, bears, peccaries (pig-like mammals), proboscideans (elephant-like mammals), sloths, squirrels, camels, bear-dogs, and tapirs during the Miocene (fig. 9.8). During the post-Miocene (Pliocene and Pleistocene), saber-tooth cats, mastodons, mammoths, armadillos, porcupines, rodents, and many hoofed mammals occupied central Florida. Fossilized remains may become mixed together when eroded, transported, and redeposited by rivers. Sorting them all out is a great challenge to paleontologists.

The most famous fossil is the shark's tooth (fig. 9.9). Venice Beach is one of the rare places where the Hawthorne Group crops out along the Gulf of Mexico coastline and is actively eroded by waves. Consequently, fossil hunting is a popular pastime, particularly for those looking for small, black, triangular teeth. Additionally, Venice Beach is unusually dark in color due to the enhanced presence of black phosphate sand grains. However, finding an undamaged serrated tooth of the extinct great shark of its day (*C. megalodon*—meaning large tooth) is a rare event. These ~15 cm long prizes are unusual because they are easily broken when transported in rivers or by breaking waves on beaches. So they can fetch hundreds of dollars when found in pristine shape.

This ancient great shark (possibly related to today's smaller great white shark) existed from about ~22 Ma to ~5 Ma. *C. megalodon* grew up to 18 m long and weighed ~18,000 kg. It had to consume food amounting to about 2 percent of its weight/day or ~360 kg. It was a top-of-the-food-chain predator—no other creature hunted this animal (fig. 9.9). Because shark skeletons (not teeth) are

Above: Figure 9.8*A*. Florida is blessed with abundant and diverse fossils, both land and marine inhabiting creatures. The woolly mammoth and mastodon, both elephant-like land-based animals, coexisted in Florida but became extinct ~11 ka. (Source: Florida Museum of Natural History in Gainesville; photo by A. C. Hine.)

Left: Figure 9.8*B*. The giant armadillos (called pampatheres; *Holmesina septentrionalis*) were nearly 2 m in length and lived on land in the Pliocene. (Source: Florida Museum of Natural History in Gainesville; photo by A. C. Hine.)

made of cartilage, they preserve poorly, rarely making fossils. Typically, all that remains of these huge creatures are their teeth. There are countless numbers of much smaller intact shark teeth from other species, thus begging the questions: (1) what environmental conditions existed to sustain so many predators, (2) how was so much food produced to sustain them, and (3) what does any of this have to do with phosphate deposition?

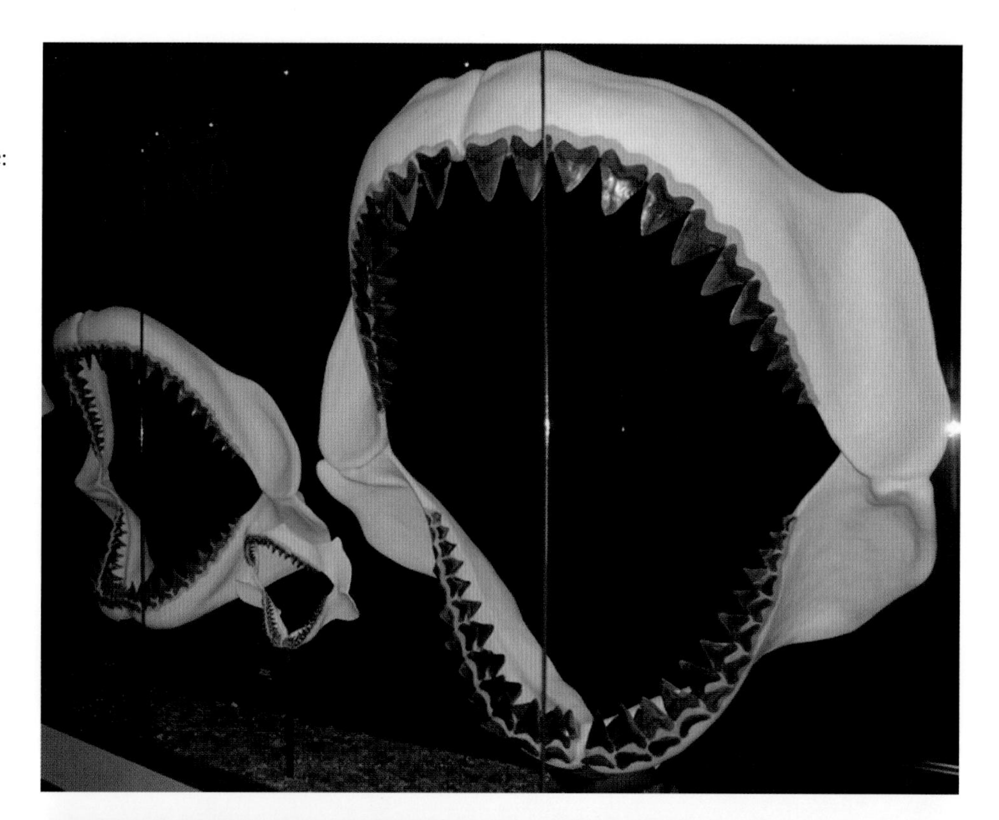

Figure 9.9*A*. Two sets of jaws of *Carcharodon megalodon*; the larger one is about 2 m high. (Source: Florida Museum of Natural History in Gainesville; photo by A. C. Hine.)

Figure 9.9*B*. Painting of ~18 m long *Carcharodon megalodon* feeding on ~8 m long Eobalaenoptera whales. This image graphically illustrates the fearsome nature of these top marine predators. No such shark exists in the modern ocean. (Image produced by Karen Carr; used with permission from Karen Carr and Ryan Barber of the Virginia Museum of Natural History.)

There are two misconceptions that need to be addressed concerning Florida phosphate deposits. First, phosphate or phosphorite deposits are not massive bone beds consisting of dinosaur fossils or other large fossil remains. (There were no dinosaurs in the Miocene in the first place!) There are numerous smaller fossils associated with phosphate deposits for reasons explained below. Most of these fossils are from marine creatures. However, few fossils are actually extracted and sent to the chemical plants for processing due to their phosphate content. The sediments that constitute the phosphate ore selected for processing are generally sand-sized, black, semi-rounded particles. The second misconception is that Florida's phosphate is not derived from guano from extensive avian colonies. There is phosphate derived by bird guano, particularly on western Pacific islands such as Christmas and Nauru, but not in Florida.

Source of Phosphorous

Since phosphorous (P) is needed for life, it makes sense that a source of P is organic matter itself. The surface waters of the ocean are filled with life in certain areas, particularly where *nutrients* are brought to the surface via *upwelling* and where sunlight can stimulate *photosynthesis*. Nutrients are required to sustain life. They consist of molecular nitrogen (N_2) and associated compounds nitrate (NO^{-}_{3-}), nitrite (NO_{2-}^{2}), and ammonia (NH_3), phosphorous in dissolved form (PO_{4-}^{3}), and silica (SiO_2), not quartz but a more soluble phase (opalline). Along with these nutrients, iron (Fe), copper (Cu), manganese (Mn), zinc (Zn), potassium (K), and cobalt (Co) are required for life.

Microscopic plants called *phytoplankton* grow in enormous abundance in upwelling zones, areas with elevated nutrient levels at the surface. The total organic mass of these plants is far greater than the sum of the mass of big fish in the ocean. This profusion of life is called *primary productivity* (total quantity of *carbon fixed* by plants) and forms the base of the *food chain* in the ocean. Small *planktonic* animals called *zooplankton* consume the phytoplankton, small fish consume the zooplankton—and so it goes on up the food chain (actually, a food web) until we reach the large predators. These steps are called *trophic levels* (trophic means nutrition). The large predators require an enormous amount of food, and therefore it is difficult to sustain large numbers of them as compared with smaller fish. So, in the Miocene oceans, and along other continental shelves where upwelling occurred, there must have been abundant quantities of food to keep these large predators fed.

The ocean is a wonderfully efficient recycling machine: planktonic organic matter that is not consumed at the surface begins to sink but rarely reaches the seafloor. This organic matter is consumed by living organisms that live deeper or is *oxidized* and dissolved, making it available for further use in the ocean. Under extreme circumstances of very high primary productivity in relatively

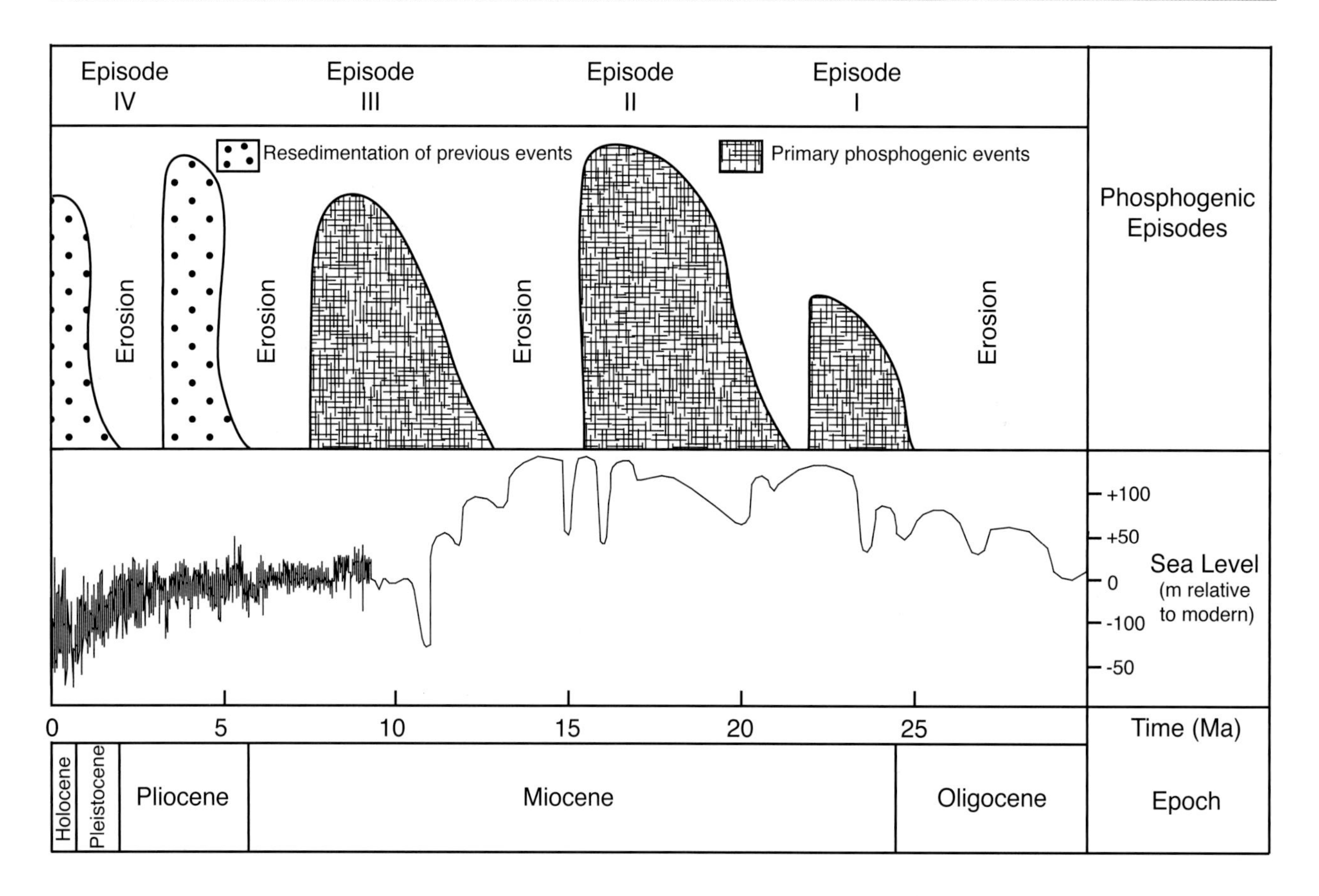

Figure 9.10. Three episodes of primary phosphate deposition (phosphogenesis) during the Miocene and reworking of these deposits later in geologic time. These data are from North Carolina, but generally are representative of the entire SE United States, including Florida. Note that the last episodes represent a period of reworking of phosphate sediments formed during one or more of the earlier three periods. Each phosphate event occurred during elevated sea level. (Modified from Riggs et al. 2000; used by permission from the Society for Sedimentary Geology.)

shallow water (shelf depths, <200 m), this organic matter can reach the seafloor whereupon it becomes buried and preserved in the sediments. Such unusual circumstances occurred a number of times in peninsular Florida and actually along the entire southeast margin of the United States during Miocene time. The water during these primary phosphate-producing events was probably only 100 m deep and most likely shallower (fig. 9.10).

Circumstances Converge to Form Phosphate: Sea Level Fluctuations

Since we see so many marine fossils in central Florida, sea level must have been higher than present. During the peak of the Middle Miocene sea level highstand, maybe as much as 100 m of water covered south-central Florida, thus linking up the Gulf of Mexico with the northern Straits of Florida (fig. 9.11*A*). So, much of peninsular Florida was submerged multiple times. Since sea level did fluctuate, portions of central Florida became shallower and, at times, were emergent, allowing rivers to flow overland to estuaries and coastlines.

C. megalodon and other sharks lived in this open shallow ocean while manatees, rays, and alligators lived in the shallower waters. But during at least three sea level highstand events in the Miocene, central Florida was a broad, shallow, and relatively warm ocean—probably warmer than today. According to a geochemical dating technique (*isotopes* of strontium), these three sea level events occurred at (1) ~23–22 Ma, (2) ~18–16 Ma, and (3) ~13–7 Ma. During the ~5–3 Myr interval in the Pliocene, there was substantial reworking of these earlier deposits by rivers and streams.

It was during long periods between these sea level events that Florida was high and dry, providing an environment for terrestrial creatures as mentioned earlier. Sea level lowstand events and tectonic activity generating many closely spaced islands in the Caribbean allowed for a Great American Biotic Interchange between land animals of the Americas. This exchange became even more vigorous once the Isthmus of Panama closed, forming a stable land bridge several million years ago (fig. 6.3*D*). Thus the sea level history of repeatedly flooding and exposing the land created one of the great fossil hunting localities, mixing the remains of abundant land and marine organisms.

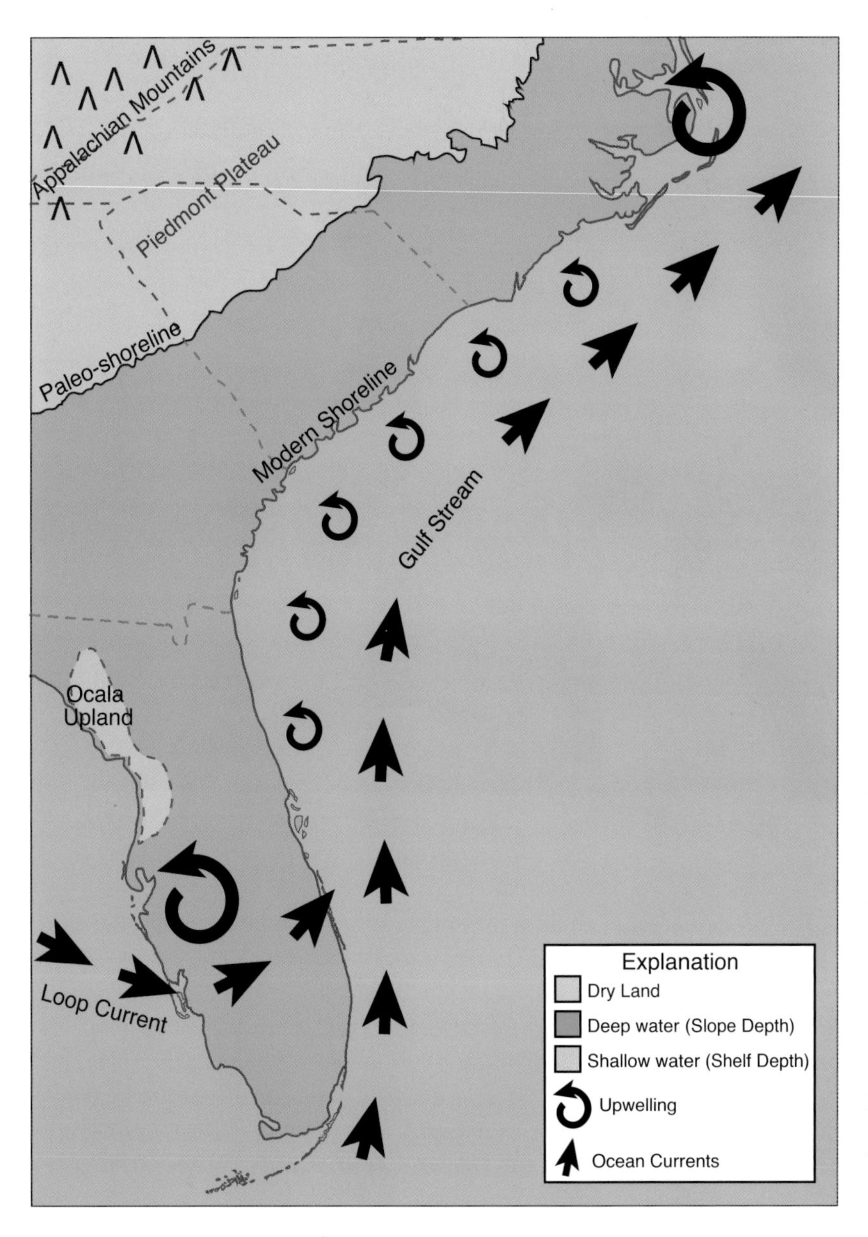

Figure 9.11*A*. Paleogeographic map of SE United States ~18 Ma during the Miocene with interpretation that the Loop Current flowed across peninsular Florida. This flow generated persistent upwelling due to the paleo Loop Current being diverted by the seafloor bathymetry (topographic steering). Topographic steering occurred not only in Florida but also in North Carolina, where large phosphate deposits are found as well. (Modified from Popenoe 1990; used by permission from the Cambridge University Press.)

Circumstances Converge to Form Phosphate: Strong Currents and Upwelling

One of the most important physical components of the modern Gulf of Mexico is a filament of moving water called the Loop Current (fig. 9.11*B*). This is a component of the *western boundary current,* the well-known Gulf Stream that flows along the east side of northern South America and North America and heads toward Europe from Cape Hatteras, North Carolina. Scotland, for example, is at the same latitude as Siberia, but the climate of the two areas is vastly different—subtropical plants can be grown outdoors in SW Scotland due to the warmth introduced by the Gulf Stream, while Siberia is one of the coldest places on Earth.

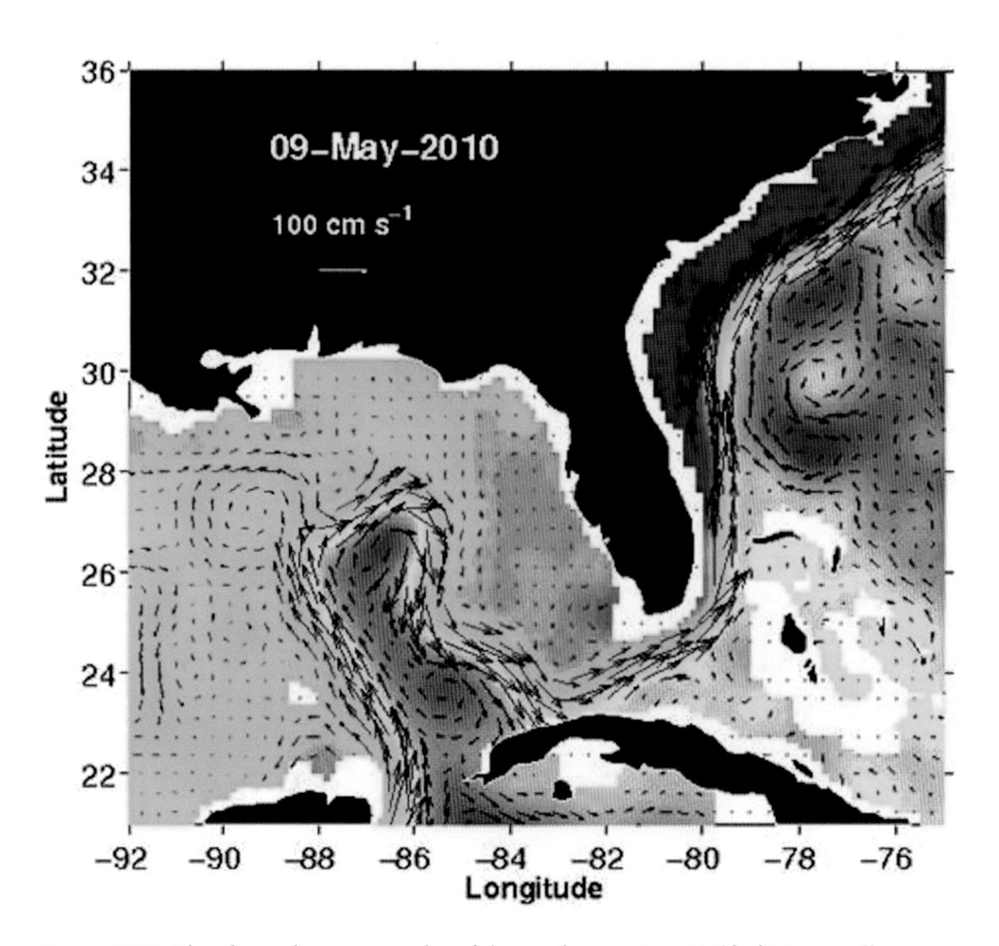

Figure 9.11*B*. The physical oceanography of the modern eastern Gulf of Mexico, illustrating satellite-based sea surface temperature (warm colors of warmer water showing the Loop Current) and velocity vectors based on satellite sea surface height measurements. Note that the Loop Current enters and leaves the Gulf of Mexico, eventually merging with the Gulf Stream. Geologists propose that the Loop Current flowed across peninsular Florida when sea level was higher ~18 Ma, producing upwelling, organic matter production, and eventually generation of phosphate-rich sediment beneath the persistent upwelling zones. (Courtesy of Drs. R. H. Weisberg and Lianyuan Zheng, College of Marine Science, University of South Florida.)

Modern satellite imagery, oceanographic data, and mathematical modeling clearly show the complexities of Loop Current behavior. This current remains well offshore but has important modern-day interactions with the water masses on the west Florida shelf that affect deeper water (60–80 m) coral-reef growth. However, geologists hypothesize that when sea level rose ~50–100 m higher in the Middle Miocene, the Loop Current moved eastward up onto peninsular Florida, flowed across the platform and directly into the northern Straits of Florida, following a much more northerly flow path than today. Equally as important, this paleo Loop Current had to flow around the topographic high posed by central peninsular Florida, forming a bend or dogleg in its flow pattern.

This bend in the current produced persistent upwelling, which brought nutrients closer to the sea surface, thus stimulating primary productivity. As the current was deflected by bathymetry, water from below replaced the water that is removed by the deflection. This is called *topographic steering* and results from the interaction of oceanographic currents with bathymetric high areas of the seafloor. As sea level rose or fell, such topographic steering was probably enhanced, reduced, or eliminated altogether.

When topographic steering was at a maximum, upwelling was not only persistent but enhanced. The result was a proliferation of food (organic matter) in the surface waters, producing a very fertile ocean, which, in turn, supported a robust food web including large numbers of fish. This enhanced the number of predators: the great sharks. Additionally, the convergence of ideal conditions, such as a robust food web, lack of competing predators, lack of disease, and low mortality of young, favored gigantism, allowing for these sharks to become very large. There were advantages to becoming very large, allowing the survivors to pass on their genes to their offspring. And perhaps the smaller sharks were consumed by their larger cousins, contributing to a natural selection process that ultimately favored a gene pool of gigantic sharks.

Most important to the phosphate story, a significant amount of organic matter reached the seafloor and became buried to form the phosphate. So how do we go from buried organic matter to phosphate grains?

Finally, the Making of Phosphate

Once buried, the organic matter is broken down by a chemical reaction called *sulfate reduction*, which is stimulated by *bacterial microbial activity*. This transformation or mineralization from organic matter to phosphorous-rich sediments occurs in a low-oxygen environment that may exist in sediments within ~10 cm of the seafloor.

One might ask, why can't we just take the mined sediments rich in carbonate-fluorapatite and put that directly into a bag of fertilizer? In this natural mineral form, the phosphorous is tightly chemically bound. To release it from these bonds, the sediments are treated with sulfuric acid, which strips away the P but produces the unusable phosphogypsum. The radioactive uranium tied to the carbonate-fluorapatite is transferred to the phosphogypsum, making the by-product slightly radioactive, posing difficulties in using it for other purposes.

The carbonate-fluorapatite crystals form inside the molds of shells (*foraminifera, gastropods*), are concentrated into *fecal pellets* by burrowing worms and other organisms, form crusts, and fill *interstitial* voids between other sedimentary particles. There are a few organisms that actually secrete carbonate-fluorapatite to form their *exoskeleton,* such as *brachiopods.* However, skeletal growth by *benthic organisms* is a minor process in producing phosphate-rich sediments as compared with the chemical reaction occurring within the shallow subsurface in the sedimentary cover just below the seafloor. Phosphate-rich muds formed on and within the seafloor form crusts that are broken up into sand and gravel-sized particles (called *intraclasts*). These phosphate-rich muds are also consumed by numerous burrowing organisms forming sand-sized rounded fecal pellets that harden. It is these sand-sized phosphate grains that are mostly sought after by the mining operators.

A distinctive feature of many phosphate deposits is that the sediments rich in carbonate-fluorapatite are reworked many times on the seafloor and later during sea level lowstands when they are *re-sedimented* in rivers. Such deposits may have been reworked by both marine and non-marine processes. Much of this reworking eliminates the organic matter that has not been consumed in this process, as well as washing out any fine-grained sediment. Also, some of the dark grains enriched in organic matter are leached or weathered, turning them white. The final ore body may consist of sand-sized grains of black phosphate and quartz sand, producing a "salt and pepper" appearance, and it appears quite different than the original sedimentary deposit formed on the seafloor. The beaches around Venice, Florida, are gray due to the black phosphate being directly eroded from the Miocene age sedimentary deposits that crop out along that stretch of coastline.

Phosphatization Not Only in Florida

As fig. 9.11*A* indicates, the phosphatization events were not confined to Florida but extended northward along the southeastern U.S. continental margin to Cape Hatteras, North Carolina. Florida and North Carolina are linked as one very large system, so we should expect that these deposits, along this large sector of the continental margin, are contemporaneous. As sea level was rising, the benthic influence of the western boundary currents (Loop Current in Florida; Gulf Stream in North Carolina) moved upslope. Where these bottom hugging currents encountered topographic highs on the seafloor, the currents were deflected, forming a zone of persistent upwelling. Within these zones primary productivity was enhanced, the ocean locally became very fertile, large predators evolved, and organic matter was buried within the seafloor. Chemical reactions within the sediments allowed carbonate-fluorapatite minerals to form. However, when sea level rose above a certain point, the topographic steering became less important, thus shutting down the phosphatization event.

Naturally, when sea level was too low, the area where phosphate is found was exposed to the atmosphere, subjecting these sediments to weathering and reworking. A certain window in time and space had to have been present to turn the phosphate factory on and off. That window does not exist today, so there are no phosphate sediments forming anywhere along the SE U.S. continental margin.

Erosion in the Ocean

Surrounding the shallow flanks of the Florida Platform in water depths ranging from 250 to 500 m are prominent eroded surfaces called marine unconformities (figs. 8.12, 9.12). These surfaces form the Pourtales and Miami Terraces as well as an unnamed 10 km wide, 100 km long rock surface that crops out along the west Florida slope. These surfaces represent a period of accelerated flow by the Loop Current, which eroded into the slope lying seaward of the continental shelf. Seismic reflection data clearly illustrate truncated surfaces indicative of

Figure 9.12*A*. Bathymetric map of the southern Florida Platform featuring the Pourtales Terrace—a bathymetric feature eroded by an accelerated Loop Current–Florida Current system during the Middle Miocene, probably when sea level was lower. (Source: Lidz et al. 2006; used by permission from the *Journal of Coastal Research*.)

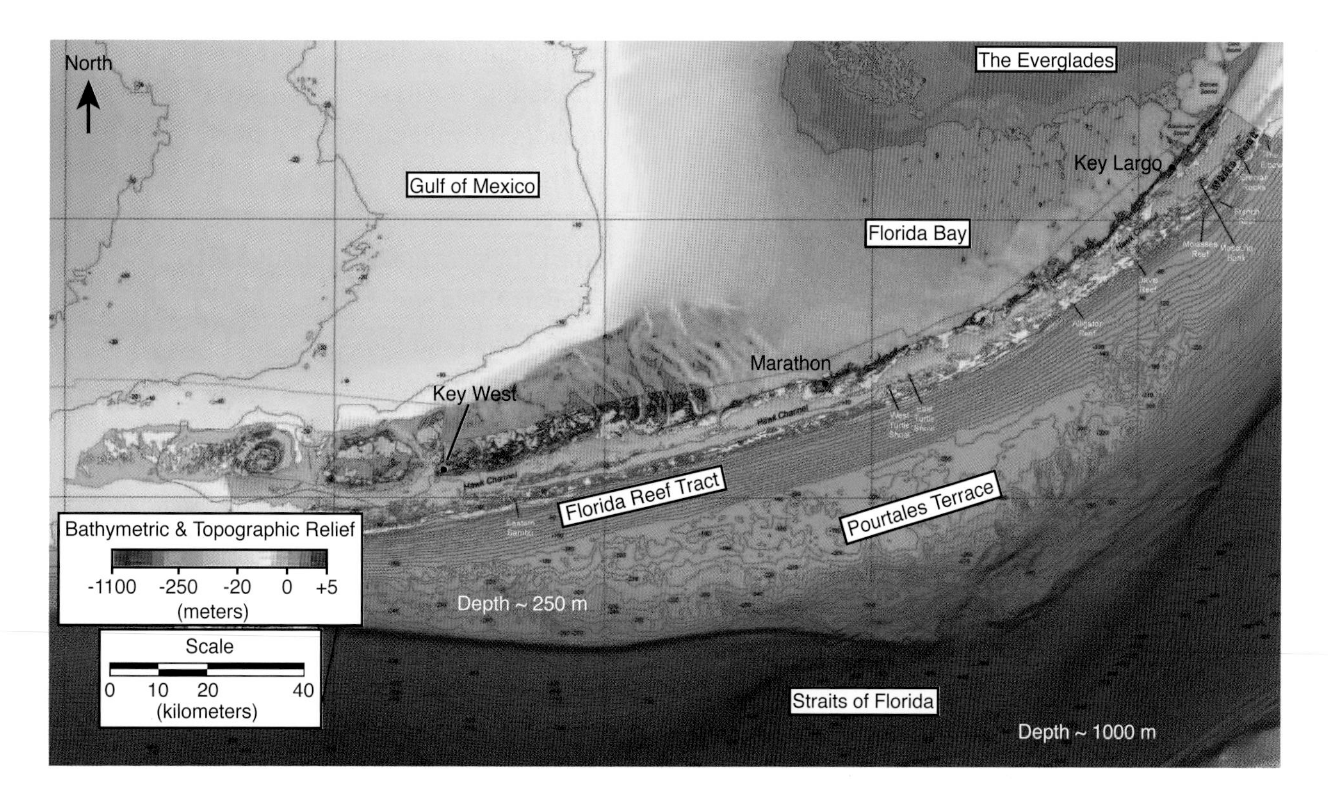

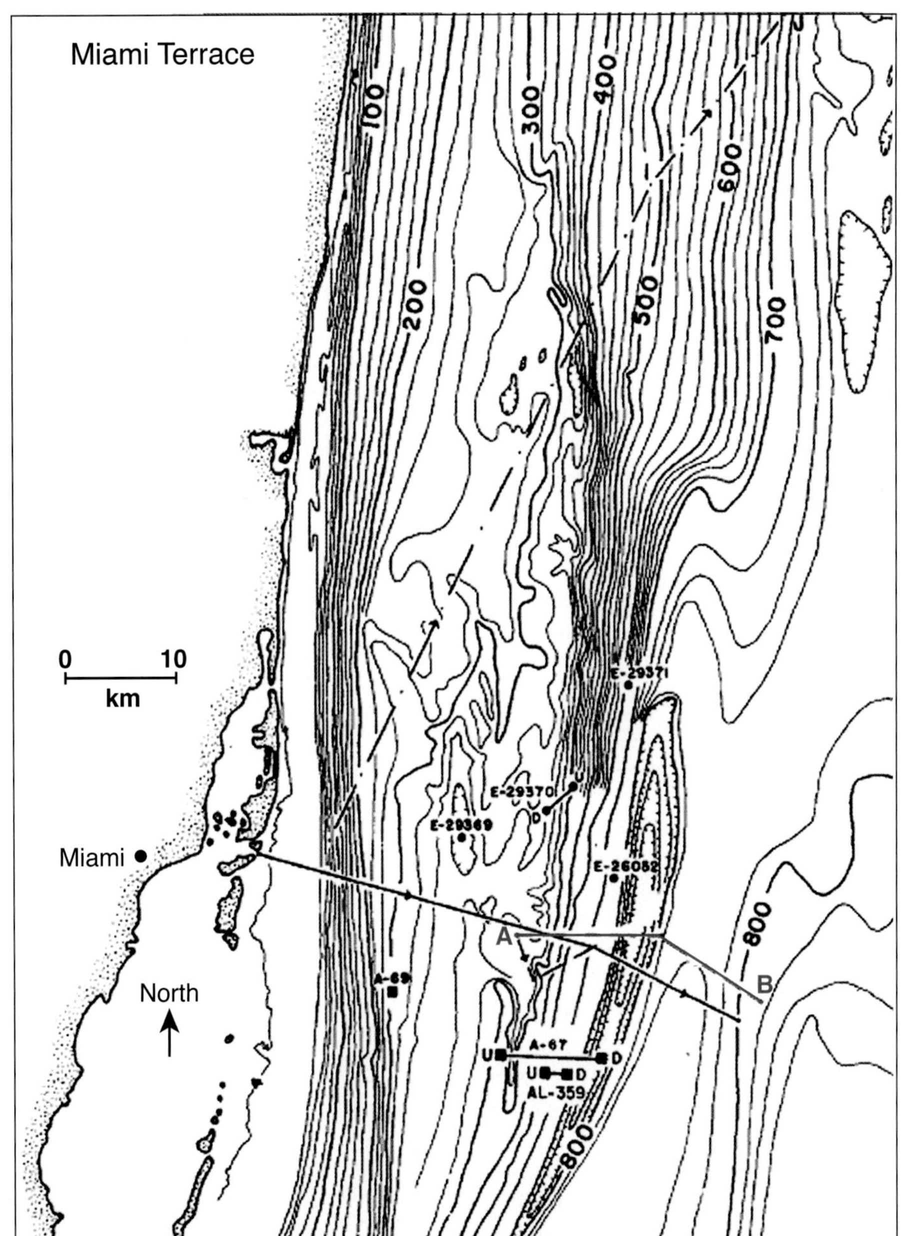

Left: Figure 9.12*B*. Bathymetric map of Miami Terrace just off Miami, located at approximately the same depth as the Pourtales Terrace and formed during the same time by accelerated Florida Current during the Middle Miocene. (Source: Mullins and Neumann 1979; used by permission from Elsevier.)

Below: Figure 9.12*C*. Seismic line across Miami Terrace illustrating the very hard (phosphatized crust) seismically impenetrable upper portion of the terrace and the truncated dipping seismic reflectors of the lower portion of the terrace indicating erosion. Numerous fossils have been dredged from the top of the Miami Terrace. (Source: Mullins and Neumann 1979; used by permission from Elsevier.)

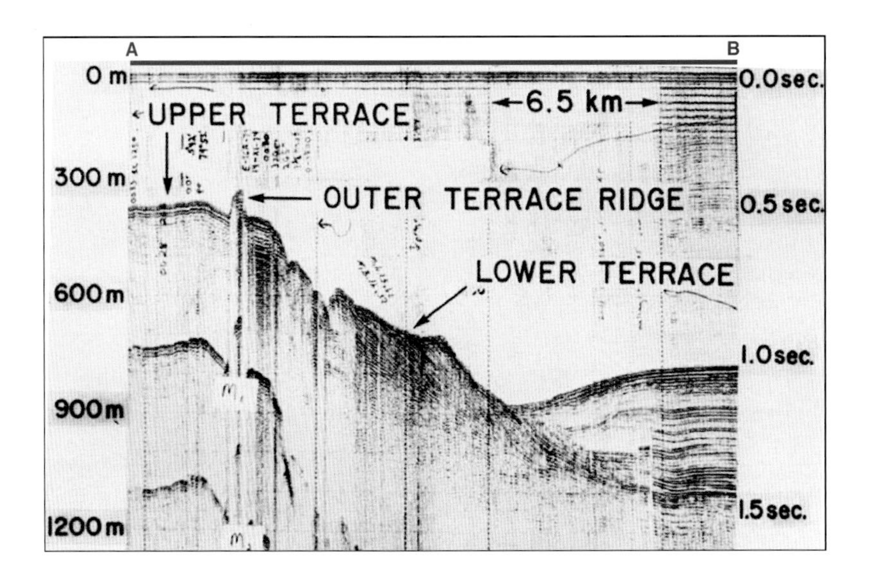

erosion. These erosion zones became hardened by phosphatization, forming a durable black crust. Portions of this crust underwent numerous episodes of erosion and re-cementation, forming a complex, multigenerational rock type. Since these surfaces were current swept, no fine-grained sediments could accumulate, allowing only large bones and teeth from marine animals to be deposited. Rock dredging from the Miami Terrace retrieved numerous fossilized bones of dugongs, shark teeth, and beaks from rare, extinct whales.

Erosion of the upper flanks of the Florida Platform began in the Middle Miocene and may have been related to tectonic events that occurred in the western Caribbean Sea. Geological oceanographers have proposed that a bathymetric feature called the Nicaraguan Rise (series of carbonate platforms and banks), which extends eastward from Nicaragua out into the Caribbean Sea, was much more prominent in the early Miocene (fig. 9.13). Due to changes in tectonic activity between the north margin of the Caribbean Plate and the south margin of the North American Plate, the Nicaraguan Rise was stretched and foundered as a result. This foundering formed several large, deep channels. These channels allowed the Caribbean Current to flow more to the north, stimulating the flow through the Yucatan Channel and accelerating the Loop and Florida Current (same filament of moving water). Previously, the Caribbean Current flowed to the west out into the eastern Pacific Ocean, since the Central American land bridge (Isthmus of Panama) had not yet been formed. So this accelerated flow around the Florida Platform eroded the margin, creating these terraces and phosphatized surfaces.

Additionally, when sea level was high, this accelerated flow stimulated upwelling on submerged central peninsular Florida, further enhancing phosphate sediment production. So an added ingredient to the Florida phosphate recipe is this tectonic event that occurred far away in the lower Caribbean Sea.

The Global Stage

Finally, phosphatization events may be linked to and may cause global climate change. During periods of extended sea level highstand worldwide with continental shelves at their maximum width due to this flooding, the extensive and rapid burial of organic matter within the seafloor may affect climate by removing carbon dioxide (CO_2) from the atmosphere. Organic matter consists primarily of carbon, and if the carbon in the ocean is removed, it has to be replaced by removing carbon from the atmosphere in the form of CO_2.

As is well known, CO_2 is an important greenhouse gas in that its presence in the atmosphere prevents incoming radiation from the sun to be radiated back out into space, allowing the Earth to heat up. Human agricultural and industrial activity produces much CO_2, as well as other greenhouse gases, sparking the

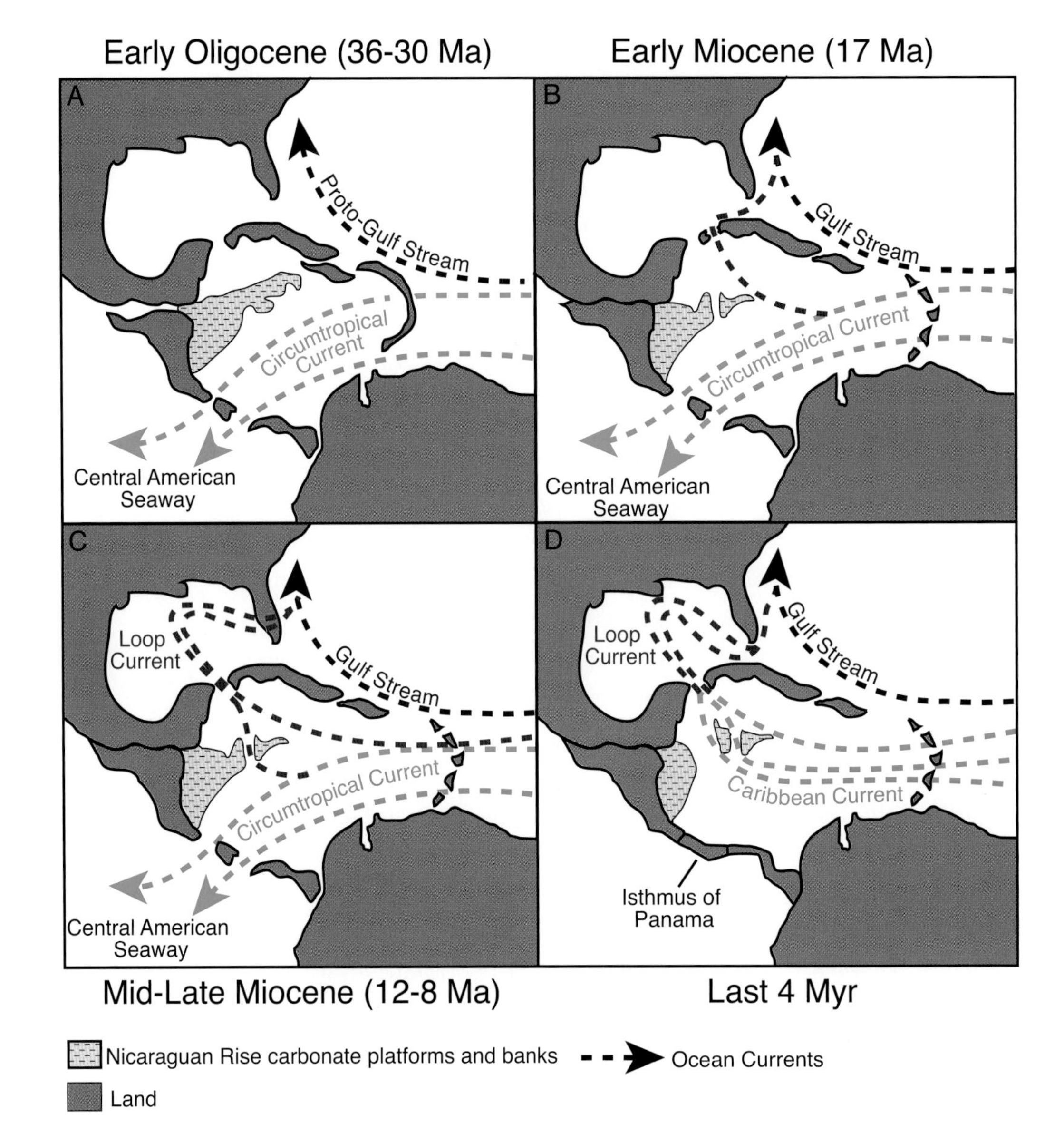

Figure 9.13. *A*. The opening up of a gateway by the drowning and collapse of a shallow-water carbonate bank system and the eventual closing of another gateway (Central American Seaway) formed the modern Isthmus of Panama. Both tectonic events allowed for the creation and acceleration of a strong northward current (Yucatan-Loop-Florida Current) which joins the Antilles Current to form the Gulf Stream. This Miocene development of this current system played an important role in the Florida phosphate system and the erosion that formed the Pourtales and Miami Terraces. The existence of an extensive shallow-water carbonate platform (Nicaraguan Rise) prevented flow into the Gulf of Mexico. Water flowed from the Caribbean to the eastern Pacific Ocean through an open seaway (Central American Seaway). *B*. The shallow water carbonate platform starts to collapse, allowing more water to flow northward. *C*. Larger, deeper channels form within the carbonate platform, allowing the establishment of the Loop Current. During sea level highstands, peninsular Florida was flooded, allowing the Loop Current to pass over the top of it, forming phosphate in west-central Florida. *D*. Closure of the Central American Seaway by formation of the Isthmus of Panama further accelerated the Loop Current, stimulating erosion of Pourtales and Miami Terraces probably during sea level lowstands. (Source: Droxler et al. 1998; used by permission from Oxford University Press.)

debate that global warming is an anthropogenic effect. The scientific community is reaching a consensus that global climate change is anthropogenic.

However, if CO_2 is removed, we can expect global cooling to occur. So some geological oceanographers have suggested that when extensive phosphatization has occurred, this event has been followed by global cooling, glacial ice development, and sea level lowering. The major sea level fall after the Middle Miocene sea level highstand is consistent with this idea. So it is possible that Florida's phosphate-producing episode along with that of the southeast United States (major phosphate deposits in North Carolina, for example) and other selected areas worldwide affected the global climate and produced a major sea level fall. This fall, in turn, played a major role in karst dissolution and collapse on the Florida Platform, creating Tampa Bay and Charlotte Harbor, as explained in chapter 7. Geologically, this is wonderfully interlinked. Local events affect global events, which, in turn, affect new local events. But it is all speculative and a working hypothesis.

So we can see now that "Erosion in the Ocean, Marine Fertility, Huge Sharks, and the Florida Phosphate Story (~22 Ma to ~5 Ma)" (the title of this chapter) is a compelling geologic narrative because at no time before this period and at no time since has this unusual convergence of events occurred in or around Florida.

Essential Points to Know

1. The mining of phosphate in central Florida is an important industry that provides fertilizer to grow food. Florida produces 30 percent of the world's phosphate. The mining operations also pose significant environmental challenges such as strip-mining, radon gas, and the disposal of phosphogypsum waste.

2. Phosphate-rich sediments were deposited mostly in the Miocene during distinct episodes resulting from an unusual convergence of environmental circumstances. These circumstances are sea level being significantly higher than today, currents producing upwelling of nutrients and stimulating primary production of organic matter, some of which settled to the seafloor; microbially stimulated chemical reactions within the sediment on the seafloor to produce phosphate minerals; reworking of these small phosphate mineral crystals by burrowing organisms and physical transport, creating the sand-sized phosphatic grains sought by the mining industry.

3. Deposition of phosphate-rich sediments occurred not only in Florida but along the SE United States continental margin extending to North Carolina, where phosphate is mined as well.

4. Associated with the high primary productivity was an enormous abundance and diversity of marine organisms forming a food web on top of which

perched one of the largest predators that ever lived—the great shark, *Carcharodon megalodon*. This fertile coastal ocean eventually produced an enormous fossil record, making Florida one of the most popular fossil hunting localities anywhere. During a period of sea level lowstand, Florida was populated with many different land animals, most of whom are now extinct. The remains of these terrestrial creatures complicate and enrich the fossil assemblages.

5. The Miocene was a period of enhanced western boundary current activity that eroded and produced the Pourtales and Miami Terraces and other exposed erosional surfaces along the west Florida margin.

6. It is possible that phosphatization, occurring worldwide during the Miocene, withdrew CO_2 from the atmosphere, affecting climate and sea level.

Essential Terms to Know

bacterial microbial activity: Microorganisms consisting of bacteria (instead of other microbes such as viruses) altering organic matter via their own metabolic processes to facilitate certain chemical reactions.

benthic organisms or *benthos:* Animals and plants that live on, in, or near the seafloor. They live in or near marine sedimentary environments, from tidal pools along the coastal zone, out to the continental shelf, and down to the abyssal depths.

brachiopod: A type of benthos fixed (sessile) on the seafloor similar to bivalves (clams, etc.) but physiologically quite different. They secrete calcium carbonate or calcium phosphate exoskeletons or shells. Most species are extinct.

carbon fixation: Any process through which gaseous carbon dioxide is converted into a solid compound. It mostly refers to the processes found in organisms that produce their own food, usually driven by photosynthesis, whereby carbon dioxide is changed into sugars (organic matter).

chemical bonds: Physical process responsible for the attractive interactions between atoms and molecules; the sharing or transfer of electrons between the participating atoms. Molecules and crystals and most of the physical environment around us are held together by chemical bonds, which dictate the structure of matter.

diagenesis: Any chemical, physical, or biological change undergone by a sediment after its initial deposition.

economic grade: Rocks and sediments that have been deemed valuable enough, generally by industrially defined standards, to merit mining.

erosional remnant: A portion of a geologic feature remaining after rock and sedimentary material have been removed by natural processes—the mesas and buttes in the SW United States are excellent examples of erosional remnants.

exoskeleton: External skeleton that supports and protects an animal's body (e.g., shell), in contrast to the internal endoskeleton of a mammal.

fecal pellet: Excrement of invertebrates occurring in sediments and as fossils in sedimentary rocks.

food chain or *food web:* Eating relationships between species within an ecosystem or a particular living place. Organisms are connected to the organisms they consume by lines representing the direction of organism or energy transfer (e.g., baleen whales consume plankton for food and the energy they need to live).

foraminifera or *forams:* Single-celled animals called protists with a shell and generally composed of calcium carbonate.

gastropod: The class *Gastropoda* forms a major part of the phylum Mollusca. Gastropods are more commonly known as snails and slugs.

interstitial: Pore space between sedimentary particles.

intraclast: Fragment of lithified or partly lithified sediment, derived from the erosion of nearby sediment and redeposited within the same area.

isotopes: Isotopes of an element have nuclei with the same number of protons (the same atomic number, and thus essentially identical chemical properties) but different numbers of neutrons. Therefore, isotopes have different mass numbers, which represent the total number of protons plus neutrons.

nutrients: Chemical compounds or elements that an organism needs to live and grow or use in metabolism, which must be taken in from its environment. Nitrogen, phosphorous, and iron are examples.

overburden: Generally unusable sediments overlying a stratigraphic unit that is the target of mining operations due to its economic value.

oxidation: The loss of electrons or gain of oxygen—increase in oxidation state by a molecule, atom, or ion. Carbon (C) is oxidized when it forms CO_2.

phosphogypsum: The primary by-product of the wet-acid process for producing phosphoric acid from phosphate rock. It is largely calcium sulfate. Gypsum is the common name for hydrated calcium sulfate, a common building material.

photosynthesis: A process that converts carbon dioxide and water into organic compounds, using the energy from sunlight and releasing oxygen.

phytoplankton: Generally marine plants (microscopic algae) that obtain energy through photosynthesis and must live in the well-lit surface layer (photic zone) of the ocean. Phytoplankton account for half of all photosynthetic activity on Earth. Thus phytoplankton are responsible for much of the oxygen present in the atmosphere. They form the basis for the vast majority of oceanic food webs and many freshwater food webs. When present in high enough numbers, phytoplankton may appear as a green discoloration of the water due to the presence of chlorophyll within their cells.

plankton: Drifting organisms (nonswimming animals, plants, or microbes) that inhabit the surface layer of the ocean.

primary productivity: Production of organic compounds from atmospheric or aquatic carbon dioxide, principally through the process of photosynthesis. All life on Earth relies directly or indirectly on primary production. The organisms responsible for primary production are known as primary producers or autotrophs, and they form the base of the food web. Phytoplankton in the ocean are primary producers.

radioactive decay: The process in which an unstable atomic nucleus loses energy by emitting particles and radiation. This decay, or loss of energy, results in an atom of one type, called the parent nuclide, transforming into an atom of a different type, called the daughter nuclide.

radionuclide: An atom with an unstable nucleus. The radionuclide undergoes radioactive decay and emits a gamma ray and/or subatomic particles.

reduction: Process of gaining electrons or losing oxygen (decrease in oxidation state) by a molecule, atom, or ion. For example, reduction of carbon by hydrogen yields methane (CH_4). See oxidation.

resedimentation: Eroding sediments from their original environment and depositing them again—many times in a different setting than that of the original depositional environment, e.g., marine sediments being eroded by streams and redeposited in freshwater environments.

sea level highstand and lowstand: Variations of the level of the global ocean. Highstands are periods when sea level is elevated sufficiently to flood the margins of continents; lowstands are periods when sea level leaves much of the continental margin subaerially exposed.

stratigraphic unit: A distinct vertical section of sediments or sedimentary rocks defined by some prominent characteristic or suite of characteristics (grain size, mineralogy, fossil content) or physical boundaries.

terrestrial ecology: Study of land-based living organisms (including those in freshwater) and their interactions with their environments in the broadest sense.

topographic steering: Deflection of currents due to bottom topography; topographic steering may induce upwelling or topographically associated upwelling.

trophic level: The feeding position in a food chain such as primary producers, herbivores, or primary carnivores. Green plants form the first trophic level, the producers. Herbivores form the second trophic level, while carnivores form the third and even the fourth trophic levels.

upwelling: An oceanographic phenomenon that involves motion of dense, cooler, and usually nutrient-rich water toward the ocean surface, replacing the warmer, usually nutrient-depleted surface water.

western boundary current: Warm, deep, narrow, and fast-flowing currents that form on the west side of ocean basins due to intensification. They carry warm water from the tropics poleward. The Gulf Stream is an excellent example.

zooplankton: Small animals such as jellyfish and krill living in seawater whose complete life cycle is planktonic. They form a critically important food source for larger animals in the ocean.

Keywords

Phosphogypsum stacks, phosphate, carbonate-fluorapatite, fertilizer, draglines, strip-mining, land reclamation, shark teeth, Hawthorne Group, primary productivity, paleo Loop Current, topographic steering

Essential References to Know

Bryan, J. R., T. M. Scott, and G. H. Means. *Roadside Geology of Florida.* Missoula, MT: Mountain Press, 2008.

Cartmell, B. C. *Let's Go Fossil Shark Tooth Hunting: A Guide for Identifying Sharks and Where and How to Find Their Superbly Formed Fossilized Teeth.* Venice, FL: Natural Science Research, 1978.

Compton, J. S. "Origin and Paleoceanographic Significance of Florida's Phosphorite Deposits." In *The Geology of Florida*, ed. A. F. Randazzo and D. S. Jones, 195–216. Gainesville: University Press of Florida, 1997.

Cutler, A. *The Seashell on the Mountaintop: A Story of Science Sainthood and the Humble Genius Who Discovered a New History of the Earth.* 2nd ed. New York: Dutton, 2003.

Droxler, A. W., K. C. Burke, A. D. Cunningham, A. C. Hine, E. Rosencrantz, D. S. Duncan, P. Hallock, and E. Robinson. "Caribbean Constraints on Circulation between Atlantic and Pacific Oceans over the Past 40 Million Years." In *Tectonic Boundary Conditions for Climate Reconstructions*, ed. T. J. Crowley and K. C. Burke, 169–91. New York: Oxford University Press, 1998.

Lidz, B. H., C. D. Reich, R. L. Peterson, and E. A. Shinn. "New Maps, New Information: Coral Reefs of the Florida Keys." *Journal of Coastal Research* 22, no. 2 (2006): 260–82. doi: 10.2112/05a-0023.1.

McFadden, B. J. "Fossil Mammals of Florida." In *The Geology of Florida*, ed. A. F. Randazzo and D. S. Jones, 118–37. Gainesville: University Press of Florida, 1997.

Mormino, G. R. *Land of Sunshine, State of Dreams: A Social History of Modern Florida.* Gainesville: University Press of Florida, 2005.

Mullins, H. T., and A. C. Neumann. "Geology of the Miami Terrace and Its Paleo-Oceanographic Implications." *Marine Geology* 30, no. 3–4 (1979): 205–32. doi: 10.1016/0025-3227(79)90016-1.

Popenoe, P. "Paleo-oceanography and Paleogeography of the Miocene of the Southeastern United States." In *Phosphate Deposits of the World*, ed. W. C. Burnett and S. R. Riggs, 352–80. New York: Cambridge University Press, 1990.

Riggs, S. R. "Petrology of the Tertiary Phosphorite System of Florida." *Economic Geology* 74, no. 2 (1979): 195–220. doi: 10.2113/gsecongeo.74.2.195.

———. "Phosphorite Sedimentation in Florida: A Model Phosphogenic System." *Economic Geology* 74, no. 2 (1979): 285–314. doi: 10.2113/gsecongeo.74.2.285.

———. "Paleoceanographic Model of Neogene Phosphorite Deposition, U.S. Atlantic Continental Margin." *Science* 223, no. 4632 (1984): 123–31. doi: 10.1126/science.223.4632.123.

Riggs, S. R., S. Snyder, D. Ames, and P. Stille. "Chronostratigraphy of the Upper Cenozoic Phosphorites on the North Carolina Continental Margin and the Oceanographic Implications for Phosphogenesis." *Marine Authigenesis: From Global to Microbial* 66, SEPM Special Publication (2000), 369–85. doi: 10.2110/pec.00.66.0369.

The Finish Line in Sight

10

Approaching Modern Florida and the Emergence of South Florida (~2.5 Ma to ~10 ka)

> Mother, mother ocean, I have heard you call,
>
> . . . .
>
> You've seen it all, you've seen it all.
>
> Jimmy Buffet, "Mother, Mother Ocean"

Preparing for the Anthropocene: Emergence of South Florida

As we saw in chapter 8, Florida received its quartz sand from a long-distance sand transport pathway extending >1,000 km from the southern Appalachian Mountains and to the Straits of Florida. This influx onto the Florida Platform started ~30 Ma. The last part of this influx occurred between 5–3 Ma in the Pliocene and brought significant quantities of quartz-rich sand and gravel to south Florida. The resulting 150 m thick package of sediment built a gently seaward, south-sloping ramp that ultimately provided a shallow-water seafloor to support the next phase of carbonate sediment production (fig. 10.1).

This change from a quartz sand (siliciclastic) dominated sedimentary system to a carbonate sedimentary system was the most significant geologic event in south Florida during the past 3–2.5 Ma. This fundamental depositional change emplaced the Pleistocene limestone formations that created the present topography in south Florida, which now supports all subsequent human activity and infrastructure in this part of the state. This topography includes the broad, elevated ridge (Atlantic Coastal Ridge; fig. 10.2) that immediately underlies Dade, Broward, and Duval Counties, the limestone floor of the Everglades, and the Florida Keys with its associated coral reefs and other environments. Without the extra few meters of elevation provided by the Atlantic Coastal Ridge, the massive presence of the 5.4 million people inhabiting SE Florida never would have occurred. A small rise in elevation due to a past geologic event makes the difference between an alligator-infested swamp and a foundation for a thriving, densely populated mosaic of cities, suburbs, superhighways, shopping malls, and skyscrapers. What geologic events converged to make SE Florida suitable for such prolific human activity?

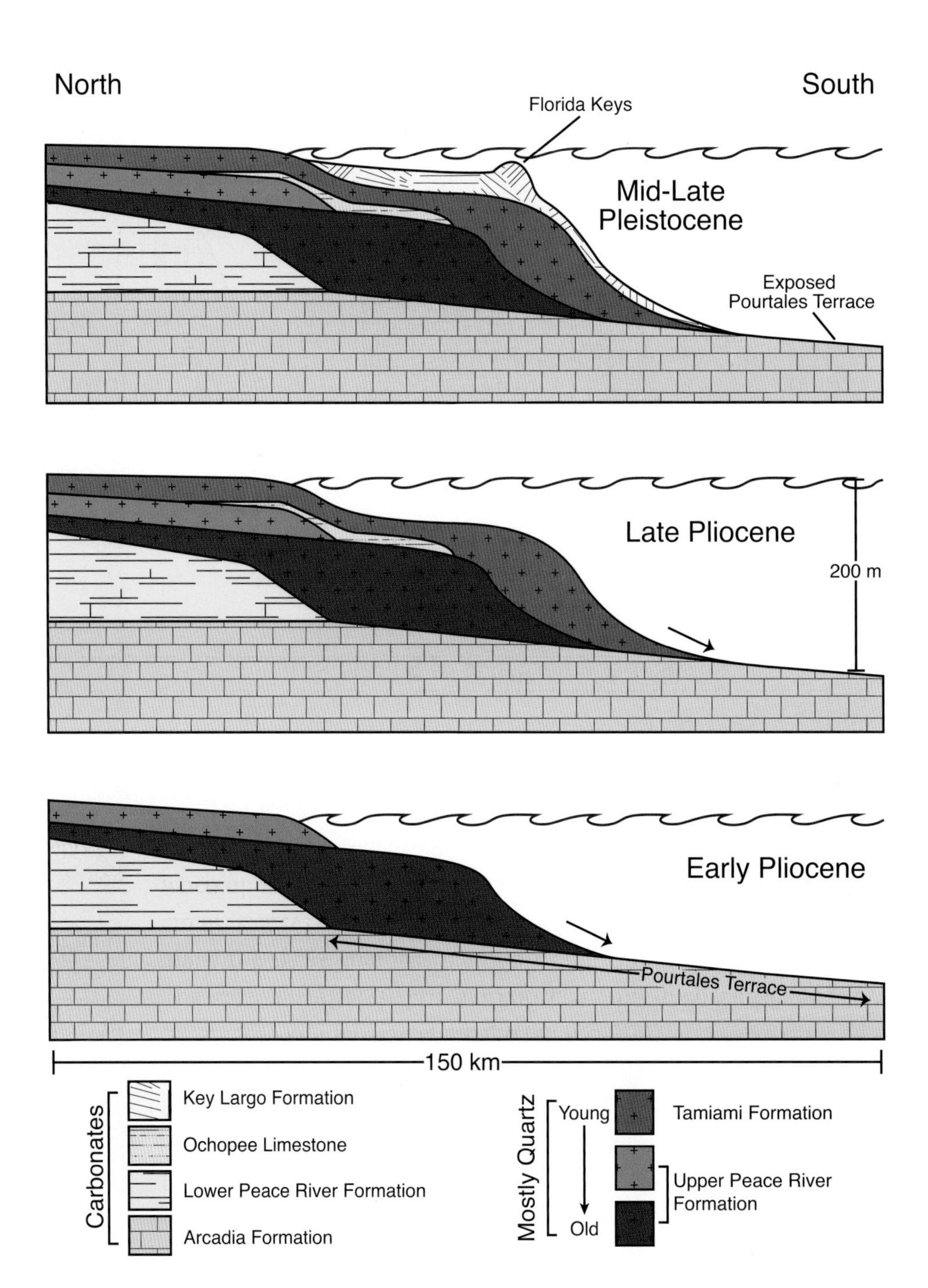

Figure 10.1. The aggradation of a ramp extending southward built up by quartz-rich sediments being introduced from the north from the Pliocene to middle Pleistocene. Eventually, the influx of these siliciclastic sediments is reduced and the ramp becomes shallow enough so that carbonate sediment returns, particularly during sea level highstand in the mid-Pleistocene, building the foundation for the modern Florida Keys and associated marine carbonate depositional environments such as the coral reefs. (Modified from Cunningham et al. 2003; used by permission from the Society of Sedimentary Geology.)

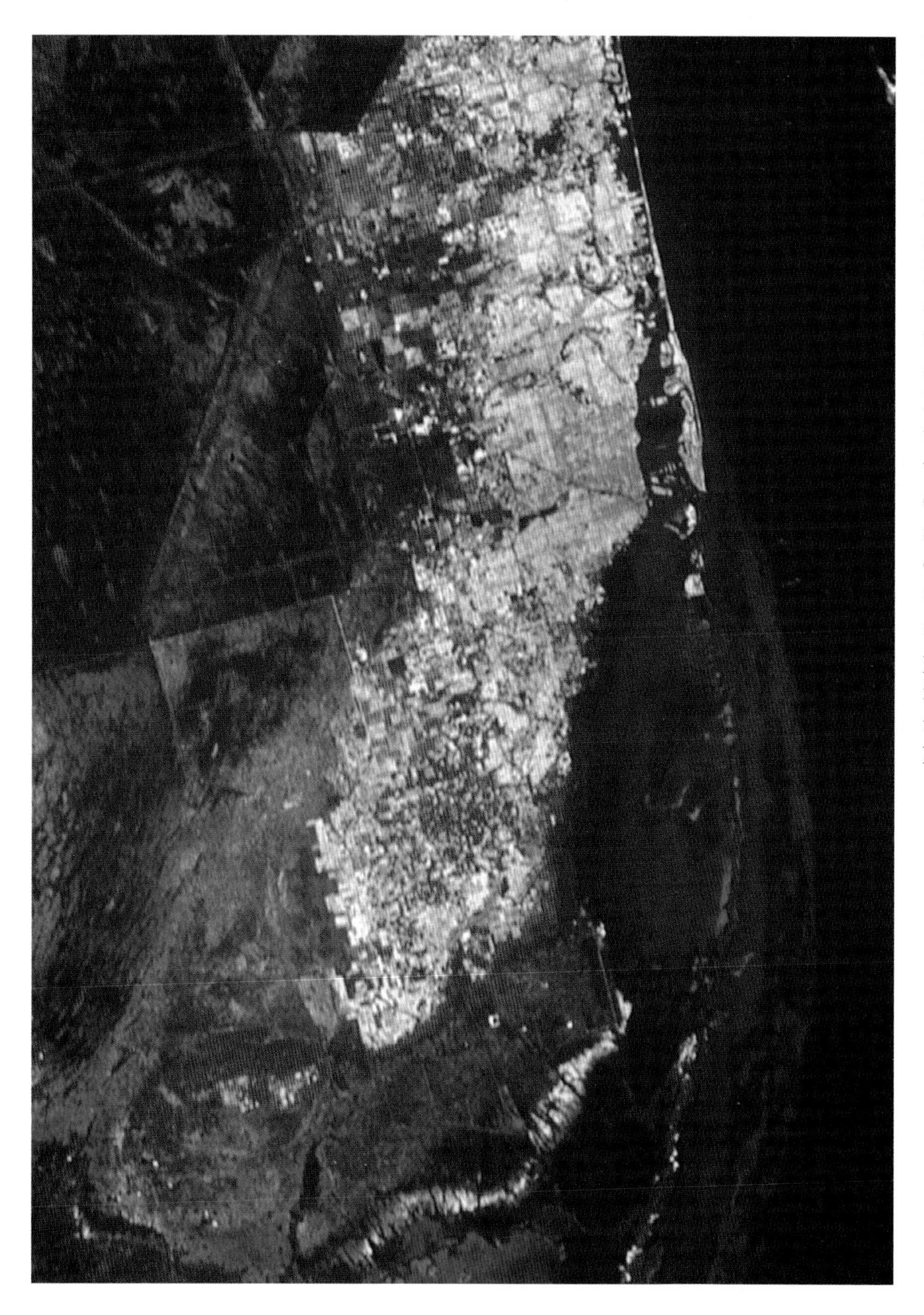

Figure 10.2. Image of south Florida illustrating the elevated Atlantic Coastal Ridge supporting millions of people from north of Ft. Lauderdale to south of Miami, the Everglades, Florida Bay, and the upper and lower Florida Keys. Only a few meters of higher elevation (accumulated Pleistocene sediment) than the Everglades to the west provide the difference between supporting a significant human habitat or a significant alligator (birdlife, etc.) habitat. (Source: Digital orthophoto quarter quadrangle imagery; http://gisdata.usgs.gov/metadata/doqq.htm.)

Quartz-Carbonate Transition

As the quartz sand and gravel-rich river deltas *prograded* down the ramp in south-central Florida, being driven by long-term sea level fall during the Late Pliocene (~3 Ma), the substrate in the area of the Keys aggraded and the seas became shallower than before. With the advent of higher frequency sea level cycles during the very Late Pliocene/Early Pleistocene (see sea level curves on page ii), the siliciclastic influx ceased and carbonate sedimentation returned.

The earliest date from carbonate sedimentary material beneath the Keys is about 370 ka, but most likely carbonate sediments appeared well before this date, marking a transition from one type of sedimentary regime to another that occurred between ~3 Ma and 0.370 Ma (~370 ka).

High-frequency sea level fluctuations, the possible reduction in intense local thunderstorm activity flooding rivers transporting sands and gravels, and the overall regional depletion of siliciclastic sediments all may have shut down river delta progradation. The cessation of the siliciclastic influx and the availability of broad shallow-water seas, particularly during periods of high sea level, probably triggered the return of carbonate sedimentation.

This production of carbonate sediments and coral reefs provided the geologic framework and the foundation for modern south Florida—what we see and enjoy today as the Everglades, Florida Bay, the Keys, and the coral reefs.

Icehouse Earth

So our narrative about the geologic history of Florida and the major events of the geologic past that formed the Sunshine State enters a final era that paleoclimate scientists call the *icehouse* or icehouse Earth. It seems somewhat counterintuitive that the "land of sunshine's" last act would occur in a period of geologic time where great ice sheets waxed and waned on the Earth's surface. How could this be?

Approximately 2.75 Ma, huge continental *ice sheets* started to accumulate in the northern hemisphere. The largest were the Laurentide and the Fenno-Scandinavian Ice Sheets covering northern North America and northern Europe, respectively (fig. 10.3). Additionally, ice sheets covered Greenland and Iceland and *alpine glaciers* filled most high mountain valleys worldwide.

Climate scientists have determined that the Earth has moved from a warm, greenhouse Earth to an icehouse Earth over the past 50 Myr with progressive cooling at both poles. Most scientists seem to favor a lowering of CO_2 over that time, but the process or processes to do that remain elusive. One method is to slow the rate of seafloor spreading and associated volcanic activity that provides a source of CO_2. There is evidence of slowed plate motion. With less CO_2 being introduced into the global system, it seems logical that global climate would cool as well. Another idea is that increased weathering of exposed rock draws down CO_2 as part of the chemical reaction:

$$CaSiO_4 + CO_2 \rightarrow H_2O + \text{clay minerals}$$

($CaSiO_4$ is general representation of Ca-rich rocks on continents exposed to weathering; it would include rocks containing feldspars and other Ca-rich minerals.)

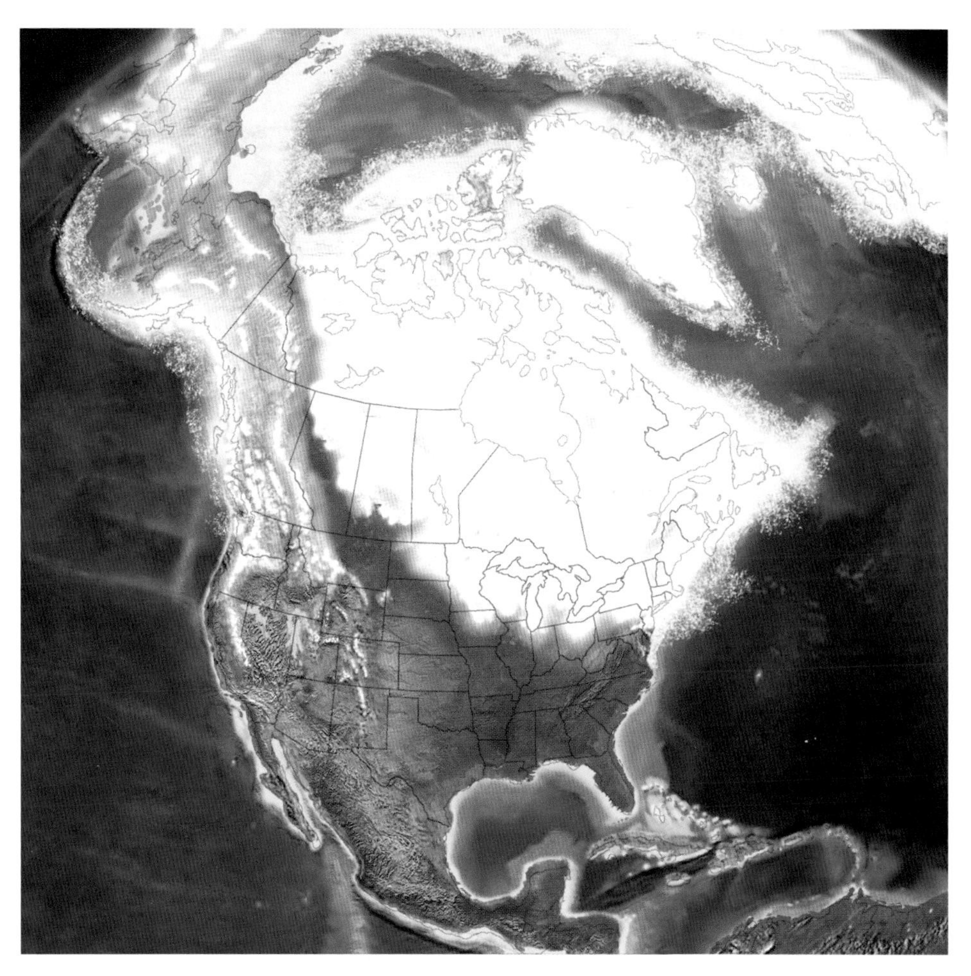

Figure 10.3. Image of North America during Last Glacial Maximum, ~20 ka. Note that northern North America is covered with the Laurentide Ice Sheet, which might have been up to 3 km thick. With so much ice sequestered on North America and northern Europe, the world's sea level fell ~125 m lower than today. (Source: Ron Blakey, Colorado Plateau Geosystems; used by permission.)

With the formation of the Himalaya Mountains and the even more extensive Himalayan-Tibetan Plateau as a result of the tectonic collision between India and southern Asia continental masses, much rock was exposed to weathering—certainly within the past 10 Myr. Increased weathering of other mountain ranges as well would reduce global CO_2. So at approximately 2.75 Ma, the northern hemisphere cooled enough to develop large ice sheets leading to the glacial and interglacial events and the resulting sea level lowstands and highstands. The causes of this 50 Myr greenhouse/icehouse transition and the appearance of northern hemisphere ice sheets at 2.75 Ma still provide fertile ground for research.

Within the past 2.75 Myr, during peak glaciations, these ice sheets extended for thousands of kilometers and were several kilometers thick. Due to oscillations in the Earth's orbit around the Sun and oscillations of the Earth's axis of rotation, heat energy on the Earth's surface from the Sun (*solar insolation*) fluctuated on 100 kyr, 41 kyr, and 23 kyr time scales (see fig. 1.5 *A–C*). These periods constructively interfered to create ice ages (glacial events) and intervening interglacial periods. These are known as *Milankovitch cycles* as described in chapter 1.

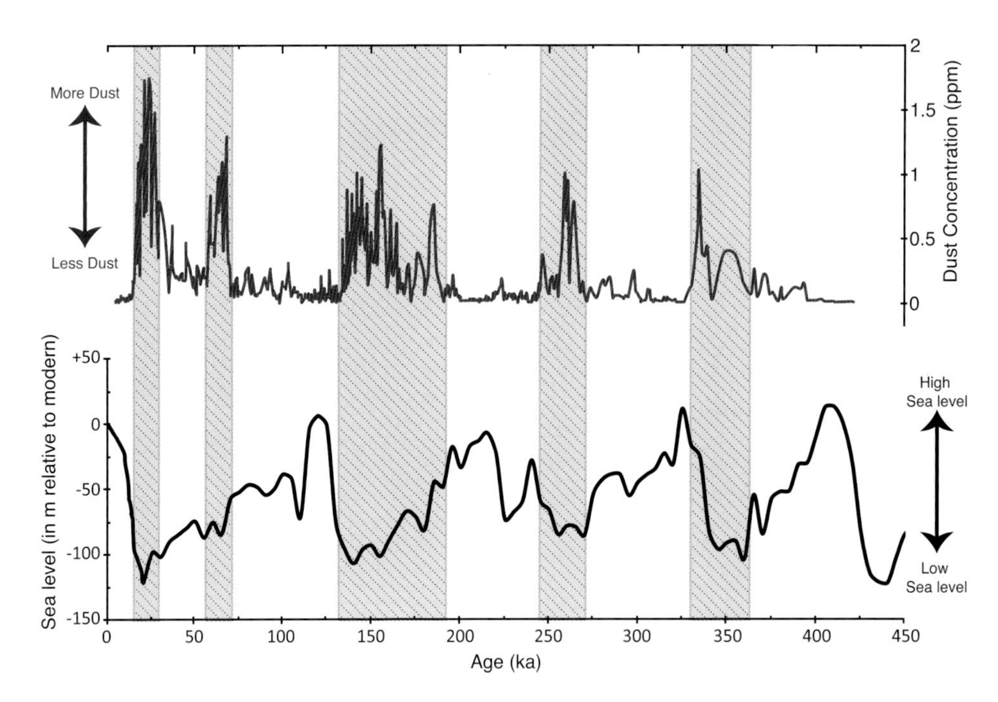

Figure 10.4. Sea level curve showing Milankovitch 100 ka sea level cycles for the past 450 kyr. The data reveal short periods of sea level highstand (~<10 kyr) during which carbonate sedimentary rocks such as tidal bars and coral reefs were deposited along the south Florida margin. These short peaks were followed by long periods (~90 kyr) of subaerial exposure during which the caliche crusts and hardened soil horizons formed. Fine dust particles blown across the Atlantic Ocean carrying iron-rich minerals were deposited during these long periods (see upper curve; identified by panels of diagonal lines), adding to the reddish color seen in fig. 10.6. (Source: Multer et al. 2002; used by permission from Springer.)

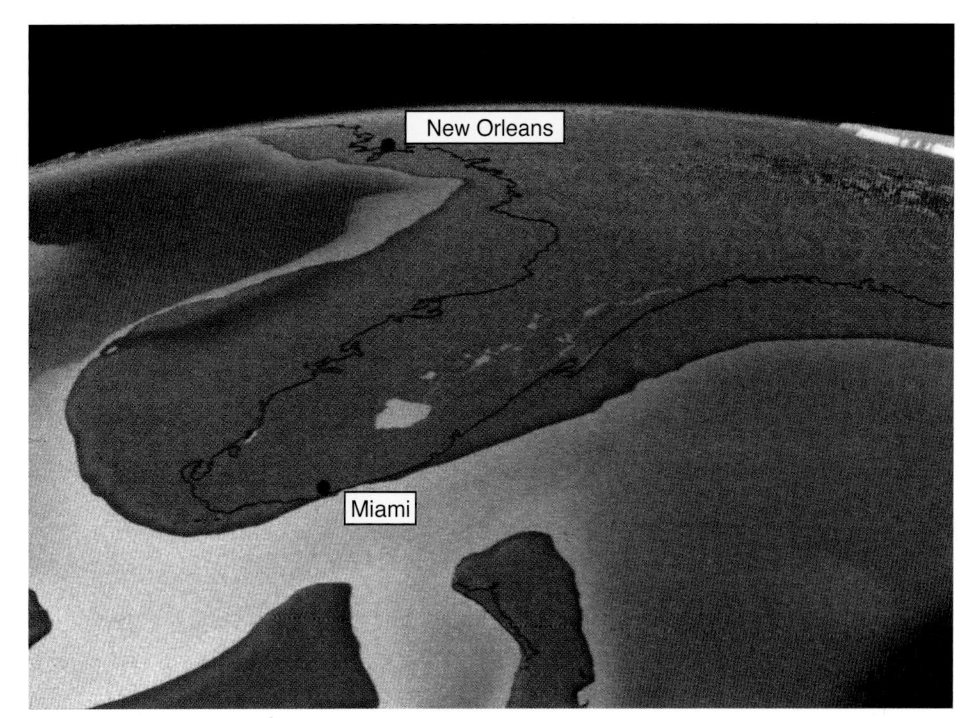

Figure 10.5. Artistic rendering of the subaerially exposed Florida Platform during the Last Glacial Maximum (~20 ka; maximum extent of the Laurentide Ice Sheet in fig. 10.3) when sea level was approximately 125 m lower than today. The green areas surrounding the state of Florida and the exposed Bahama Banks supported terrestrial plants, animals, and eventually pre-Columbian humans. (Source: *Geotimes*, now *Earth Magazine;* used by permission.)

For the past 1 Myr or so, the 100 kyr cyclic signal had been dominant (fig. 10.4). Climate cycles, and therefore sea level cycles, fluctuated ~121 m in amplitude over that time frame. Consequently, there have been roughly 10 100 kyr sea level cycles (one highstand and one lowstand) in the past 1 Myr. Some highstands were higher than others, and some lowstands were lower than others, but not by much. That last major sea level highstand (interglacial period), for example, was ~125 ka, and it was ~6 m higher than today's sea level highstand.

The Earth is in an interglacial period at the moment, but at just 20 ka, the northern hemisphere ice sheets were fully developed, forcing sea level to fall ~121 m lower than what we see today—the Last Glacial Maximum (fig. 10.5). These ice sheets disappeared in near catastrophic fashion, so by about ~8 ka, they were nearly completely gone (not including Greenland). It was these multiple sea level highstands of the past that built the Florida Keys and helped shape Florida's modern coastal system.

The Florida Keys

The carbonate environment of south Florida commencing some time before ~350–400 ka is defined largely by the Key Largo Limestone, divided into at least five *stratigraphic units.* Each unit or layer of rock was formed during a sea level highstand, and each is capped by a reddish, soil-stone crust (also called *caliche,* duricrust, and paleosol). These soil horizons formed during *subaerial exposure,* trapping iron-rich atmospheric mineral dust originating from Africa and carried across the Atlantic Ocean by the Trade Winds (fig. 10.6). Such surfaces were also etched by slightly corrosive rainwater forming *karstic* solution features during the long, extensive sea level lowstands.

Each rock unit was formed at the peak interglacial event each spaced ~100 kyr apart when the top of the south Florida Platform was submerged by marine water for a relatively short time, perhaps <10 kyr, allowing for additional carbonate sediments and coral reefs to form—another stratigraphic unit. During most of the cycle (>90 kyr) sea level was below the top of the platform, exposing the carbonate sediments to subaerial cementation. This exposure formed the limestone and the soil-stone crust capping each unit with karst processes etching the top of each unit.

The earliest Key Largo Limestone unit consisted of skeletal sands forming in deeper water on the undulating bathymetry of the siliciclastic unit below, probably at ~1 Ma. Patchy coral reefs started to form on the topographic high areas. By ~350–300 ka, coral reefs had fully developed on karst-formed, topographically high areas. The next sea level highstand at ~225–190 ka did not match the coral reef productivity that occurred during the previous interglacial highstand.

By the ~125–80 ka sea level highstand, the modern Keys began to take shape. The coral reefs formed parallel to the shelf edge, which dropped off into the

Figure 10.6*A*. Outcrop showing reddish caliche/duricrust/paleosol in Florida Keys on top of Pleistocene age limestone. The reddish color is from iron introduced by dust blown over from Africa. (Photo by A. C. Hine.)

Figure 10.6*B*. Hand sample that has been slabbed or cut by a rock saw showing reddish layering of this diagenetically altered limestone that had been subaerially exposed for >80 kyr. (Photo courtesy of Dr. E. A. Shinn.)

deep Straits of Florida. The highest point of this highstand was at ~125 ka with sea level being ~6 m higher than present-day sea level. Patch reefs probably grew close to the sea surface at this time. When sea level fell, these reefs were subaerially exposed and received their soil-stone, caliche cover. When the last ice age ended at ~18 ka, sea level rose rapidly to today's present level, which is about 6 m lower than when the previous highstand's coral reefs were formed. Hence these 125 ka reefs are now islands, form the upper Florida Keys, and are composed of reef rock. These islands are oriented parallel to the shelf edge (fig. 10.7). So, most likely, the modern upper Keys were at one time a line of patch reefs situated behind a deeper, more massive *forereef* located further seaward (to east/southeast). This 125 ka forereef is now submerged and buried under a very thin, modern (Holocene) coral reef in about 10 m water depth. We know it is there because it can be seen in geophysical images (fig. 10.8).

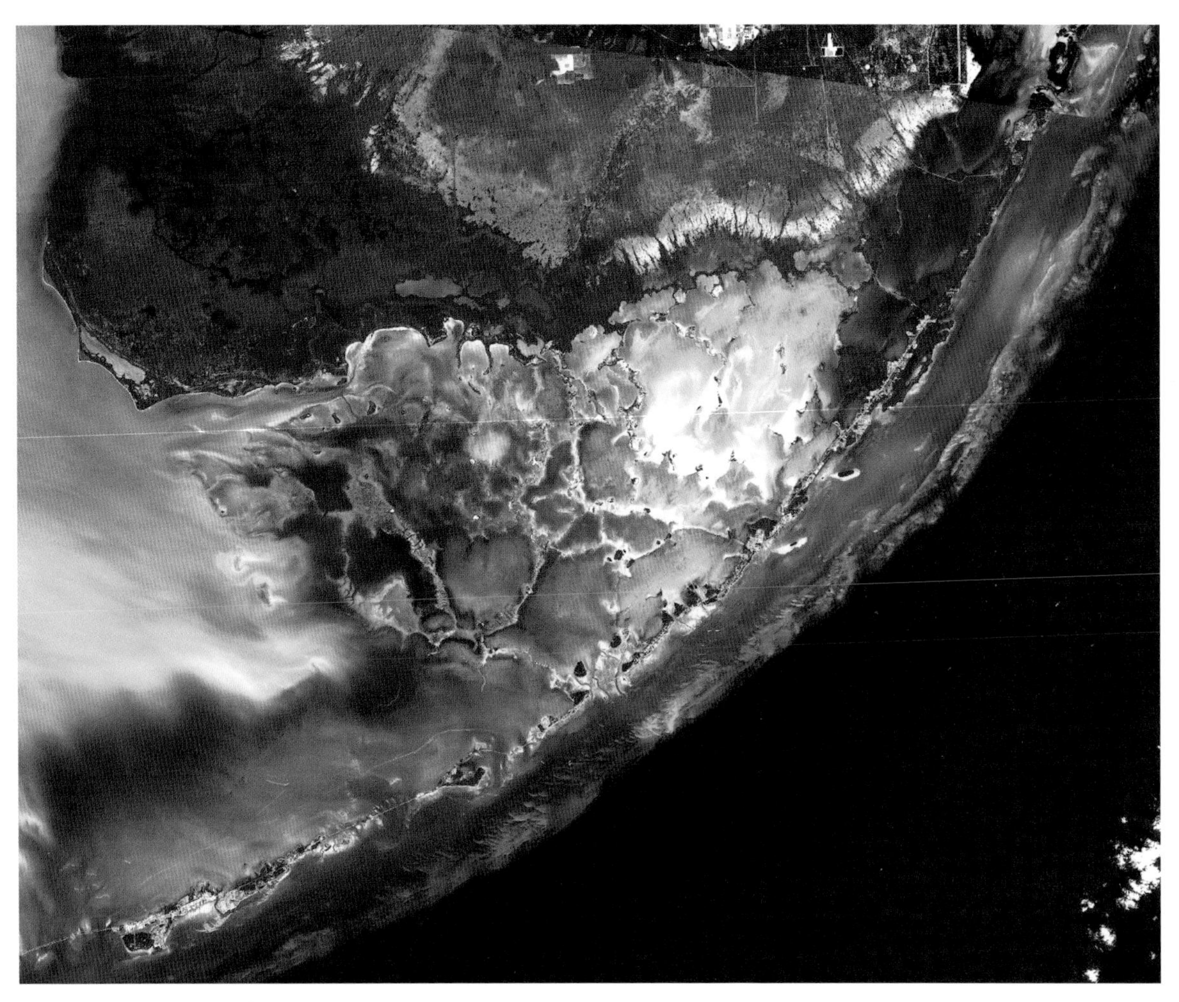

Figure 10.7. Vertical image of south Florida showing linear, parallel-to-shelf-edge orientation of the upper Florida Keys islands. The upper Keys are former coral reefs formed ~125 ka during the last highstand of sea level (+6 m). These islands consist of the Key Largo Formation. (Source: Digital orthophoto quarter quadrangle imagery; http://gisdata.usgs.gov/metadata/doqq.htm.)

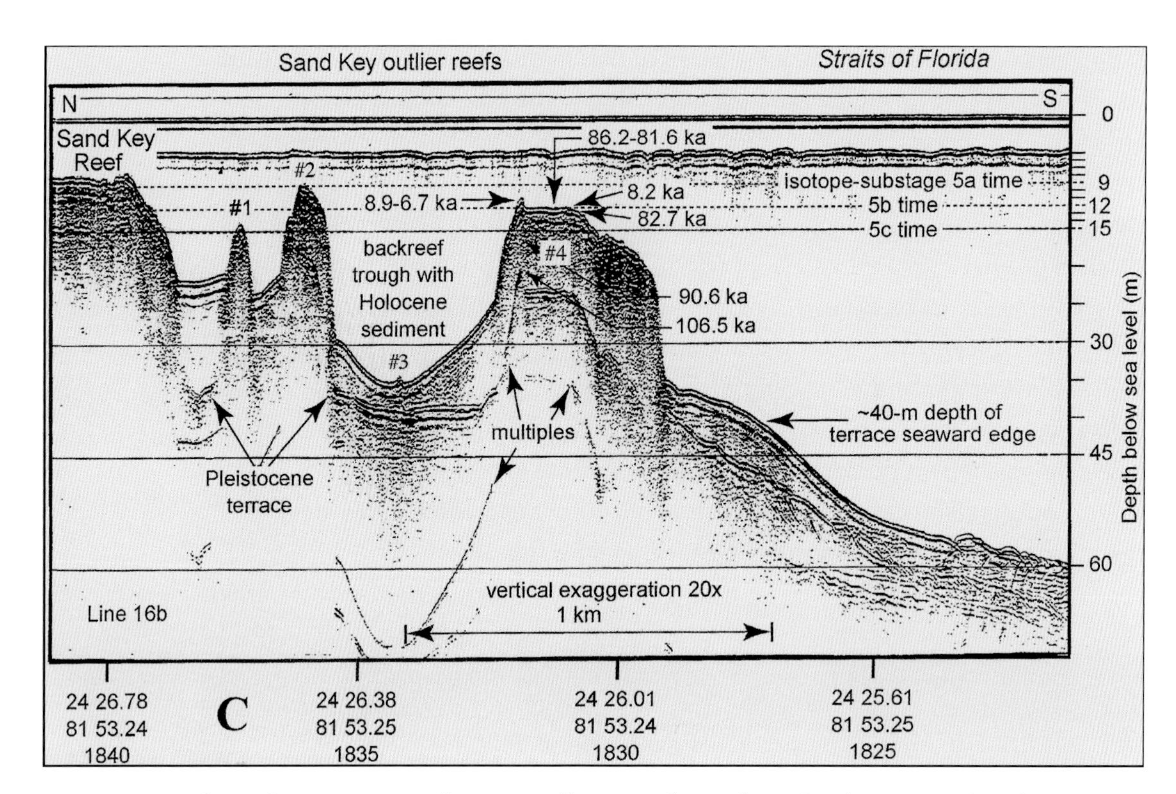

Figure 10.8. High-resolution seismic reflection profile across the modern Florida Keys coral reef tract near Sand Key just south of Key West. Profile shows multiple reefs that have developed since 106 ka (numbers on profiles are radiometric dates in thousands of years). Note that the youngest reef growth (~6.7 ka) is very thin. Most of the reefs seen today off the Keys are very thin geologically; most of the reef edifice was built during early periods of the most recent sea level highstand. (Source: Lidz et al. 1997; used by permission from the *Journal of Coastal Research.*)

The reef rock (called "keystone" by the public) of the upper Keys was mined extensively at Windley Key—now the Windley Key Fossil Reef State Park (http://www.floridastateparks.org/windleykey), where excavations into these fossil reefs reveal corals in growth position forming one of the best coral reef outcrops in the United States (fig. 10.9). These reef rocks, although obtained elsewhere now, are still used for many purposes.

Tidal Sand Bodies

While the Pleistocene coral reefs were growing along the shelf edge in what was to become the upper Keys, further to the southwest, vigorous tidal currents swept on and off the platform margin (fig. 10.10). These strong flows transported carbonate sediments along the shallow seafloor, rolling them over and over/back and forth. Few coral reefs formed here. This agitation caused the individual carbonate grains such as small broken skeletal fragments or hardened fecal pellets from lime mud to become coated with very thin layers of calcium carbonate (*aragonite*). Through time, dozens of these thin limestone layers accumulated,

Figure 10.9*A*. Fossil head coral embedded in Key Largo Formation that was once mined in the Windley Key rock quarry. Much of this rock is seen as facing stone of older buildings in Florida. (Photo by A. C. Hine.)

Figure 10.9*B*. Large coral in outcrop exposed during cutting of a canal across one of the upper Keys. The linear, parallel-to-shelf upper Keys were former coral reefs (actually a back reef), and radiometric dates indicate that they formed during the last sea level highstand at ~125 ka. (Photo by A. C. Hine.)

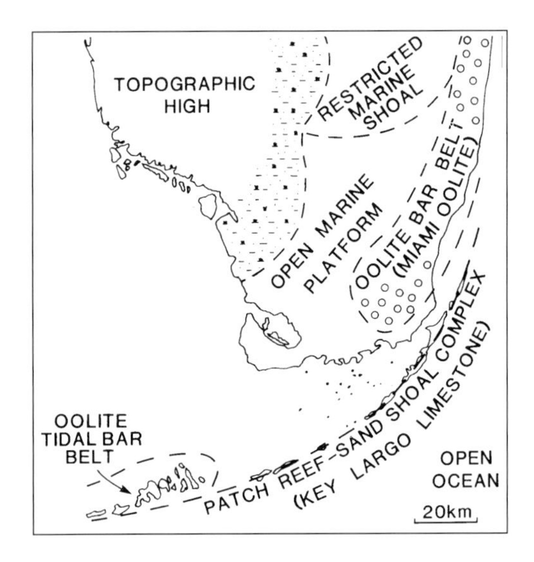

Left: Figure 10.10*A*. Facies map of southeastern Florida including the Keys which shows the distribution of the Pleistocene age (~125 ka) Key Largo Formation and the Miami Limestone. (Source: Tucker and Wright 1990; used by permission from Blackwell Scientific Publications.)

Below: Figure 10.10*B*. Vertical image of lower Florida Keys showing linear islands oriented perpendicular-to-shelf edge, having been formed by strong tidal currents passing on and off the shelf margin. The rocks forming these islands are part of the Miami Limestone and consist mostly of cemented ooid sand grains. (Source: Digital orthophoto quarter quadrangle imagery; http://gisdata.usgs.gov/metadata/doqq.htm.)

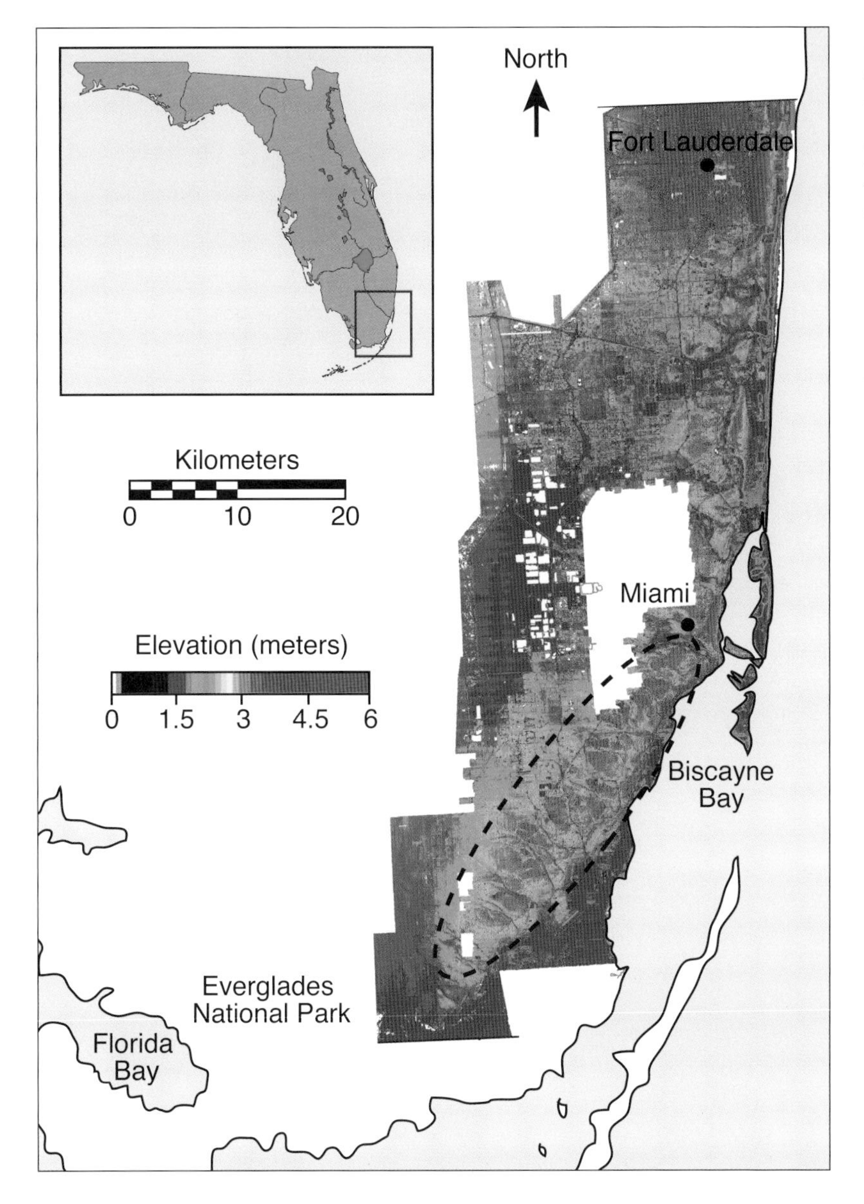

Figure 10.10C. Colorized LIDAR (light detection and ranging) image showing topography of densely populated Miami area, revealing similar topography as lower Keys islands. Both areas are part of the Miami Limestone, which surrounded the coral reef–dominated Key Largo Formation. Strong tidal flows coursed on and off the shelf margin at ~125 ka when sea level was ~6 m higher than today. (Used by permission from Dr. Dean Whitman, Florida Atlantic University.)

eventually forming very well rounded, spherical grains called *ooids* (like tiny snowballs rolling around growing larger and more spherical; fig. 4.5).

These ooid grains accumulated in enormous numbers to form 5–10 m thick, 10 km long linear tidal sand shoals that were oriented perpendicular to the bank edge and parallel to the tidal currents (fig. 10.11). The tops of these shoals were separated by channels 3–10 m deep and were probably awash at low tide. However, when sea level fell, leaving these sand ridges subaerially exposed for ~80 kyr, they, too, became cemented to form limestone and developed a hardened, soil-stone crust and karst-etched surfaces. A similar tide-dominated environment occurred in the Miami area and built up the Atlantic Coastal Ridge, which supports the 5.4 million people now inhabiting that area.

Instead of reef-rock, this ooid-rich limestone is called an *ooid grainstone* and forms the Miami Limestone (contains Miami Oolite). The Key Largo Limestone

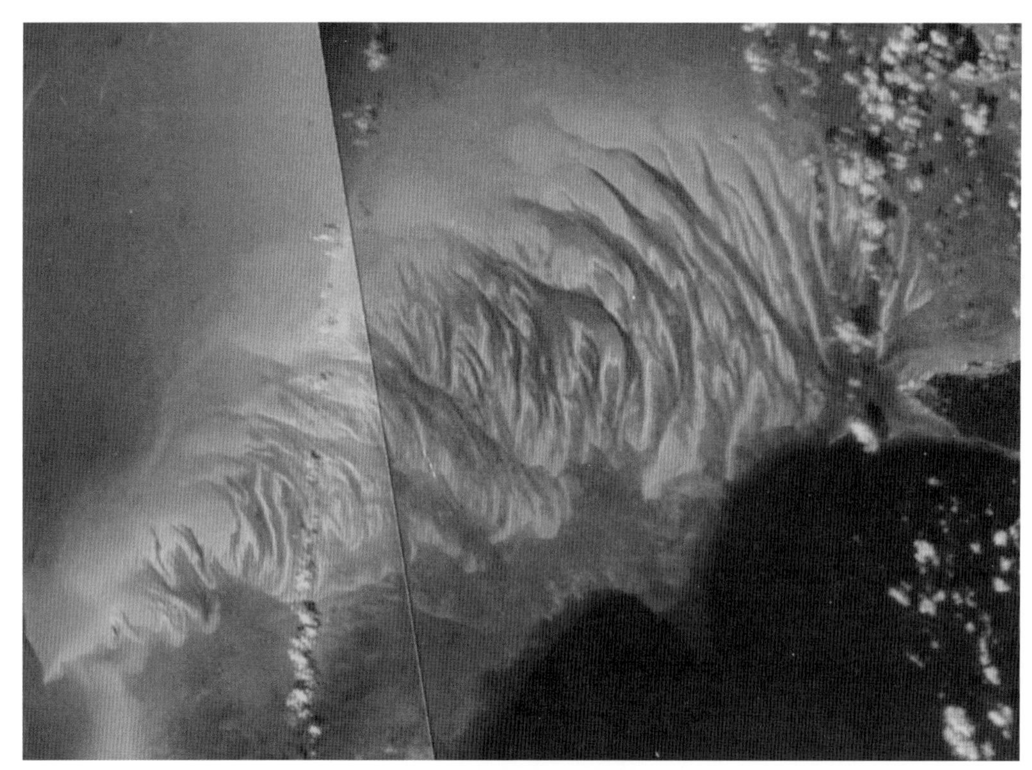

Figure 10.11*A*. Space image of modern tidal-dominated ooid bar belts in the Bahamas roughly similar to Pleistocene lower Florida Keys and Miami area (fig. 10.10). (Source: USGS EROS Data Center.)

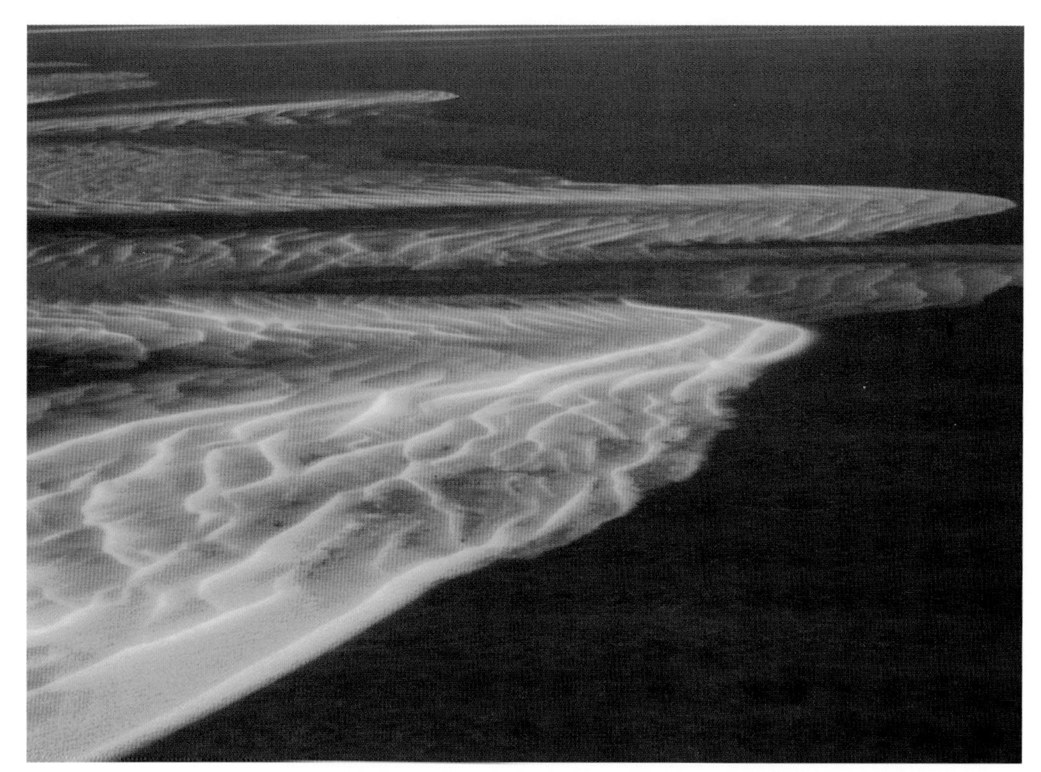

Figure 10.11*B*. Aerial photo of sedimentary bedforms (sand waves or underwater dunes) indicative of strong water flow onto the Bahama Banks. Location is the Lily Bank ooid shoal on Little Bahama Bank. (Photo by A. C. Hine.)

Figure 10.11*C*. Outcrop Miami Limestone showing cross-bedding primary sedimentary structures in the rock resulting from migration of sand waves shown in fig. 10.11*B*. (Photo by A. C. Hine.)

and the Miami Limestone formed contemporaneously side by side—coral reefs in the center with tide-dominated shoals on either side all along the SE Florida Platform margin. One limestone rock unit developed from coral reef *depositional environment,* and the other developed from a tidal bar depositional environment. Some of the building stone seen in Key West and Miami is from the Miami Limestone. It does not contain corals like the Key Largo Formation, but consists of cemented laminations and thin layers of ooid grains deposited by migrating submarine dunes or sand waves.

Paleo-shorelines and Coral Reefs: The Finishing Touches

Once sea level had fallen below the edge of the south Florida shelf at ~80 ka, south Florida remained subaerially exposed until sea level reflooded the platform at ~10 ka. As the Earth approached the last ice age, which peaked at ~26.5–19 ka, sea level had fallen ~121 m. The west Florida coastline was about 150 km further to the west. Had Florida been a state back then, it would have had twice the area that it has today. For sure, what is now the seafloor once supported forests, grasslands, lakes, streams, and animals—a terrestrial ecosystem as complex as today's—and the first humans. Native Americans most likely lived along coastlines and estuaries that now lie beneath 100 m of water. Most likely there are submerged middens and other pre-Columbian artifacts strewn across the shelf.

A recent discovery by geological oceanographers studying the Florida shelf is a series of at least four shorelines, now drowned, that were constructed during sea level rise as the last great ice sheets rapidly melted. The shorelines must have formed during brief periods when melting ceased, allowing beaches and dunes to form. Where these paleo-shorelines are composed of carbonate sand, they are well preserved and were not destroyed during ensuing sea level rise when ice sheets resumed melting. Carbonate sediments rapidly cement together, sometimes in a matter of a few decades, forming limestone and resisting erosion by the rising seas. These linear ridges on the seafloor can be easily seen as imaged (fig. 10.12*A*). Where conditions are well suited to allow corals to form, these paleo-shorelines provide a hard substrate suitable for reef development (figs. 10.12*B* and 10.12*C*). One of the significant discoveries on the SW Florida shelf was the deepest light-dependent coral reef in the United States situated upon

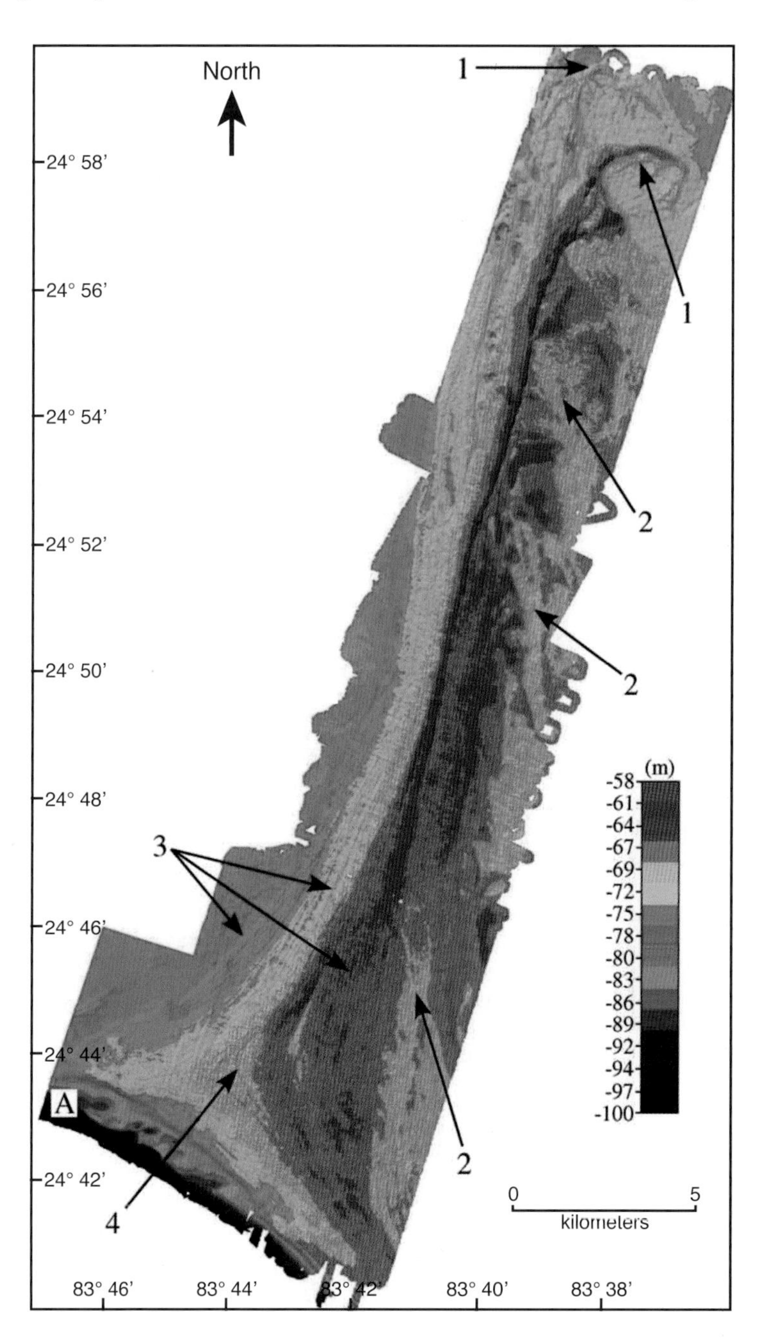

Figure 10.12*A*. Multibeam image of southern Pulley Ridge drowned barrier island, which now supports the deepest, light-dependent coral reef in continental U.S. waters. This carbonate shoreline formed at ~14 ka when the rapid rise of sea level, resulting from the melting of the world's ice sheets, stopped for probably a few hundred years. Note the barrier island geomorphological features such as former tidal inlets and channels, beach ridges, and recurved spits. (Source: Jarrett et al. 2005; used by permission from Elsevier.)

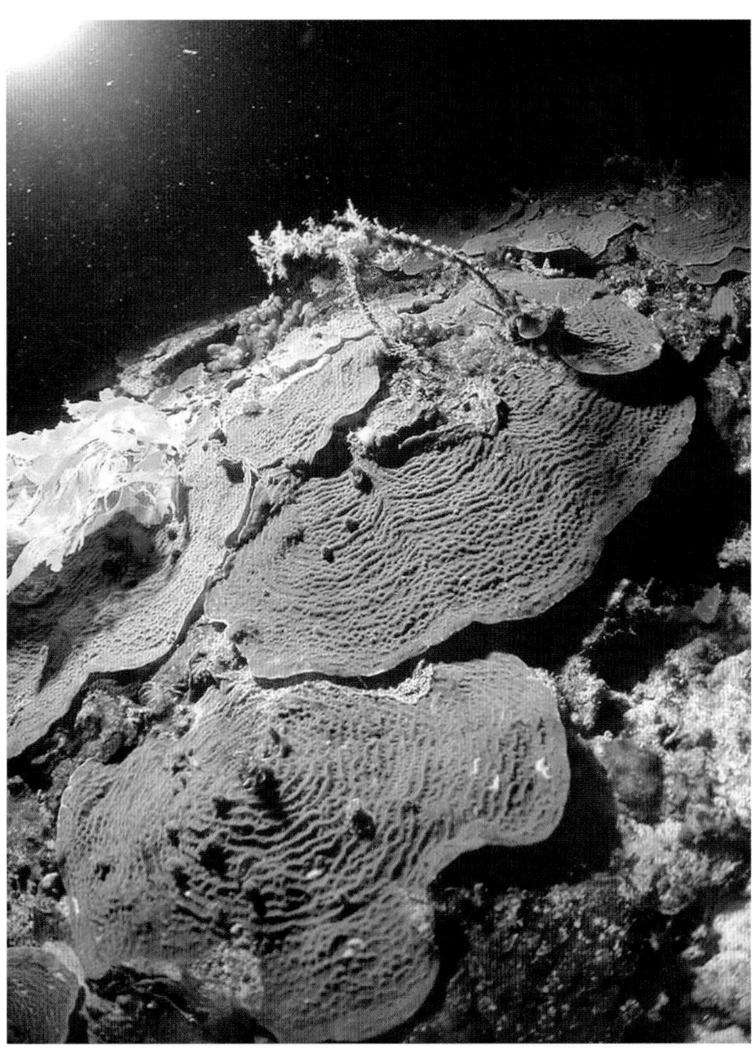

Left: Figure 10.12*B*. Bottom photo at Pulley Ridge revealing corals, encrusting red algae and green fleshy algae growing on top of the ancient barrier island. (Used by permission from Dr. Robert Halley, USGS.)

Below: Figure 10.12*C*. Bottom photo of Pulley Ridge revealing similar features as shown in fig. 10.12B. (Used by permission from Dr. Robert Halley, USGS.)

Figure 10.13. Vertical image of Cape Canaveral area showing "multiple" Cape Canaverals—capes behind capes all formed during previous sea level highstands. The modern Cape Canaveral was probably formed within the past few thousand years. The cape just to the west inside the modern cape was probably formed during the last sea level highstand ~125 ka. (Source: Digital orthophoto quarter quadrangle imagery; http://gisdata.usgs.gov/metadata/doqq.htm.)

these paleo-shorelines in 65–75 m water depth—southern Pulley Ridge. At the same depth but in another location, radiocarbon dating of the paleo-shorelines indicates that they formed at ~14 ka.

Similar, former sea level *highstand* shorelines are even more common than *lowstand* shorelines because they are emerged and can easily be seen on land. Satellite imagery indicates that a cape-like feature formed at ~125 ka that was the precursor to today's Cape Canaveral (fig. 10.13). In Miami, a famous outcrop of limestone named the Silver Bluff is a remnant of a former highstand of sea level (fig. 10.14*A*). A modern example of waves breaking against a similar cliffed limestone shoreline from the Bahamas is shown in fig. 10.14*B*. Indeed, a drowned example lying in ~70 m of water off Key West is shown in fig. 10.14*C*.

The entire modern coastline of Florida has formed as a result of slow sea level rise over the past few thousand years, including the flooding of the karst-induced basins that are now the Tampa Bay and Charlotte Harbor estuaries (see chapter 7).

Former elevated coastlines are ubiquitous, and during these elevated sea level events, Florida's coastline must have appeared quite different than it is today.

Figure 10.14*A*. Ancient erosional cliffed shoreline called Silver Bluff in Miami Limestone indicating past wave activity, but at ~6 m higher when sea level was elevated at ~125 ka. (Photo by A. C. Hine.)

Figure 10.14*B*. Modern, erosional cliffed shoreline of carbonate island in the Bahamas. (Photo courtesy of Dr. S. D. Locker.)

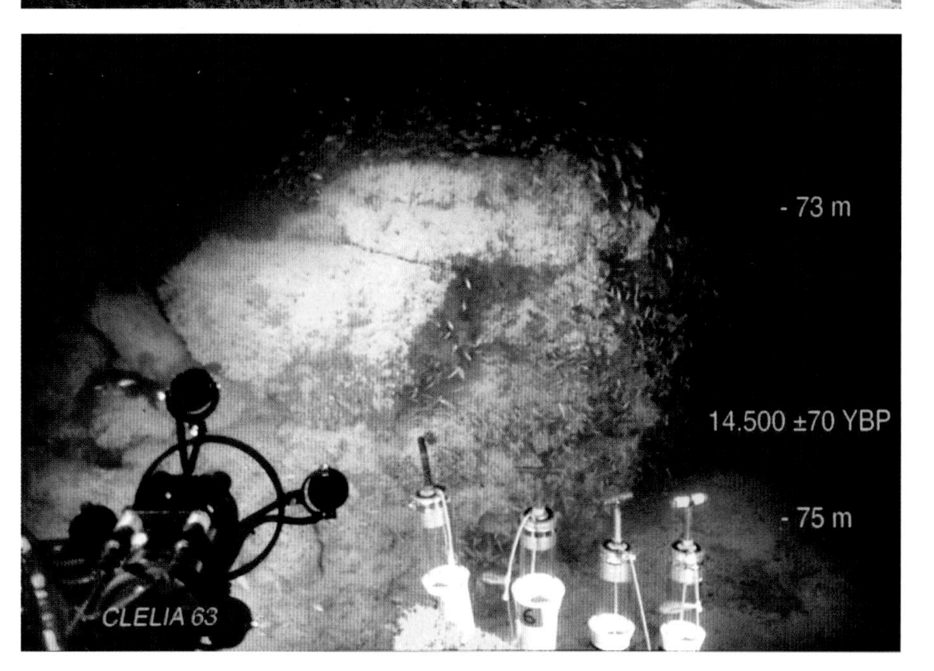

Figure 10.14*C*. Underwater photo of similar environment as shown in figs. 10.14*A* and 10.14*B*, but in 70 m water depth when sea level was lower at ~14 ka. This cliffed paleo-shoreline lies south of Key West. (Photo courtesy of Dr. S. D. Locker.)

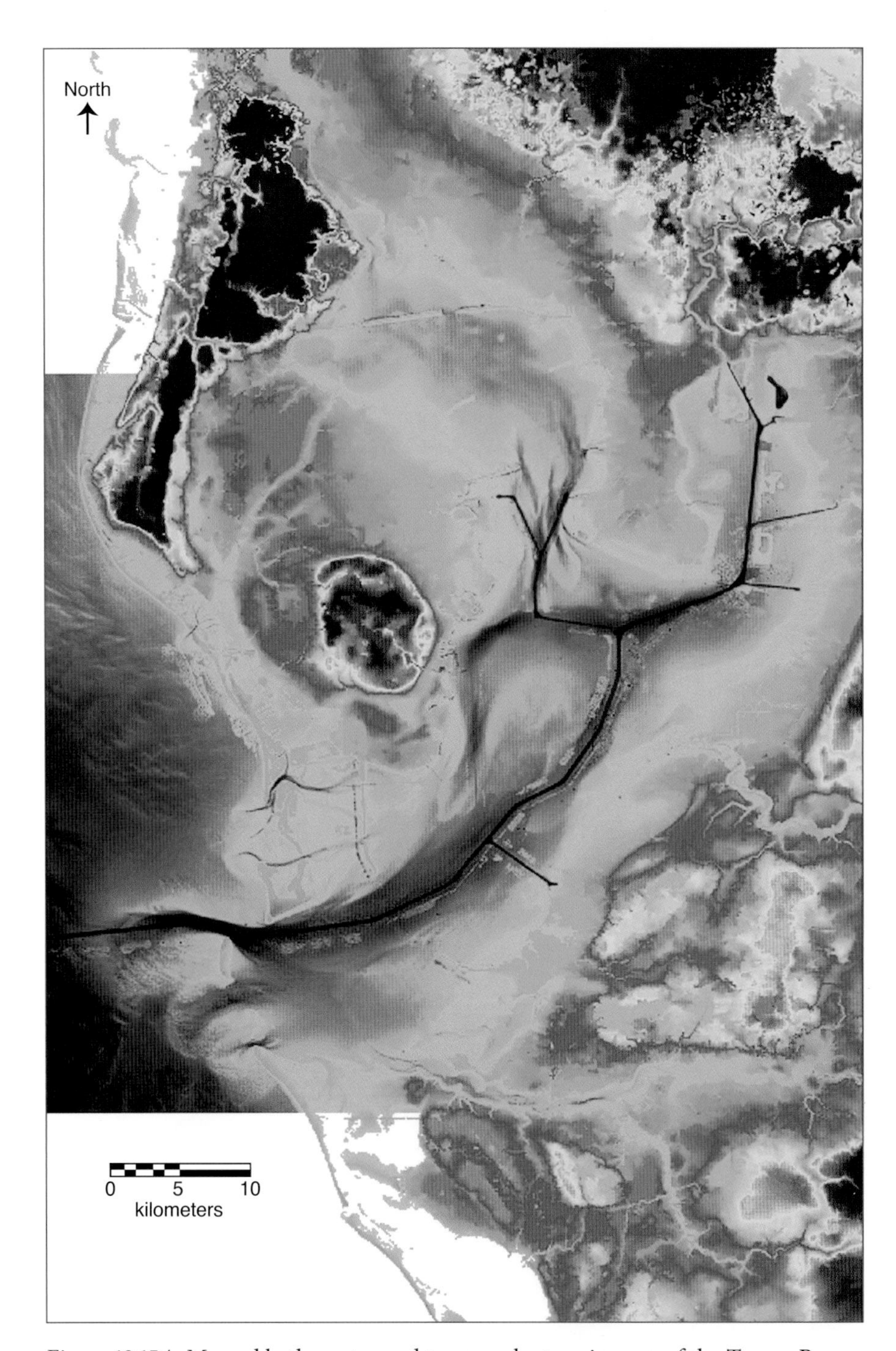

Figure 10.15*A*. Merged bathymetry and topography terrain map of the Tampa Bay Basin. These state-of-the-art images illustrate the elevation continuum between underwater and terrestrial morphology. The sharp color differences generally represent paleo-shorelines or paleo sea level stillstands. Note that Pinellas County consisted of two large islands (St. Petersburg occupies one) that existed when sea level was 6 m higher about at ~125 ka. (Source: USGS EROS Data Center.)

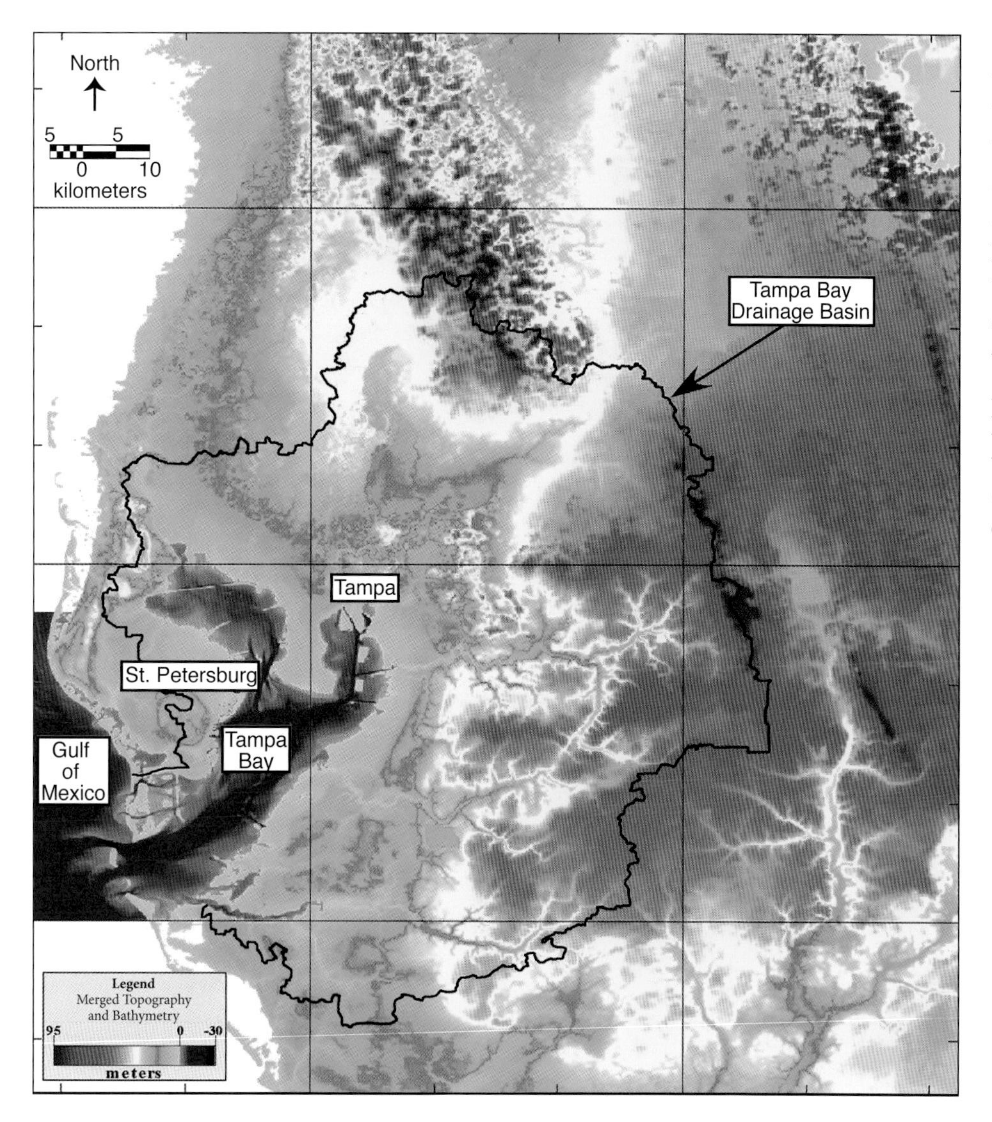

Figure 10.15*B*. The entire Tampa Bay drainage basin. Note rivers flowing into Tampa Bay. Entire green area was underwater during the last sea level highstand, making the Tampa Bay area a large, open marine embayment with a number of islands. Also note numerous circular features that represent thousands of sinkholes. (Source: USGS EROS Data Center.)

Fig. 10.15*A* clearly shows that Pinellas County must have been two large but separate islands during the last interglacial event occurring some 125 ka. Finally, the distribution of Florida's rivers became established since the major sea level highstand in the mid-Miocene, when the phosphate deposits were formed. But much of the drainage was probably formed in the past 2.75 Myr as a result of the climate changes associated with numerous glacial/interglacial oscillations (fig. 10.15*B*).

The End, Fini, That's All Folks!

So we have reached the end of Florida's geological history. Of course, the most recent events of the late Holocene (past 10 kyr) through today continue to shape the Florida Platform. Geology never stands still. As long as there are winds, waves, currents, weather and climate, biological activity, chemical reactions, and motions within the Earth, geology is present. Perhaps the most significant recent geologic agent, never seen on Earth before, is the collective activity of human civilization. One of the great challenges we and future generations face is to understand the extent to which our advanced agriculture and our industrial-based civilization become agents of geologic change.

It was the sum total of all the geologic events extending back to the origin of Florida's basement rocks in deep time once located at the South Pole (~650 Ma) that defined the template upon which the modern geology of Florida rests.

Essential Points to Know

1. The southward prograding river delta system migrating down the seaward-sloping ramp forming southernmost Florida 5–3 Ma built up a surface on which carbonate sedimentation would eventually dominate.

2. This switch from siliciclastic sedimentation to carbonate sedimentation was a key event as it led to the development of topography that defines south Florida. This topography includes the Atlantic Coastal Ridge, which supports the millions of people living in the Miami area, the floor of the Everglades, and the islands that form the Florida Keys.

3. Multiple sea level highstands resulting from climate-driven sea level cycles built the Florida Keys. These islands consist of limestone from cemented, late Pleistocene coral reefs and tidal sand bodies.

4. These sea level fluctuations also left behind paleo-shorelines, now submerged on the west Florida shelf and emerged on land.

5. A portion of the submerged paleo-shorelines now supports the deepest light-dependent coral reef in the United States, a significant new discovery.

6. The emerged paleo-shorelines have controlled the geomorphology of the modern coastal system.

7. The 100 kyr long sea level cycles of the past 1 Myr were the latest events instrumental in shaping the geology of modern Florida. It was the sum total of all the geologic events extending back to the origin of Florida's basement rocks in deep time (~1 Ga) that defined the template upon which the modern geology of Florida rests.

Essential Terms to Know

aggradation: Term used in geology for the increase in surface elevation due to the deposition of sediment, meaning that it builds vertically, as opposed to progradation, which builds horizontally. Aggradation occurs in areas in which the supply of sediment is greater than the amount of material that the system is able to transport horizontally.

albedo: The extent to which an object diffusely reflects light from the Sun. Albedos of typical materials in visible light range from 90 percent for fresh snow to 4 percent for charcoal, one of the darkest substances.

alpine glaciers: Mountain glaciers that form at high elevations where temperatures remain cold enough during the summer to keep the previous winter's snow from melting, allowing snow and ice to accumulate. The ice migrates downslope, cuts valleys, transports rocky debris, and creates rivers and streams from meltwater. Alpine glaciers terminate where downslope ice transport is offset by melting.

aragonite: A carbonate mineral, one of the two common, naturally occurring types (polymorphs) of calcium carbonate ($CaCO_3$). The other type is the mineral calcite. They have similar crystal chemistry but a different internal arrangement of elements, making them different minerals.

biogenic: Produced by life processes. It may be either constituents or secretions of plants or animals.

caliche: A hardened deposit of calcium carbonate at the surface or within soil. This calcium carbonate cements together other materials, including gravel, sand, clay, and silt, forming a crust, commonly hardened soil on top of a carbonate rock unit. See duricrust.

duricrust: A thin hard layer on or near the surface of soil, usually a few millimeters to a few centimeters thick, forms on top of carbonate rock units when exposed to air. It can be composed of any water soluble mineral in contrast to caliche.

forereef: The main seaward portion of a reef that starts from the shallow reef crest and extends downslope and seaward into deeper water.

glacial and interglacial events: Generally cycles of forming large ice sheets (glacial events, global cooling) and melting/disappearance of ice sheet (interglacial events, global warming).

glacio-eustacy: Global changes in sea level that result from waxing and waning of ice sheets forming on continental masses.

ice sheet: A mass of glacier ice that covers surrounding terrain and is greater than 50,000 km^2. The only current ice sheets are in Antarctica and Greenland; during the last glacial period at Last Glacial Maximum (LGM), the Laurentide ice sheet covered much of Canada and North America, the Fenno-Scandinavian ice sheet covered northern Europe, and the Patagonian Ice Sheet covered southern South America.

icehouse: Period in Earth's past when climate was locked into a series of glacial and interglacial cycles. We are currently under icehouse conditions (but in an interglacial at the moment) that started at ~2.75 Ma.

karst: A landscape or surface shaped by the dissolution of a layer or layers of soluble bedrock, usually carbonate rock such as limestone or dolomite.

Marine isotope stage (MIS) or *marine oxygen-isotope (O) stages:* Alternating warm and cool periods in the Earth's paleoclimate, deduced from oxygen isotope data reflecting temperature curves derived from deep-sea core samples. Each stage represents a glacial or interglacial. Interglacials are odd-numbered, glacials are even-numbered, one for each stage, starting from the present and working backward in time. For example, the Holocene is MIS 1, or O-stage 1, or just stage 1. The previous major interglacial was MIS 5, or O stage 5, or just stage 5.

Milankovitch cycles: Climate cycles generating glacial and interglacial events on time scales driven by eccentricity of the Earth's orbit around the Sun and variations of the Earth's rotation on its axis—all altering incoming solar radiation striking the Earth's surface and changing climate. Milankovitch was an early twentieth-century Serbian engineer and mathematician.

ooids/ooid grainstone: Small (<2 mm), spheroidal, "coated" (layered) sedimentary grains, usually composed of calcium carbonate, but sometimes made up of iron- or phosphate-based minerals. Ooids usually form on the seafloor, most commonly in shallow tropical seas. These ooid grains can be cemented together to form a sedimentary rock called an oolitic (not entirely composed of ooids) or ooid grainstone.

paleogeography: Geography, distribution of landforms and oceans in the past.

paleo-shorelines: Beaches, dunes, and other coastal features that formed in the past, generally when sea level (or lake level) was either higher or lower than present. Shorelines that consist of carbonate sands are commonly cemented to form limestone. The limestone resists erosion and enhances preservation.

paleosol: Ancient and fossil soil.

positive feedback: A process that generates another process which adds to or further stimulates the initial process. For example, global warming might produce large swamps and bogs that produce methane, a greenhouse gas that contributes to further global warming.

progradation: Lateral filling in of an area such as a river delta prograding into a lake basin or a carbonate platform prograding into deeper water—filling in of a basin by laterally introducing sediments—horizontal filling rather than vertically filling.

solar insolation: A measure of solar radiation energy received on a given surface area in a given time. The name comes from the words *incident solar radiation*. It is commonly expressed as average irradiance in watts per square meter (W/m^2) or kilowatt-hours per square meter per day ($kW{\cdot}h/(m^2{\cdot}day)$) (or hours/day).

subaerial exposure: Rock or sediments exposed to air due to sea level fall or tectonic uplift.

Keywords

Atlantic Coastal Ridge, depositional change, icehouse Earth, sea level cycles, Florida Keys, caliche, African dust, Key Largo Formation, Miami Limestone, ooids

Essential References to Know

Cunningham, K. J., S. D. Locker, A. C. Hine, D. Bukry, J. A. Barron, and L. A. Guertin. "Surface-Geophysical Characterization of Ground-Water Systems of the Caloosahatchee River Basin, Southern Florida." In *Water-Resources Investigations Report* 76. Tallahassee: U.S. Geological Survey, 2001.

Cunningham, K. J., S. D. Locker, A. C. Hine, D. Bukry, J. A. Barron, and L. A. Guertin. "Interplay of Late Cenozoic Siliciclastic Supply and Carbonate Response on the Southeast Florida Platform." *Journal of Sedimentary Research* 73, no. 1 (2003): 31–46. doi: 10.1306/062402730031.

Enos, P., and R. D. Perkins. *Quaternary Sedimentation in South Florida.* Vol. Memoir 147. Boulder: Geological Society of America, 1977.

Halley, R. B., and C. C. Evans. *The Miami Limestone: A Guide to Selected Outcrops and Their Interpretation.* Miami: Miami Geological Society, 1983.

Hine, A. C., R. B. Halley, S. D. Locker, B. D. Jarrett, W. C. Jaap, D. J. Mallinson, K. T. Ciembronowicz, N. B. Ogden, B. T. Donahue, and D. F. Naar. "Coral Reefs, Present and Past, on the West Florida Shelf and Platform Margin." In *Coral Reefs of the USA*, ed. B. M. Riegl and R. E. Dodge, 127–73. New York: Springer, 2008.

Hoffmeister, J. E., K. W. Stockman, and H. G. Multer. "Miami Limestone of Florida and Its Recent Bahamian Counterpart." *Geological Society of America Bulletin* 78, no. 2 (1967): 175–90. doi: 10.1130/0016-7606(1967)78[175:mlofai]2.0.co;2.

Hoffmeister, J. E., K. W. Stockman, and H. G. Multer. "Miami Limestone of Florida and Its Recent Bahamian Counterpart." *Geological Society of America Bulletin* 78, no. 2 (1967): 175–90. doi: 10.1130/0016-7606(1967)78[175:mlofai]2.0.co;2.

Jarrett, B. D., A. C. Hine, R. B. Halley, D. F. Naar, S. D. Locker, A. C. Neumann, D. Twichell, C. Hu, B. T. Donahue, W. C. Jaap, D. Palandro, and K. Ciembronowicz. "Strange Bedfellows: A Deep-Water Hermatypic Coral Reef Superimposed on a Drowned Barrier Island: Southern Pulley Ridge, SW Florida Platform Margin." *Marine Geology* 214, no. 4 (2005): 295–307. doi: 10.1016/j.margeo.2004.11.012.

Lidz, B. H. "Pleistocene Corals of the Florida Keys: Architects of Imposing Reefs—Why?" *Journal of Coastal Research* 22, no. 4 (2006): 750–59. doi: 10.2112/06-0634.1.

Lidz, B. H., E. A. Shinn, A. C. Hine, and S. D. Locker. "Contrasts within an Outlier-Reef System: Evidence for Differential Quaternary Evolution, South Florida Windward Margin, USA." *Journal of Coastal Research* 13, no. 3 (1997): 711–31.

Lidz, B. H., C. D. Reich, R. L. Peterson, and E. A. Shinn. "New Maps, New Information: Coral Reefs of the Florida Keys." *Journal of Coastal Research* 22, no. 2 (2006): 260–82. doi: 10.2112/05a-0023.1.

Locker, S. D., A. C. Hine, L. P. Tedesco, and E. A. Shinn. "Magnitude and Timing of Episodic Sea Level Rise during the Last Deglaciation." *Geology* 24, no. 9 (1996): 827–30. doi: 10.1130/0091-7613(1996)024<0827:matoes>2.3.co;2.

Mallinson, D., A. Hine, P. Hallock, S. Locker, E. Shinn, D. Naar, B. Donahue, and D. Weaver. "Development of Small Carbonate Banks on the South Florida Platform Margin: Response to Sea Level and Climate Change." *Marine Geology* 199, no. 1–2 (2003): 45–63. doi: 10.1016/s0025-3227(03)00141-5.

Multer, H. G., E. Gischler, J. Lundberg, K. R. Simmons, and E. A. Shinn. "Key Largo Limestone Revisited: Pleistocene Shelf-Edge Facies, Florida Keys, USA." *Facies* 46, no. 1 (2002): 229–72. doi: 10.1007/BF02668083.

Ruddiman, W. F. *Earth's Climate: Past and Future.* 2nd ed. New York: W. H. Freeman, 2008.

Shinn, E. A. "The Geology of the Florida Keys." *Oceanus* 31, no. 1 (1988): 46–53.

Tucker, M. E., and P. V. Wright. *Carbonate Sedimentology.* Oxford: Blackwell Science, 1990.

Whitman, D., K. Zhang, S. P. Leatherman, and W. Robertson. "Airborne Laser Topographic Mapping: Application to Hurricane Storm Surge Hazards." In *Earth Science in the City: A Reader*, ed. G. Heiken, R. Fakundiny, and J. F. Sutter, 363–78. Washington: American Geophysical Union, 2003.

Wilson, R. C. L., S. A. Drury, and J. L. Chapman. *The Great Ice Age: Climate Change and Life.* New York: Routledge and the Open University, 2000.

Zhang, K., S.-C. Chen, D. Whitman, M.-L. Shyu, J. Yan, and C. Zhang. "A Progressive Morphological Filter for Removing Nonground Measurements from Airborne Lidar Data." *IEEE Transactions on Geoscience and Remote Sensing* 41, no. 4 (2003): 872–82. doi: 10.1109/TGRS.2003.810682.

Epilogue

What's Next in Our Understanding of Florida's Geologic Past?

I have presented Florida's top ten hits, so to speak, in the previous ten chapters—not that there is anything magical about the number ten—it just turned out that way. What I have not mentioned is that much of what we know about Florida's past is based on a most highly fragmented and grossly incomplete data set from the past. Such limitations stoke the imaginations of earth systems scientists to generate the scientific narrative that fills in the gaps. So it is when dealing with the geologic past—we use what nature provides as best we can, and we use our imaginations guided by the best scientific principles to occupy the voids. What data and techniques we have at hand are extremely limited for creating images and for taking samples from the subsurface—that is, after all, where the geology lies.

Although there are many boreholes that have penetrated Florida's surface, they are only a handful of those required to develop a complete narrative. For example, as was mentioned in chapter 5, the huge meteorite impact in nearby Yucatan that defined the K/Pg boundary appeared to have little effect in Florida. How could that be? This dinosaur-killing event happened not far away. Most likely the shallow seas that covered the Florida Platform at the time were hugely affected. But our limited core data through strata of that age reveal nothing. Alternatively, maybe those sediments and rocks were removed by erosion, erasing evidence of one of the seminal geologic events of history—at least within the Florida Platform. Such records of events can be lost forever. The narrative of Florida's geologic past will be what it is now, subject to interpretation by different scientists, until more data can be obtained. Many gaps are to be filled by evidence found elsewhere and extrapolated to Florida.

PhD or MS committee members like to ask graduate students, "If you had unlimited funds, what more would you do to address the problem you tackled in your dissertation or thesis?" Allow me to provide my own answer. A colleague of mine once said that "we will only understand the Earth better once we are able to image it better." Images provided by various aspects of remote sensing (light, sound, magnetic and gravity variations, etc.) are powerful tools. One only has to marvel at the imagery provided by the Hubble telescope and point to how that has elevated our understanding of the cosmos. The traffic jam of Earth-orbiting satellites looking downward at our planet has revolutionized weather forecasting, recognizing large-scale aquifer depletion, mapping mineral resources, distribution of forests, and wetlands, and furthering our understanding of the ocean's surface—and now with decades of data from space, we can quantify trends. Think about how GPS and Google Earth have affected our daily lives. They did not exist a mere 25 years ago.

A more mundane form of remote sensing is seismic reflection profiling—the key exploration tool of the oil industry. This requires generating a large acoustic pulse (i.e., a big bang—no, not that Big Bang!) into the Earth or ocean. As this pulse travels into the sediments and rocks, energy is reflected off geologic strata and returned to the surface where it is converted to an electric current, which is then digitized and processed to make images by computers. The images of the Earth's sedimentary cover and crust produced by these geophysicists have enormously elevated our understanding of Earth history and Earth processes. So if I had unlimited funds, I would shoot more seismic reflection profiles across both the submerged and emerged portions of the Florida Platform. Based on interpretations of these subsurface images and 3D visualizations, I would strategically drill holes obtaining rock cores to verify the interpretations and provide important paleo-environmental data and age dates. Timing is everything.

Being able to place events in proper chronological sequence and to determine their duration and rates of processes involved is essential. From this, we could much more fully understand the drowning of the west Florida margin, or the forming of the southern Straits of Florida resulting from the Antillean Orogeny, or connecting onshore to offshore geology. Seismic profiling and drilling would provide a much more thorough understanding of Florida's deep past.

Now for the Present

Looking back at the map of the Florida Platform in chapter 1, a critical point to recall is that 50 percent of it is underwater. Imagine a territory as large of the state of Florida with no airports, no highways, no restaurants, no hotels, no gasoline stations—none of the infrastructure that land-based geologists can use to get around and to study surface or subsurface (mobile drilling rigs on trucks) geology. Examining geologic features on land is relatively easy—you can walk right up to them for the most part! I know, I know, I have colleagues who work in remote and dangerous places on land. I have done that type of work as well, and it is not easy! On land there are very detailed maps, aerial photos, and space images to work from. Generally, we do not have these data products of the seafloor when we go to sea.

Now imagine a Florida without any of these human infrastructural conveniences. To make matters even more difficult, the surficial geology (i.e., the seabed) is separated from you by <1 m to >1,800 m of water. The only way to get around is by research vessels that generally move ~10 knots through the water taking days (>24 hrs continuous) of steaming to go from Pensacola to Miami, for example, only in good weather—and such vessels commonly cost >$10,000/ day. Additionally, modern scientific instrumentation to examine the seafloor and its subsurface is enormously expensive to purchase, expensive to operate, becomes obsolete sooner than it should, and can easily be lost. There are no finely detailed maps, bathymetric or otherwise, to guide one's path. As a consequence, we know considerably less about the geology of the modern seafloor than we do about the modern surface of the emerged Florida Platform.

With my bank account of unlimited funds, I would start to generate seafloor maps showing the centimeter-scale bathymetry, the distribution of coral reefs, paleo-shorelines, hardbottom communities, springs, sand deposits, rocky outcrops that provide habitat for fish, drowned river valleys, and many other geologic features that control life in the ocean and may provide minerals that can be retrieved from the seabed. Such maps are essential for environment impact for burying pipelines, cables, and providing information on substrate for potential energy uses. And, as has been shown recently, such maps can be used as baselines for disasters such as the BP/*Deepwater Horizon* explosion and subsequent hydrocarbon release. Simply put, the modern geology of the seafloor remains in the intellectual Dark Ages.

What about the Future?

As mentioned in chapter 1, we could consider ourselves to be in the Anthropocene—a new period of geologic time whereby human activity has become a dominant geologic agent. Although still controversial, the majority membership of the earth systems scientific community is now convinced that humans have initiated a grand, uncontrolled experiment by introducing greenhouse gases into the atmosphere at rates that far exceed any known past natural processes. Even if the skeptics turn out to be clairvoyant and the human effect is not that significant, our climate is still changing at unprecedented rates. So predicting these changes is vital to humans to produce food on land and from the ocean, to effectively deal with disease, to prepare for potentially more violent and extreme meteorological phenomena, and even to plan a strategic retreat along low-lying coastlines threatened by a rising sea level. "Evacuate south Florida now!" could be a bumper sticker in 75 years.

Coupled with predicting rates of processes that are important to human time scales is understanding the recent geologic past. Extracting climate records from sediments stored in Florida's lakes and in the deep-sea sediments surrounding the Florida Platform is critical to establishing a baseline. Such a baseline of data would allow us to predict rates of natural processes so that they could be deconvolved or separated from those of the Anthropocene. With whatever funds I might have left in my imaginary bank account, I would generate high-resolution, paleoclimate records to assist climate modelers in predicting Florida's climate so we can make appropriate preparations. Without those predictions and the necessary preparations to meet whatever nature has planned for us, the quality of human life could significantly regress. The geologic past introduced in these ten chapters has taught us that the Earth will go on with or without us—it is our choice if we want to see how the Anthropocene and beyond will turn out.

A Word about Offshore Oil Drilling

On April 20, 2010, the British Petroleum–leased drilling platform, named the *Deepwater Horizon,* owned by TransOcean, Inc., located just off the Mississippi River delta in ~1,500 m water depth, exploded and sank, killing 11 workers. During the ensuing 86 days at flow rates of up to 62,000 barrels of oil/day, approximately 4.1 million barrels of oil were released into the Gulf of Mexico, thus perpetrating one of the nation's largest environmental disasters. Since hydrocarbons are products of geology, knowledge of Florida's geology and how that is tied to the Gulf of Mexico's geologic history are critically important to Floridians. Florida sits adjacent to one of the Earth's hydrocarbon megaprovinces—perhaps second only in reserves to the great Middle Eastern Arabian to Iranian megaprovince. To date, the state of Florida (exposed part of the Florida Platform) has not produced much oil or gas (some, but not much). Fig. E.1 shows how little oil and gas have been extracted from Florida compared with the rest of the Gulf of Mexico and surrounding land. The Florida Platform's basement rock, carbonate platform structure, and geologic history apparently did not provide the basic ingredients such as source rock, reservoir rock, impermeable seals, traps, and thermal history, to be an eastern extension of the Gulf of Mexico hydrocarbon megaprovince.

However, a number of thoughts come to mind for the reader's consideration. First, there may be significantly important reserves along the deeply submerged, outer western margin of the Florida Platform extending east toward the shallow west Florida shelf. Indeed, the oil industry continues to obtain seismic reflection data as part of its ongoing exploration activity in the eastern Gulf of Mexico. Second, there appear to be enormous hydrocarbon reserves lying beneath the seafloor in much deeper water than the 1,500 m deep BP *Deepwater Horizon* blowout site (actually called Macondo 252). The oil industry considers ultra-deep-water drilling to start at ~1,800 m water depth, making the BP site in relatively shallow water. The maximum depth of the Gulf of Mexico is ~3,400 m water depth, and it is possible that hydrocarbons ultimately could be extracted

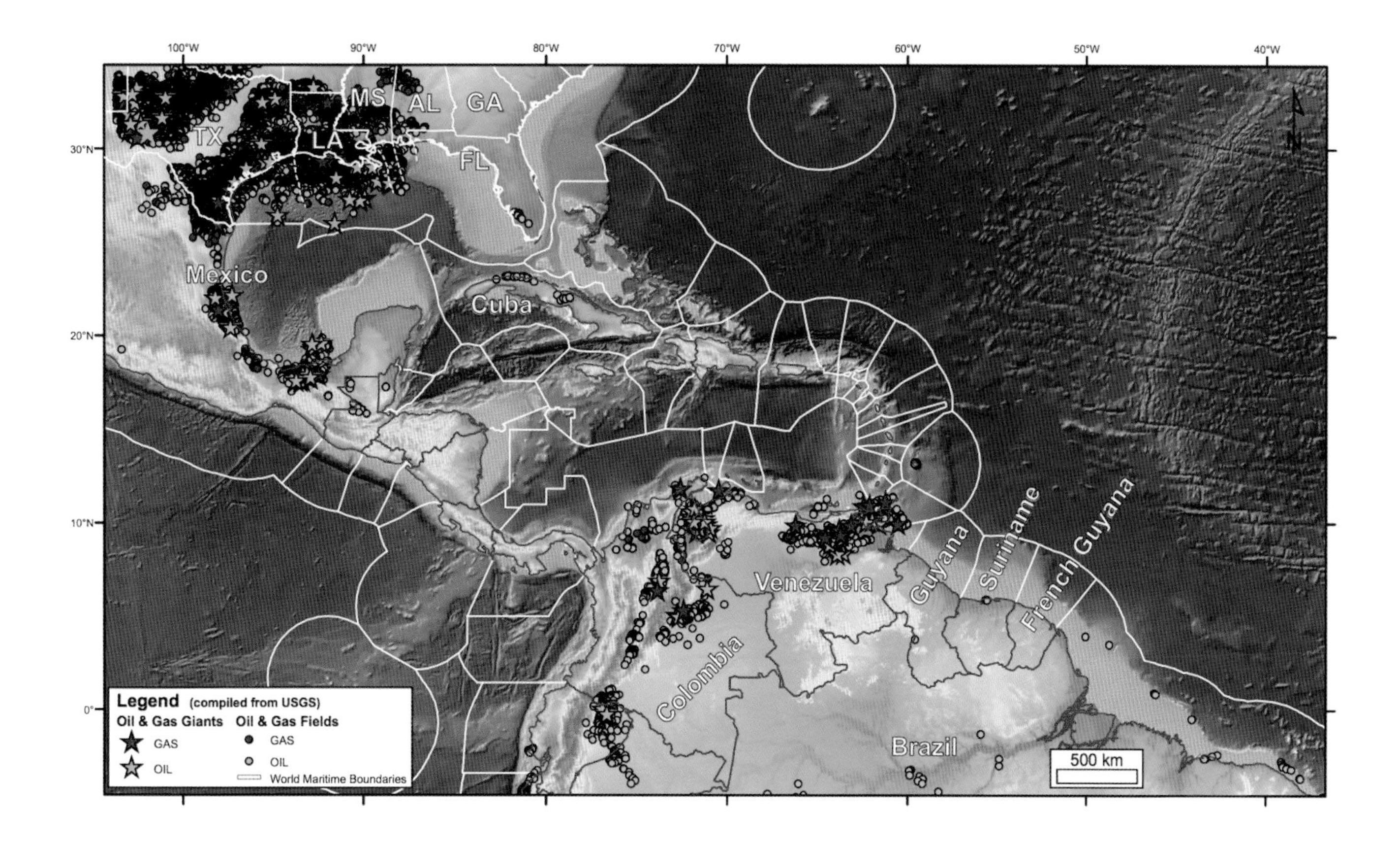

Figure E.1 Oil and gas fields in the Gulf of Mexico, the Caribbean, and northern South American regions. Note the almost complete absence of such fields on the Florida Platform. (Source: Mann and Escalona 2011; used by permission from the Society of Exploration Geophysicists.)

from those water depths—more than twice as deep as the Macondo 252 site! Third, it was apparent from the BP event that Florida was simply fortunate that the Loop Current and wind fields during the event kept oil from reaching peninsular Florida's coastline and possibly entering large estuaries such as Tampa Bay and Charlotte Harbor. Indeed, a worst-case scenario would have allowed surface oil to have been entrained by the Loop Current and transported around Florida and up the southeast coast of the United States. Fourth, the U.S. domestic production of oil and gas is about 6.3 million barrels/day or about one-third of the 19 million barrels that are required to run our economy every day. Simply put, we have been importing ~66 percent of our hydrocarbon energy needs. The number could increase to 75 percent in the not too distant future. Fifth, the United States will be dependent upon oil and gas for many decades to come.

What do these five observations mean? First, regardless of any moratorium short of preventing all oil and gas extraction from the entire Gulf of Mexico, Florida is exposed to risk of the consequences of a BP-type event or worse. Second, with the multi-decadal time scale needed to introduce new energy sources such as nuclear power, we will require hydrocarbon extraction from the Gulf of Mexico for dozens of years—including the new drilling in the southern Straits of Florida off northern Cuba! Third, technology should be in place to greatly reduce the risk of another BP-style blowout. Should such an event occur, the technology to contain the blowout rapidly and shut it down within hours or one or two days should be on standby. Fourth, the present moratorium preventing offshore drilling could probably be reduced in scope without further imperiling Florida's coastal environments, given that the technology just mentioned is credible. If substantial and really significant hydrocarbon resources are found 150 km west of St. Petersburg, for example, oil drilling and extraction could boost our domestic supply and still protect the environment. For sure, no one wants to gaze out from Florida's spectacular beaches and see an oil platform. But the possibility of industry activity 150 km offshore, if it could significantly boost our domestic supply, could be closely reexamined—and maybe nixed, but reexamined nevertheless.

Regardless of how one feels about the issue of offshore oil drilling, the enormous issue of having affordable and available energy for the future requires a 50- to 70-year plan. Now is the time to develop such a plan and to decouple it from the time frame of 2–6 years we find ourselves locked into by our political system. See John Hofmeister's book on *Why We Hate the Oil Companies.*

Finally, a Word about Climate Change

One of the great scientific issues of the day is what our climate, both regionally and globally, will be like by the year 2100 and beyond. Certainly, our climate in Florida will change—floods and droughts may occur, thus affecting availability of water to agriculture and human habitation. Storms may become more frequent and more powerful—no one knows for sure, and there does not appear to be solid evidence for that at the moment. However, it seems that there is scientific consensus that sea level will rise and could rise significantly, perhaps up to 1.8 m (~6 ft) by 2100, maybe even more. One highly accomplished climate scientist told me in September 2011 that a 5 m (~16.5 ft) rise by 2100 is possible! But most scientific experts in the field do not predict such a catastrophic rate. However, enhanced sea level rises, perhaps coupled with changing tropical storm activity, could enormously change how we invest and maintain our infrastructure along Florida's coastline and very low-lying areas. But there is hardly consensus at this point about the magnitude of sea level rise—other than sea level will rise faster than it is rising today.

Some planners have talked eventually about a strategic retreat from the barrier islands. Such a notion seems drastic and needlessly worrisome at the moment. If the projections made now by scientists seem to hold up or are seen to become more dire in the next decade or so, then we need to begin to formulate a 50- to 70-year *strategic retreat plan* to go along with our 50- to 70-year *energy plan*. Both plans will require a paradigm shift in how we think about the future.

For the moment, we simply need to figure out how we can move from our shackles of what John Hofmeister calls "political time" to longer term planning that makes sense for the multiple generations to come.

Essential References to Know

Hofmeister, J. *Why We Hate the Oil Companies: Straight Talk from an Energy Insider.* New York: Palgrave Macmillan, 2010.

Mann, P., and A. Escalona. "Major Hydrocarbon Plays in the Mexican Sector of the Gulf of Mexico, the Caribbean, and Northern South America." In *Society of Exploration Geophysicists Annual Meeting*, 2328–33. San Antonio, TX, 2011.

Index

Page numbers in *italics* refer to illustrations; page numbers in **bold** refer to definitions.

Acadian Orogeny, 29
Acidity: defined, **129**; in mogote formation, 93, 96, 122; in submarine canyon formations, 82, *82*, *83*; *specific acids*: carbonic (in acid rain), 115–16, *116*; phosphoric, 157; sulfuric, 165, 173–74. *See also* Alkalinity
Active margins: defined, **104**; formation of, 96–98, *98*; passive margin colliding with, 78, *85*, 96, 99, *100*, 101, *101*, *102*, 103, 104. *See also* Caribbean Plate; Greater Antilles Volcanic Arc
Africa: dust originating from, 93, *95*, *190*, 191, *192*; rift basins of, 38–39; small section still attached to North America, 40–41, 43, 44, *45*. *See also* Gondwana
Aggradation, *186*, **206**
Alabama Promontory, 25, *27*
Alaskan topography, 2–3, 16
Albedos, **207**. *See also* Sun
Aleutian Trench, 16
Algal blooms, 166
Alkalinity, 115, **129**. *See also* Acidity
Allegheny Orogeny, *27*, 29–31, *30*
Alligators, *3*, 132, 166, 170
Alpine glaciers, 188, **207**. *See also* Glacial and interglacial events
Alvarez, Walter, 86–87
Anions, 60, **67**
Anoxia and anoxic events: defined, **88**; described, 74–76. *See also* Ocean anoxic events
Antarctica, *75*, 145. *See also* Gondwana; Ice sheets
Anthropocene: defined, **21**; new epoch of, 18; South Florida's emergence and, 185, *186*, *187*; understanding threats in, 213–14. *See also* Florida (modern); Human population
Apalachicola Island, *134*
Apalachicola River, 152
Appalachian Mountains: acid rain's effects on, 115–16; coastlines impacted by erosion of, 140, *142*, 143, *144*; elevation of, 2; exotic terrain of, 29–30; origins of, *26*, 27, *27*, 29, *30*; quartz sand grains from erosion of, 138, *139*, 140
Aquifers: defined, **129**; permeability of, 113, 118, **130**; satellite imagery of, 212; secondary porosity and, 124; surface, 116, *116*, **130**; transmissivity and, 124–25, **130**. *See also* Floridan Aquifer; Groundwater; Water
Arabia hydrocarbon megaprovince, 42, 214
Aragonite. *See* Calcium carbonate (aragonite)
Arctic Ocean, 74, *74*
Astrophysics, 23. *See also* Hubble telescope imagery
Atlantic Coastal Ridge, 185, *187*, 197
Atlantic Ocean: basement rocks of, *30*, 31; Boulder Zone and, 114, *115*; Pacific Ocean connected to, 42, 96, *178*; precursor to, 25; South Atlantic origins, 38. *See also* North Atlantic Ocean
Australia, Great Barrier Reef, 56, 63, *64*

Bacterial microbial activity, 173–74, **180**. *See also* Cyanobacteria; Vent-type communities
Bahama Banks: carbonate muds of, *57*; in last glacial event, *190*; limestone strata produced on, 143–44; recharge area and, 128; sedimentary environments of, *58*, 59; tidal-dominated formations of, *198*
Bahama Block: depth of basement rocks, 43–44; origins of, *36–37*, 38, *39*
Bahama Fracture Zone, *45*, 46
Bahama Platform: active sedimentary processes in, 108; as carbonate megaplatform, 52, *55*, 56, *57*, *58*, 63, *64*, 65, *65*; cross section of, *53*; deep seaways formed in, 77–78; Florida Platform compared with, 128; Florida Platform separated from, 18–19, 78; topographic asymmetry of, 77. *See also* Florida-Bahama Platform; Yucatan-Florida-Bahama Platform
Bahamas: carbonate cliff of, 202, *203*; caverns of, 109; exploration oil drilling in, 108–9; freshwater recharge of, 128; space image of, *58*
Baja Peninsula, 29
Banda Aceh (Indonesia): tsunami of, 97
Barrier islands: aerial views and locations of, *133*, *134*, *135*; drowned, 200, *200–201*; sediments and, 140, *142*, 143; strategic retreat plan for, 216–17
Basalt, **48**. *See also* Ocean crust
Basement rocks: completed assembly of Florida, *41*, 42; defined, **32**, **153**; essential points about, 31–32, 47–48; as exotic terrain, 29–30, **32**, 41, **48–49**; faults likely in, 42, 124; flooding of, 53; origins and migration of, 25, *25*, *26*, 27, *27*, 29–31, *30*; quartz sand grains from erosion of, 138, *139*, 140; seismic reflection profile of, *71*; in transition before breakup, 38, *39*. *See also* Bedrock
Basins: formation of, *30*, 38–41, *39*, *40*, *41*; karst-generated type, 150, 202; quartz-rich sediments in, 150. *See also* Foreland basins; Ocean basins
Bathymetry and bathymetric changes: Georgia Seaway Channel as barrier in, 40, 45; Loop Current deflected in, *171*, 173; of Miami and Pourtales terraces, 175, *175*, *176*; of Nicaraguan Rise, 177; of scalloped embayments, 78, *80*; of siliciclastic units, 191; of submarine canyons, *82*; of submerged portion of Florida Platform, *13*, 18–19; of Tampa Bay, *204–5*, 205
Bay-head deltas, **153**
Beaches: aerial view of, *3*; color of, 174; formation of, 200; of siliciclastic sediments, *136*, 137; splendor of, 132, *133*, *134*. *See also* Dunes; Quartz sand; Sanibel Island

Bedload, 143, **153**
Bedrock: coastlines and weathering of, 140, *141*, *142*, 143; crystalline, **153**. *See also* Basement rocks
Benthic organisms (benthos), 82–83, 174, **180**
Big Bend area (Ozello): aerial view and location of, *134*, 135; oyster reefs in, *121*, 122, *122*; as sediment starved, 152; surficial karst in, *116*, 119, *120*, *121*
Biogenesis, **207**
Bird guano, 169
Bivalves: nucleation sites for, *121*, 122, *122*; rudist, 56, *57*, **68**, *73*, 78
Blake Plateau Basin (including Escarpment), *53*
Block faults, 42, **48**, 124
Blue Ridge Mountains, *141*
Bone Valley Member (Middle Miocene), 159, *161*. *See also* Phosphates and phosphorous
Boulder Zone, 114, *114*, *115*. *See also* Floridan Aquifer
Boundaries. *See* K/Pg boundary; Middle Cretaceous Sequence Boundary; Sequence boundaries; Western boundary current
Bové Basin, *30*, 44
BP (British Petroleum): *Deepwater Horizon* disaster of, 5, 213, 214–15
Brachiopods, 174, **180**
Britton Hill, *2*, 2–3
Broward County: Atlantic Coastal Ridge under, 185, *187*; quartz-rich and limestone units underlying, 146, 149

Cacarajícara Formation (Cuba), 85, *85*, *86*
Caladesi Island, *133*, *142*
Calcification process, 54. *See also* Carbonate platforms
Calcite crystals, *57*
Calcium carbonate (aragonite): components of, 54; defined, **207**; saturation in relation to, **130**; tidal currents carrying, 194, *196*, 197. *See also* Caliche crust
Calcium sulfate, hydrated. *See* Phosphogypsum
Caledonian Orogeny, 29
Caliche crust: defined, **207**; described, *190*, 191, *192*, 193
Caloosahatchee River, *144*, 146, *148*, *149*
Campeche Bank, 77
Canyons. *See* Submarine canyons
Cape Canaveral, 202, *202*
Cape Florida, 151
Cape Romano, 151
Carbonate factory: described, 56, *57*, *58*, *59*, 59–60; essential points about, 66–67; halted in drowning platform, 76–77; restarting of, 185, *186*, *187*, 187–88; in shallower water, 151; threats to, 144. *See also* Carbonate platforms; Carbonate sedimentation
Carbonate-fluorapatite, 163–64, 173–74. *See also* Phosphates and phosphorous
Carbonate muds, 56, *57*, 59
Carbonate platform margins: collision of active and passive, 78, *85*, 96, 99, *100*, 101, *101*, *102*, 103, 104; erosion and drowning as reshaping, 70, 76–77, 78–79, *79*, *80*, 81, *81*, 200, *200–201*, 202; reef-dominated, 59, *59*. *See also* Carbonate platforms
Carbonate platforms: basic ingredients, 52, *53*; essential points about, 66–67; evaporites on, 60, *61*; formation of, 53–54, *54*, *55*, 56; holes in, *109*, 109–10, *110*; sedimentary environments on, 56, *57*, *58*, *59*, 59–60; seismic reflection profile of, *71*, 72; siliciclastic sediments covering, 53, *54*, *55*, *62*, 63, *64*, 65, *65*; submarine canyons of, 81–82, *82*, *83*; tectonic subsidence and vertical building of, *62*, 62–63, *64*, *65*, 65–66. *See also* Bahama Platform; Carbonate platform margins; Florida-Bahama Platform; Florida Platform; Guadalupe Mountains; Karst topography; Yucatan-Florida-Bahama Platform
Carbonate sedimentation: active areas of, 108, 110, *111*, *112*, 113; block faults controlling, 42; building up of, *62*, 62–63, *64*, *65*, 65–66; changes from shallow- to deep-water, 72–73, 76–77; components in, 56, *57*; current limit of, 66; processes of, 56, 59–60; as resistant to erosion, 200, *200–201*; restarting of, 185, *186*, *187*, 187–88; shallow-water deposition of, 53–54, *54*; shells made of, *133*; transported by tidal currents, 194, *196–97*, 197, *198–99*, 199. *See also* Caliche crust; Carbonate platform margins; Carbonate platforms
Carbon dioxide: addition vs. subtraction of, 177, 179; current higher levels of, 18, 179, 213–14; Early Cretaceous level of, *73*, 73–74; lowered level in Icehouse Earth, 188. *See also* Climate change; Global warming; Greenhouse Earth
Carbon fixation, **180**. *See also* Primary productivity
Carbonic acid, 115–16, *116*
Carcharodon megalodon (ancient great shark), 166–67, *168*, 171, 179–80
Caribbean Basin: changing current of, 177; complex geology of, 96; Great American Biotic Interchange via, 171
Caribbean Current, 177
Caribbean Plate: collision margin of, 46; essential points about, 104; Florida Platform in relation to, *19*; formation of, 97–98, *98*; migration direction of, 99; subduction of, 96–97
Caribbean Sea: formation and widening of, 41–42, 96; origins of proto Caribbean, 35, *36–37*, 38, *39*, *41*, 42, 45; paleogeographic reconstruction of, 97, *98*
Carlsbad Caverns (New Mexico), *108*, 110, *111*
Cations, 60, **67**
Caverns: active formation and collapsing of, 110, *112*, 113; Boulder Zone and, 114, *114*, *115*; example of, *3*; exploration of, 109, *109*; spring systems linked to, 109, *110*; weakened in sea level lowstands, 127–28; in western Cuba, *95*. *See also* Carlsbad Caverns
Cementation: of animal and plant remains, 135, 143–44; of carbonate sediments, 65; defined, **67**, **153**; numerous episodes of, 177; speed of, 60; subaerial type of, 191. *See also* Subaerial exposure
Cenotes, 83, **88**. *See also* Sinkholes
Cenozoic: carbonate stratigraphic succession in, 66; fauna of, 166. *See also* Miocene; Oligocene
Central American Seaway, *178*
Central Florida Phosphate District, *158*. *See also* Phosphates and phosphorous
Challenger Salt, 42
Charlotte Harbor: as karst-generated basin, 150, 202; limited river discharge in, 152; origins of, 125, *126*, 127, 179; sediments of, 127
Chemical (electrical) bonds, 163–64, **180**
Chemical weathering: defined, **153**; limits of, 122; red clay soils as evidence of, 140, *141*; during sea level lowstands, 191; warmer climates as stimulating, 145

Chemosynthesis, 82–83
Chicxulub Crater (Yucatan Peninsula), 83–87, *84*, *85*, *86*, 88, 211
Climate: currents' impact on, 172–73; Gulf of Mexico's role in, 43; interglacial period of (current), 191; Milankovitch cycles' role in, 14; of Pliocene, 150–51; temperature differences between poles and equator, 74–75, *75*. *See also* Currents; Glacial and interglacial events; Global warming; Greenhouse Earth; Icehouse earth; Sea level changes and cycles; Sun; Winds
Climate change: as human induced, 18, 179, 213–14; phosphatization and, 177, 179; planning response to, 213–14, 216–17. *See also* Global warming
Coastlines: suspended load of sediments on, 140, *142*, 143, **154**; varieties of, 132, *134*; weathering as important to, 140, *141*, *142*, 143, *144*. *See also* Beaches; Quartz sand
Coccolithophorids, 73
Cocos Plate, *19*
Collision margins: defined, **48**; passive margin transformed into, 46; uplift of, 99, *100*, 101, *101*, *102*, 103. *See also* Active margins; Cuba; Greater Antilles Volcanic Arc; Passive margins
Conjugate continental margins: collision of, 25, *26*, 27, *27*; defined, **32**, **48**; origins of, 35; tectonic subsidence of, *62*, 62–63
Continental shelves and slopes: breaks of, 71, *72*; defined, **21**; extended by sediments, 143; structural boundaries of, 12, *13*
Continents (or continental fragments): age of, 24; defined, **32**; rifted zones within, 35; suturing of, 29–31, *30*. *See also* Margins; Plate tectonics; Supercontinent cycle; Supercontinents; *and specific continents*
Coral reefs: current's impact on, 173; as environment indicators, 56; final touches in emergence of Florida, 199–200, *200–201*, 202, *202*, *203*, *204–5*, 205; formation of, 191, *193*, 193–94, *195*; foundation for, 185, *186*, 188; tide-dominated shoals along, 199
Cotton Valley Formation, 53
Coya Granite, 44
Cretaceous (K): land areas in, *55*, 56; ocean anoxic events in (middle), 74–76; sea level highstands and warm temperatures in, 53, 63–64, 74, *74*. *See also* Greenhouse Earth; K/Pg boundary; Middle Cretaceous Sequence Boundary; Tethys Ocean
Crystalline bedrock, **153**
Crystal River, *116*, 119
Cuba: essential points about, 103–4; evidence of K/Pg boundary extinction event in, *85*, 85–86, *86*; hydrocarbon accumulation off coast of, 103; mogotes of western, 93, *94–95*, 96, 119, 122; uplift of, in collision of margins, 99, *100*, 101, *101*, *102*, 103. *See also* Greater Antilles Volcanic Arc
Cuban Exclusive Economic Zone, 103
Currents: climate impacted by, 172–73; factors in, 74–75, *75*; tectonic events and acceleration of, *178*; *specific*: Caribbean, 177; Loop, *171*, *172*, 172–73; Yucatan-Loop-Florida, *178*. *See also* Gulf Stream; Paleo Loop Current; Tidal currents; Topographic steering; Western boundary current; Winds
Cyanobacteria, **67**

Dade County: Atlantic Coastal Ridge under, 185, *187*; quartz-rich and limestone units underlying, 146, 149
Dating methods: deep borings and geophysical remote sensing, 43–44, 52, 211; geologic time, 5–7; imaging techniques, 211–12; isotopes of strontium in, 170–71, **181**
Davis, R. A., Jr., 93
Debris flows: defined, **88**, **104**; deposits of K/Pg boundary extinction event, *85*, 85–86, *86*; in margin collisions, 99. *See also* Sediment gravity flow deposits
Deepwater Horizon disaster, 5, 213, 214–15
Deltas: bay-head, defined, **153**; buried, *149*, 149–50; migration halted, 150–51. *See also* Estuaries and bays
Denali (Mount McKinley, AK), 2–3, 16
Density: of continental vs. oceanic crust, 29; of fresh- vs. seawater, 118, *119*
Depositional environment, 199
De Soto Canyon, 40
Devil Spring system, *110*
Diagenesis, **180**. *See also* Carbonate platforms; Limestone
Diapirs, 43, **48**, *71*
Diepolder Cave, 109, *109*
Dinosaurs: extinction event for, 83–87, *84*, *85*, *86*, 88, 211; size of draglines compared with, 159; three-toed, 39
Dissolution tectonics: Boulder Zone, 114, *114*, *115*; deep subsurface processes in, 122–25, *123*, *124*, *125*, *126*, *127*, 127–28, 179; defined, **129**; essential points about, 128–29; Florida and Bahama platforms compared, 128. *See also* Floridan Aquifer; Karst topography
Distally steepened ramp, 71, *72*, **88**
Dolomite: Boulder Zone development in, 114, *114*, *115*; defined, **129–30**; limestone transformed into, 60
Downlapping, *148*, **153**
Draglines for phosphate mining, 159, *160*, *161*, *162*, 163
Drowning: carbonate platform margins reshaped in, 70, 76–77, 78–79, *79*, *80*, 81, *81*; defined, **21**; essential points about, 87–88; isthmus formed in, *178*; of paleo-shorelines, 199–200, *200–201*, 202, *202*, *203*, *204–5*, 205. *See also* Sea level highstands
Dunes: formation of, 200; of Lake Wales Ridge, 146, *147*; splendor of, 132, *133*, *134*. *See also* Beaches; Quartz sand
Duricrust, 191, *192*, **207**
Duval County: Atlantic Coastal Ridge under, 185, *187*

Earth: age of, 5–7; cycles of, 27–29, *28*; early development of, 23–24; highly mobile crust of, 123; rotation and orbit variations of, 14, *15*. *See also* Basement rocks; Climate; Continents; Greenhouse Earth; Icehouse Earth; Ocean(s); Plate tectonics
Earthquakes: tsunamis from, 85–86, 97. *See also* Microseisms; Plate tectonics; Volcanic eruptions
Earth systems science: components of, 1–2; consensus on climate change in, 213–14, 216–17; interconnectivity in, 7; questions remaining in, 211–12
Eccentricity orbital cycle, 14, *15*
Ecology. *See* Terrestrial ecology
Economic grade, **181**
Economy and economic development: in early years of state, 132; exploration and offshore oil drilling in, 108–9, 214–16; phosphate's role in, 159, 163;

rock valuation in, **181**. *See also* Beaches; Oil and gas production; Strip-mining, phosphate
Electricity, 159, *160*
Elevation (Florida), *2*, 2–3, *144*, 146, 157, *158*, 159, 185
Embayments: origins of, 25, *27*, 78, *80*; submarine canyons formed from, 81–82, *82*, *83*; in supercontinent cycle, 35
Energy planning, 214–17
Environment. *See* Terrestrial ecology
Eocene: limited chemical weathering of rocks in, 122; rocks exposed in, 128; surficial karst and rocks of, *116*, 119, *120*, *121*; warmer climates of, 145
Epicenters, 124, **130**
Erosion: essential points about, 87–88; in ocean, 175, *175–76*, 177, *178*; remnants of, defined, **181**; warmer climates and, 145; weathering process in, 140, *141*; West Florida Escarpment reshaped by drowning and, 70, 76–77, 78–79, *79*, *80*, 81, *81*. *See also* Quartz sand
Escarpments, **21**, *53*. *See also* West Florida Escarpment
Estuaries and bays: geology underlying, 20; as marshes vs. lagoons, 140, *142*, 143; sea level changes and sediments of, 127, *127*. *See also* Charlotte Harbor; Deltas; Rivers and streams; Tampa Bay
Ethanol production, 166
Europe: ice sheets of, 188–89, *189*, *190*, 191
Eustatic movements: coining of, 70. *See also* Sea level changes and cycles
Evaporite minerals: on carbonate platforms, 60, *61*; defined, **48**, **67–68**; deposited in Gulf of Mexico, 41–43. *See also* Salts
Everest, Mount, 138, *139*
Everglades: alligators of, *3*; buried river deltas under, *149*, 149–50; sea level rise potential for, 18; underlying structure of, *144*, 146, *148*, 149, 185, *187*. *See also* Pourtales Terrace
Exoskeletons: carbonate-fluorapatite secreted in formation of, 174; defined, **181**; shells of, *133*, 135, *136*
Exotic terrain: defined, **32**, **48–49**; left behind in rifting, 41; in suturing process, 29–30. *See also* Volcanic island arcs
Exposure. *See* Subaerial exposure
Extensional forces, 42, **49**. *See also* Rifting and rift system
Extinction event at K/Pg boundary, 83–87, *84*, *85*, *86*, 88
Exuma Sound (seaway), 78

Facies, sedimentary, 56, 59, 60, **68**
Faults: block type, 42, **48**, 124; thrust type, **105**; transform type, 41–42, 46, **50**; transmissivity due to, 124–25. *See also* Epicenter; Microseisms; Tectonic plates
Fauna: fossils of, 166–67, *167*, *168*; Georgia Seaway land bridge for, 145
Fecal pellets: carbonate sedimentation and, 56, *57*, 174; defined, **68**, **180**; tidal currents carrying, 194, 197
Feldspars, 140, **153**
Fenno-Scandinavian Ice Sheet, 188
Ferrel cell, 75, **89**. *See also* Winds
Fertilizer: chemical processing of, 165–66; exports of, 159, 163; phosphorous in, *164*, 164–65. *See also* Phosphates and phosphorous
Flexural loading, 101, **105**
Florida (modern): appeal of, 3, 132, *133*, *134*; defined, **10–11**; elevation of, *2*, 2–3, *144*, 146, 157, *158*, 159, 185; essential points about, 205–6; future of, 211–17; latitude of, 25; local and global impacts on, 5; quartz-carbonate transition and, 187–88; tidal sand bodies of, 194, *196–97*, 197, *198–99*, 199. *See also* Beaches; Carbonate platforms; Coral reefs; Dissolution tectonics; Drowning; Economy and economic development; Florida Keys; Food chain; Phosphates and phosphorous; Plate tectonics; Quartz sand; Sea level changes and cycles; South Florida
Florida-Bahama Platform: bathymetric features of submerged portion, *13*, 18–19; in breakup of Pangea, *36–37*, 38, *39*; carbonate platform of, 52–67 (*see also* Carbonate platforms); completed assembly of basement rocks of, *41*, 42; cross-section, *13*; deep seaways formed in, 77–78; defined, **12**; effects of colliding margins on, 101, *101*, *102*; essential points about, 47–48; as exotic terrain, 41; origins of, *30*; Permian Basin compared with, 107–8; shape and depth of basement rocks, 43–44, *44*; structural boundaries, *12*, 45–46, *46*; topographic relief of, 16, *17*, 18. *See also* Bahama Platform; Florida Platform; Middle Cretaceous Sequence Boundary
Florida Bay, *148*
Florida Keys: beaches and coastline of, 132, 136, *136*; carbonate rocks underlying, *148*, 188; carbonate succession continuing in, 66; coral reef of, *3*; currents and orientation of, *196*; emergence of, 191, *192*, *193*, 193–94, *194*, *195*; exploration oil drilling in, 109; foundation for, 185, *186*, *187*; geology underlying, 20; Miami topography compared with, *197*; quartz-rich and limestone units underlying, 146, 149; reef rock mined in, 194, *195*. *See also* Key West
Floridan Aquifer: boulder zone of, 114, *114*, *115*; deep subsurface structures in, 122–25, *123*, *124*, *125*, *126*, *127*, 127–28, 179; essential points about, 128–29; overview of, *113*, 113–14; recharge areas of, 125, 128; surface and subsurface structures in, 115–16, *116*, *117*, 118–19, *119*, *120*, *121*, 122, *122*
Florida Panhandle, *2*, 2–3
Florida Platform: active sedimentary processes in, 108, 110, *111*, *112*, 113; Bahama Platform separated from, 18–19, 78; carbonate megaplatform of, *55*, 56; Cuba compared with, 96; Cuba's origins linked to, 103; deep seaways formed in, 77–78; defined, **21**; erosion of flanks of, 175, *175–76*, 177, *178*; essential points about, 87–88, 135; examining underwater portion of, 212–13; fracture patterns in, 123, *123*; in Icehouse Earth period, 188–89, *189*, *190*, 191; K/Pg boundary extinction event's impact on, 83–87, *84*, *85*, *86*, 88; origins of, 23–25, *25*, *26*, 27, *27*, 35, *36–37*, 38; quartz sediments introduced to, 143–46, *144*; subaerial exposure of, 128, *190*, 191, *192*; tectonic stability of, 19, *19*; topographic asymmetry of, 77. *See also* West Florida Escarpment
Florida Straits. *See* Straits of Florida
Florida Straits Block: Suwannee Basin Block in relation to, *30*, 42
Fluorine, 163. *See also* Phosphates and phosphorous
Food chain (or food web): defined, **180–81**; primary and secondary productivity in, 75; trophic levels in, 169, **182**. *See also* Phytoplankton; Zooplankton
Forams (foraminifera), 73, **88–89**, 174, **180**
Foreland basins, *100*, 101, *102*, 103, **105**
Forereefs, 193, *194*, **207**

Fossils: abundance of, 166–67, *167*, *168*, 169; conditions for land and marine mixtures of, 170–71; essential points about, 179–80; preserved in park, 194
Fracture patterns, 123, *123*, 124–25

Galapagos Islands, 35
Gastropods, 174, **181**
Geological duration, 6
Geological oceanography, 1–4
Geological Society of America, 5
Geologic Time Scale: overview, 5–7. *See also specific eras*
Geology: dynamic nature of, 70, 205–6; hierarchy in (Member-Formation-Group), 159; interconnectivity in, 7; local and global impacts on, 5; metric measurements in, 8, *8*; questions remaining in, 211–12
Georgia: chemical weathering in, 140; coastline of, *142*. *See also* Appalachian Mountains
Georgia Rift Valley (or Basin), *39*, 39–40, *40*, 45
Georgia Seaway Channel: flooding of, 53–54, *54*; formation of, 40, 45; sedimentation of, 145–46; as sediment barrier for Florida, 144–45
Giant armadillos (pampatheres, *Holmesina septentrionalis*), *167*
Gigaplatform (or gigabank), 55, 63, *64*, *65*. *See also* Carbonate platforms
Glacial and interglacial events: cycles of, 14, 188–89, *189*, *190*, 191; defined, **207**; shoreline changes in Florida, 14, *16*. *See also* Icehouse Earth; Ice sheets; Sea level changes and cycles
Glaciers. *See* Alpine glaciers; Ice sheets
Glacio-eustacy (sea level changes), **207**. *See also* Sea level changes and cycles
Global warming: atmospheric and ocean circulation slowed in, 74–76, *75*; chemical weathering and erosion increased in, 145; as human induced, 18, 179, 213–14; phosphatization and, 177, 179; planning response to, 213–14, 216–17. *See also* Sea level rise
Gneisses, 138, **153**
Gondwana: basement rocks of, 24–25; in breakup of Pangea, *36–37*; North America's collision with, 25, *26*, 27, *27*, 31; as supercontinent, 29; transition signs before breakup, 38–41, *39*, *41*
Granites, 44
Gravity anomalies, 43–44, **49**
Great American Biotic Interchange, 171
Great Bahama Bank. *See* Bahama Banks
Great Barrier Reef (Australia), 56, 63, *64*
Greater Antilles Volcanic Arc: collision with passive margin, 99, *100*, 101, *101*, *102*, 103; essential points about, 104; formation and active margin of, 96–98, *98*, 99; K/Pg boundary extinction event and, 85, *85*; water and land masses of, 93, *94*, *98*. *See also* Cuba
Greenhouse Earth: anoxic events in, 74–76; atmospheric and ocean circulation slowed in, 74–75, *75*; carbon dioxide level in, 73, *73*; defined, **89**; sea level rise and warm temperatures of, 63–64, 74, *74*
Greenhouse gases. *See* Carbon dioxide
Greenland: ice sheets of, 188–89, *189*, *190*, 191
Groundwater: Bahama vs. Floridan system of, 128; carbonate rocks affected by, 60; flushed via gravity, *124*, 125; seawater contact with, 118, *119*. *See also* Aquifers; Irrigation; Karst topography; Mixing zones
Guadaloupe Island volcanic eruption (1970s), 98
Guadalupe Mountains, 107, *108*. *See also* Carlsbad Caverns
Guaniguanico Terrain (western Cuba), 93, *94*, 101
Gulf of California, 39
Gulf of Mexico: basement rocks of, *30*, 31; extinctions of faunas of, 145 (*see also* Chicxulub Crater); formation of, 41–42; halite in, 60; mid-Cretaceous connection to Arctic Ocean, 74, *74*; modern temperature and currents of, *172*; nutrient run-offs and dead zones in, 166; as oil and gas production megaprovince, 42–43, 53, 103, *215*; origins of, 35, *36–37*, 38; restricted marine circulation in, 76; salt deposits in, 41–43; sediment gravity flow deposits in, 78–79, *80*. *See also* Carbonate platforms; Oil and gas production
Gulf Stream, 75, *171*, *178*. *See also* Western boundary current
Gypsum, *61*, 82, *82*, *83*. *See also* Evaporite minerals; Phosphogypsum

Hadley cell, 74–75, **89**. *See also* Winds
Halite, 60, *61*
Hawaiian Islands, 16, *17*, 35
Hawthorne Group, 159. *See also* Fossils
Hercynian Orogeny, 29. *See also* Mauritanide Mountains
Highway excavations, 1
Himalaya Mountains, 29, 99, 138, *139*, 189
Himalaya-Tibetan Plateau, 189
Hispaniola: as part of Caribbean Plate, 99
Hofmeister, John, 216, 217
Holmesina septentrionalis (giant armadillos or pampatheres), *167*
Holocene: Florida Keys in, 193–94; proposed end of, 18
Hot spots, 35, 38, **49**. *See also* Mantle plumes
Hubble telescope imagery, 6, 23, *24*, 211
Hudson Canyon, 81
Human population: as agent of geological change, 206; climate change induced by, 18, 179, 213–14; early territory of, 199
Hydrocarbon accumulation: factors in source-rock deposition, 76; megaprovinces of, 42–43, 53, 214; potential sites of, *100*, 103, 214–15; seepage vs., 109. *See also* Oil and gas production
Hydrostatic loading, *62*, 63
Hypoxia. *See* Anoxia and anoxic events

Iapetus Ocean, 25, 29–31, 40
Icehouse Earth: defined, **207**; described, 188–89, *189*, *190*, 191; Florida Keys and cycles of, 191, *192*, *193*, 193–94, *194*; river transport and delta migration halted in, 150–51. *See also* Glacial and interglacial events
Iceland: ice sheets covering, 188–89, *189*, *190*, 191; origins of, 35
Ice sheets: accumulation cycle of, 188–89, *189*, *190*, 191; defined, **207**; sea level fluctuations due to, 12, 14, *14*. *See also* Glacial and interglacial events; Sea level changes and cycles
Indian Ocean, 97, *139*
Indonesia: as volcanic island arc, 97
Interstitial space, 174, **181**
Intraclast fragment, 174, **181**
Invertebrates. *See* Fecal pellets
Iran hydrocarbon megaprovince, 42, 214
Iridium, 83–85
Irrigation, 110, 113
Island arc, **105**. *See also* Volcanic island arcs
Islands. *See* Barrier islands; *and specific islands*

Isostatic adjustment: defined, **130**; rebound effects, 138, 140, 145, **153**
Isotopes, 170–71, **181**. *See also* Marine isotope stage
Isthmus of Panama, 97, *98*, 171, *178*
Iturralde-Vinent, M. A., 93

Japanese Islands: tectonic setting of, 19; tsunami of, 97
Jurassic: gigaplatform of, *55*, 63, *64*, *65*; seafloor spreading in (early), 41–42; sea level rise in (late), 53. *See also* Dinosaurs

Karst Plateau (Yugoslavia), 118
Karst topography: chemical weathering of, 191; deep subsurface, 122–25, *123*, *124*, *125*, *126*, *127*, 127–28, 179; defined, **130**, **207**; essential points about, 128–29; geomorphology linked to, 118–19, 122; origins of term, 118; surface and shallow subsurface, 115–16, *116*, *117*, 118–19, *119*, *120*, *121*, 122, *122*. *See also* Carbonate platforms; Caverns; Cenotes; Mogotes; Sinkholes
Key Largo Limestone: distribution in Pleistocene, *196*; formation of, 191, *192*, *193*, 193–94, *194*; fossils in, *195*; Miami Limestone in relation to, *197*, 199
Key West: building stone of, 199; drowned shoreline off coast of, 202, *203*
Kohout, Francis, 125
Kohout convection concept, *124*, 125
K/Pg boundary (or K/T boundary): essential points about, 88; extinction event at, 83–87, *84*, *85*, *86*, 211; meteor strike as defining, 6

Lake Upland, *147*
Lake Wales Ridge, *144*, 146, *147*, *148*, 150
Land reclamation, 165. *See also* Strip-mining, phosphate
Last Glacial Maximum (LGM), *189*, *190*, 191
Laurasia: basement rocks of, 24–25; in breakup of Pangea, *36–37*; transition signs before breakup, 38–41, *39*, *41*
Laurentia (ancestral North America): origins of, *26*, *27*, *27*, 29–31, *30*
Laurentide Ice Sheet, 188, *189*, *190*, 191
Lesser Antilles Islands (Leeward and Windward Islands), 98
Light detection and ranging image (LIDAR image), *197*
Limestone: components of, 54; differences in formation times, 60; emplacement of, 185, *186*, *187*; produced in Bahama Banks, 143–44; quartz sand areas separated by, *137*, 138, *138*; in submarine canyon formations, 82, *82*, *83*; in western Cuba, 93. *See also* Calcium carbonate (aragonite); Key Largo Limestone; Miami Limestone
Lithosphere. *See* Ocean crust; Tectonic plates
Lithostatic load: in Appalachian Mountains, 140; of carbonate sediments, *62*, 62–63; defined, **154**
Long Key Formation, 146
Longshore transport: mechanism described, 146, 150, 151–52; timing of, 185, *186*
Loop Current: described, *172*, 172–73; paleogeographic map of, *171*. *See also* Paleo Loop Current
Louann Salt, 42

Macondo 252. See *Deepwater Horizon* disaster
Madagascar: as continental fragment, 29
Magnetic anomalies, 43–44, **49**
Manatees (sea cows), 118, 166, 170
Mangroves, *121*, 132, *134*, *135*
Mantle plumes, *73*, 73–74, **89**. *See also* Hot spots
Margins: phosphatization events along, 174–75, 177; transcurrent, 46, **50**. *See also* Active margins; Carbonate platform margins; Collision margins; Conjugate continental margins; Passive margins
Marine isotope stage (MIS) or marine oxygen-isotope (O) stages, **208**. *See also* Isotopes
Marine unconformity, 175, *175–76*, 177, *178*. *See also* Middle Cretaceous Sequence Boundary
Marshes. *See* Big Bend area; Everglades; Mangroves; Wetlands and marshes
Martinique Island volcanic eruption (1902), 98
Mastodons, *167*
Mauna Kea (Hawaii), 16, *17*
Mauritanide Mountains, *26*, 29, *30*
MCSB (or MCU). *See* Middle Cretaceous Sequence Boundary
Measurements, 8, *8*. *See also* Dating methods
Metamorphic rocks, 44, 138, **154**
Meteor impact. *See* K/Pg boundary
Metric measurements, 8, *8*
Miami: building stone of, 199; Silver Bluff remnant in, 202, *203*
Miami Limestone: distribution in Pleistocene, *196*; formation of, 197, 199, *199*; Key Largo Limestone in relation to, *197*, 199
Miami Terrace, 175, *176*, 177, *178*
Mica, 140, **153–54**
Microbes and microbial activity: acidic soil due to, 116; bacterial, 173–74, **180**; defined, **32**, **68**, **130**
Microseisms, 124, **130**
Middle Cretaceous Sequence Boundary (MCSB, aka Middle Cretaceous Unconformity [MCU]): drowning event of, *79*; formation of, 72–73, 77; platform profile in, *71*, *72*
Mid-ocean ridge, **32**, *62*, 62–63, 73. *See also* Ocean basins; Seafloor spreading; Vent-type communities
Milankovitch cycles, 14, 189, *190*, 191, **208**
Mineral rights, 163
Minerals: erosion of, 140; satellite imagery of, 212; *specific*: halite, 60, *61*; mica, 140, **153–54**; silicates, 53, 140, **154**; sulfates, 60, *61*, **68**, 173–74. *See also* Calcium carbonate (aragonite); Dolomite; Evaporite minerals; Phosphates and phosphorous; Salts
Mining. *See* Strip-mining, phosphate
Miocene: Bone Valley Member (Middle), 159; erosion in ocean during, 175, *175–76*, 177, *178*; fauna of, 166; quartz sand areas in, *137*, 138, *138*; sea level high- and lowstands in, 127–28, 170–73, *171*, *172*, 179; upwelling in, 169–70, *170*, 173; Yucatan-Loop-Florida Current and Gulf Stream in, *178*. *See also* Paleo Loop Current
MIS (Marine isotope stage), **208**. *See also* Isotopes
Mississippi River, 43, *71*, 79
Mitchell, Mount, 2, *139*. *See also* Appalachian Mountains
Mixing zones: defined, **130**; in Kohout convection, *124*, 125; of sea- and freshwater, 118, *119*
Mogotes: defined, **105**; in western Cuba, 93, *94–95*, 96, 119, 122

Moncada Formation (Cuba), 85, *85*
Monroe County: quartz-rich and limestone units underlying, 149
Montserrat Island volcanic eruption (1995), 98
Mountains: internal elevation and external weathering of, 70. *See also* Alpine glaciers; Erosion; Orogeny; *and specific mountain ranges*

Nana (supercontinent), 29
Native Americans, 199
Natural gas. *See* Hydrocarbon accumulation; Oil and gas production
Nazca Plate, *19*
NE/NW Providence channels, 78
New York: Hudson Canyon off coast of, 81
Nicaraguan Rise, 177, *178*
Nitrogen, *164*, 165
Norphlet Formation, 53
North America: ice sheets covering, 188–89, *189*, *190*, 191
North American Plate: Cuba as part of, 96, 99; Florida in relation to, 19, *19*; northward movement of, 42, 96; small section of Africa still attached to, 40–41, 43, 44, *45*. *See also* Florida-Bahama Platform; Florida Platform; Laurentia
North Atlantic Ocean: carbonate platform system along, *55*, 63, *64*; formation of, 41; origins of, 35, *36–37*, 38; Pacific Ocean connected to, 42, 96; rift basins in, 38–41, *39*, *40*, *41*
North Carolina: chemical weathering in, 140; continental margin of, *65*; geology of Outer Banks, 35, 38; phosphatization events along, 174–75. *See also* Appalachian Mountains
Nutrients: in anoxic events, 76; dead zones due to run-offs of, 166; defined, **89**, **181**; elements of, 169; resurfacing of, in upwelling, 76, 169–70, *170*, 173

Obduction, **105**
Obliquity orbital cycle, 14, *15*
Ocala Platform, *144*, 146
Ocean(s): appeal of, 3; erosion in, 175, *175–76*, 177, *178*; mapping floor of, 212–13; as recycling machine, 169–70, *170*; thermocline of, 75
Ocean anoxic events (OAEs): anoxia defined, **88**; carbonate sedimentation and, *73*, 77; effects of, 42; essential points about, 87; in mid-Cretaceous, 74–76
Ocean basins: age of, 24; decreased volume of, *73*, 73–74; defined, **32**, **49**; formation of new, 24–25, 35, 38, *41*; Gulf of Mexico as, 41–42; Wilson cycle in formation of, *28*, 28–29. *See also* Seafloor spreading
Ocean circulation. *See* Currents
Ocean crust: defined, **49**; essential points about, 104; formation, 25, 39, 73; high sea level linked to expanded production of, *73*, 73–74, *74*; obduction and, **105**; subduction of, 29, 97–98
Oceanic plateaus, *73*, 73–74, **89**
Offshore oil drilling: BP disaster, 5, 213, 214–15; exploratory, 108–9; opportunity and technology considered, 214–16
Oil and gas production: baseline for disaster response in, 213; megaprovinces of, 42–43, 53, 214; seismic studies for, *71*; sites of, *215*; Texas areas of, 107–8, *108*. *See also* Hydrocarbon accumulation
Oligocene: sea level lowstands in, 127–28; siliciclastic sedimentation in, 145–46
Olistostromes, *85*, 85–86, *86*, **89**. *See also* Debris flows
Ooids/ooid grainstone: in carbonate sediments, 56, *57*; cementation of, 60; defined, **68**, **208**; tidal currents and formation of, *196*, 197, *198*, 199, *199*
Orlando Ridge, *147*
Orogeny: Allegheny, *27*, 29–31, *30*; defined, **32–33**; Wilson cycle in, *28*, 28–29. *See also* Collision margins; Foreland basins; Greater Antilles Volcanic Arc; Mountains; Ocean basins; Supercontinent cycle
Osceola Granite, 44
Osceola Plain, *147*
Ouachita Embayment, 25, *27*
Ouachita Mountains, 25
Outer Banks (NC), 35, 38
Overburden sediments: defined, **181**; strip mining for phosphate under, *158*, 159, *161*, *162*, 165. *See also* Phosphates and phosphorous
Oxidation, **181**. *See also* Ocean anoxic events
Ozello/Chassahowitzka area: oyster reefs in, *121*, 122, *122*; surficial karst in, *116*, 119, *120*. *See also* Big Bend area

Pacific Ocean: Atlantic Ocean connected to, 42, 96, *178*; origins of, *36–37*
Paleoceanography, 76, **89**
Paleoclimate and paleoclimatology, **89**, **208**
Paleogene (Pg). *See* K/Pg boundary
Paleogeography: defined, **208**; of shifting continents, 24–25, *25*, *26*, 27, *27*
Paleo Loop Current: erosion associated with, 175, *175–76*, 177, *178*; phosphate formation and, 170–71, *171*; siliciclastic sediment of, *54*. *See also* Loop Current
Paleontology, 166–67, *167*, *168*, 169
Paleo-shorelines: defined, **21**, **208**; drowning of, 199–200, *200–201*, 202, *202*, *203*, *204–5*, 205. *See also* Drowning; Shorelines
Paleosol, 191, *192*, **208**
Palm Beach area, *3*
Panama, Isthmus of, 97, *98*, 171, *178*
Pangea: assembly of, 29–31, *30*; breakup of, 35, *36–37*, 38; essential points about, 31–32, 47; reconstruction of, *26*; transition signs before breakup, 38–41, *39*, *40*, *41*
Panthalassa Ocean, 31, *36–37*
Passive margins: active margin colliding with, 78, *85*, 96, 99, *100*, 101, *101*, *102*, 103, 104; along Florida, 45–46, *46*; defined, **49**, **105**; formation of, 99, *100*, 101, *101*, *102*, 103; of North American Plate, 19, *19*; transformed into collision margin, 46
Peace River, *149*
Peace River Formation, 146, *149*, 159
Peñalver Formation (Cuba), 85, *85*
Peninsular Arch, 43, *44*
Permeability, 113, 118, **130**. *See also* Porosity; Transmissivity
Permian Basin (Texas), 107–8, *108*. *See also* Carlsbad Caverns; Guadalupe Mountains
Phosphates and phosphorous: chemical processing of, 165–66; elements of, 163–64; essential points about, 179–80; formation of, 170–73, *171*, *172*; global context of, 177, 179; importance of, *164*, 164–65; making of, *171*, 173–75; mining of, 159, *160*, *161*, 163; misconceptions about, 169; remnants from mining, 157, *158*, 159, *162*; seafloor deposits of, 76; source of, 169–70, *170*
Phosphatization: areas of, 174–75, *176*; in climate change context, 177, 179

Phosphogypsum (and phosphogypsum stacks): defined, **181**; environmental concerns about, 163; production of, 157, *158*, 159, 165
Phosphoric acid, 157
Phosphorite, 164
Photic zone, 76, **89**
Photosynthesis: chemical process of, 54; defined, **68**, **90**, **181**; at ocean surface, 75, 76, 169–70, *170*, 173. *See also* Carbon fixation; Photic zone; Phytoplankton; Primary productivity
Physical weathering, 140, **154**. *See also* Erosion
Physiography, 10–11, *11*. *See also* Topographic relief
Phytoplankton: defined, **89–90**, **181–82**; production of, 75, 169–70. *See also* Primary productivity
Piedmont, 138, 140, *141*, **154**
Piercement structures (diapirs), 43, **48**, *71*
Pinellas County: sediment thickness of offshore sand, *137*, *138*. *See also* Tampa Bay
Plankton, **182**. *See also* Phytoplankton
Plate tectonics: constant mobility in, 123; defined, **33**; energy source for, 23–24; extensional forces in, 42, **49**; obduction in, **105**; Wilson cycle in, *28*, 28–29. *See also* Dissolution tectonics; Faults; Margins; Rifting and rift system; Subduction; Supercontinent cycle; Suturing process and suture zone; Tectonic movement; Tectonic plates; Volcanic island arcs
Platforms: drowning of, **21**, *79*; segmentation of, 18–19, 77–78. *See also* Carbonate platforms; Drowning; Paleo-shorelines; *and specific platforms*
Pleistocene: fauna of, 166; limestone distribution in, *196*; limestone emplacement in, 185, *186*; sea level cycles in, 187–88. *See also* Coral reefs
Pliocene: climate of, 150–51; fauna of, 166, *167*; ramp aggradation in, *186*; sea level cycles in, 187–88
Polar cell, 75
Porosity, secondary, 124. *See also* Permeability; Transmissivity
Positive feedback, **208**
Potash, *164*, 165
Pourtales Terrace: formation of, 175, *175*, 177, *178*; location of, *144*; sediment deposits on, 146, *148*, 150, 151
Precambrian, Late: rocks of, 29; Rodinia broken up in, 24–25
Precession orbital cycle, 14, *15*
Primary productivity: defined, **90**, **181**; essential points about, 179–80; at ocean surface, 75, 169–70. *See also* Phytoplankton
Productivity. *See* Primary productivity; Secondary productivity
Progradation, 150, **154**, 187–88, **208**
Promontories, 25, *27*, 35
Pteropods, 73, **90**
Puerto Rico: as part of Caribbean Plate, 99
Pulley Ridge, *200–201*, 202

Quartzites, 138, **154**
Quartz sand: abundance of, *3*, 132, *133*, *134*, 135, *135*, 137–38; change to carbonate sedimentation from, 185, *186*, *187*; characteristics of, 135–38, *136*, *137*, *138*; climate differences and, *148*, 150–51; essential points about, 152–53; introduced to Florida Platform, 143–46, *144*; source of, 138, *139*, 140; transported by longshore process, *144*, 146, *147*, *148*, *149*, 149–50; transported by tidal currents, 194, *196–97*, 197, *198–99*, 199; transport processes, summarized, 151–52, 185, *186*. *See also* Sedimentation

Radiation, solar, **208**
Radioactive decay, 23–24, 165, **182**
Radionuclides, 157, 174, **182**
Radon gas, 165
Rainfall, 115–16, *116*, 150. *See also* Aquifers; Groundwater
Ramp: aggradation of, *186*, **206**; distally steepened, 71, *72*, **88**; of West Florida Escarpment, 71, *72*. *See also* West Florida margin
Rays, 166, 170
Red Snapper Sink, *123*, 128
Reduction, **182**
Reef rocks ("keystone"), 193–94, *195*
Reefs: appeal of, *3*; biologic, as carbonate factory product, 56, *57*; carbonate platform margin dominated by, 59, *59*; nucleation sites for oysters, *121*, 122, *122*; outliers, *194*; rudistid, 56, *57*, *73*, 78. *See also* Coral reefs; Forereefs
Remote sensing, 43, 52, 83, 86, 211. *See also* Dating methods; Seismic reflection profiling
Resedimentation, **182**
Rift basins or valleys, 38–41, *39*, *40*, *41*
Rifting and rift system: defined, **33**, **49**; Gondwana and Laurasia transition before, 38–41, *39*, *41*; Rodinia broken up via, 24–25, *27*, 28–29, 31; thermal uplift in, 35, **50**; Wilson Cycle and, *28*, 28–29, **33**. *See also* Bové Basin; Florida Straits Block; Georgia Rift Valley; Plate tectonics; Supercontinent cycle; Suwannee Basin Block
Rivers and streams: color and bedload of, 143; current distribution of, 205; mountain erosion carried in, 140, *141*; shoreline erosion by, 146, *148*; small discharge of, 152; springs and, 109, *110*, 118, 143. *See also* Deltas; Estuaries and bays; *and specific rivers*
Rock: economic grade (value) of, 181; fracture patterns in, 123, *123*; three types of, 44; *specific types*: basalt, **48**; gneisses, 138, **153**; granites, 44; saprolites, 140, *141*, **154**; volcanic, **105**. *See also* Basement rocks; Carbonate platforms; Dolomite; Exotic terrain; Metamorphic rocks; Sedimentation and sedimentary rock; Weathering
Rocky Mountains, 16
Rodinia: break up of, 24–25, *27*, 28–29, 31
Rudist bivalves, 56, *57*, **68**, *73*, 78

Salt Lake (Australia): halite in, *61*
Salts: deposits of, 41–43; evaporation increasing, 60, *61*. *See also* Diapirs; Evaporite minerals
Sand: carbonate type of, 136, *136*. *See also* Quartz sand
Sand Hill Boy Scout Reservation (Brooksville), *109*
Sand Key, *137*, *138*, *194*
Sandstones, 138
Sanford High, *144*, 146
Sanibel Island, *133*, *149*, 150
Saprolites, 140, *141*, **154**
Satellite imagery, 211–12
Saturation in relation to calcium carbonate, **130**
Savannah River, *142*
Schists, 138, **154**
Sea cows (manatees), 118, 166, 170
Seafloor spreading: defined, **33**, **49–50**; ocean anoxic events linked to, 76; ocean crust expanded production in, *73*, 73–74, *74*; oceans formed by, 41–42;

rift basins formed by, 38–39; Rodinia broken up via, 25; Wilson Cycle and, *28*, 28–29, **33**. *See also* Mid-ocean ridge; Ocean basins; Vent-type communities
Sea level changes and cycles: carbonate rock variations due to, 65; coral reef formation and, 191, *193*, 193–94; delta migration and, 150; essential points about, 206; estuaries/bays impacted by, 127, *127*; frequency of, 187–88; influences on, 12, 14, *14*, *15*, *16*, 20; mixing zone changes in, 118–19; springs impacted by, 118; surficial processes impacted by, 122; terminology for, 70. *See also* Sea level highstands; Sea level lowstands; Sea level rise
Sea level highstands: Cape Canaveral area during, 202, *202*; defined, **21**, **182**; essential points about, 87; highest ever, in mid-Cretaceous, 74, *74*; longshore transport of sand in, 146, 150, 151–52; ocean anoxic events linked to, 76; ocean crust expanded production linked to, *73*, 73–74, *74*; phosphate formation in, 170–73, *171*, *172*; Silver Bluff remnant from, 202, *203*; tectonic subsidence and sedimentation in relation to, *62*, 62–63
Sea level lowstands: carbonate-fluorapatite-rich sediments reworked in, 174; caves weakened in, 127–28; defined, **21**, **182**; in glacial events, 188–89, *189*, *190*, 191; sedimentation of Georgia Seaway Channel in, 145–46; shoreline erosion by rivers in, 146, *148*. *See also* Glacial and interglacial events
Sea level rise: atmospheric heat impacting, 74, *74*; due to climate change, 18; due to melting ice sheets, 199–200, *200–201*, 202; hydrostatic loading of, *62*, 63; in Late Jurassic/Early Cretaceous, 53; planning response to, 213–14, 216–17. *See also* Drowning; Global warming; Sea level highstands
Secondary productivity, 75, **90**. *See also* Zooplankton
Sedimentary facies, 56, 59, 60, **68**
Sedimentary minerals, 163–64
Sedimentation and sedimentary rock: age of, 44; economic grade (value) of, 181; erosion and redeposition of, **182**; rivers and streams' role in, 143; sandstone as, 138; sea level rise and tectonic subsidence in relation to, *62*, 62–63. *See also* Aggradation; Carbonate platforms; Cementation; Phosphates and phosphorous; Progradation; Quartz sand; Sediments
Sediment gravity flow deposits: defined, **90**; formation of, 78–79, *80*; of K/Pg boundary extinction event, *85*, 85–86, *86*. *See also* Debris flows; Turbidites
Sediments: defined, **181**; over salt deposits in Gulf of Mexico, 42–43; stromatolites as, **68**; suspended load of, 140, *142*, 143, **154**; transported by longshore process, *144*, 146, *147*, *148*, *149*, 149–50; transported by tidal currents, 194, *196–97*, 197, *198–99*, 199; transport processes, summarized, 151–52, 185, *186*; turbidites as, 85, *85*, **90**. *See also* Olistostrome; Ooids/ooid grainstone; Overburden sediments; Quartz sand; Sedimentation and sedimentary rock; Siliciclastic sediments; Stratigraphic units
Seismic reflection profiling: of basement rocks, 43–44, *71*; buried river deltas found via, *149*, 149–50; of carbonate platforms, *71*, 72; defined, **50**, **130**, **154**; of Florida Keys coral reef tract, *194*; process and benefits of, 212; of Tampa Bay and Charlotte Harbor, 125, *125*, *126*, 127, 179
Sequence boundaries, **90**. *See also* Middle Cretaceous Sequence Boundary
Seychelles (islands), 29
Sharks, ancient (*Carcharodon megalodon*), 166–67, *168*, 171, 179–80
Shark teeth, 166–67, *168*, 177
Shells, *133*, 135, *136*. *See also* Exoskeletons
Shorelines: drowning of, 14, **21**; eroded by rivers and streams, 146, *148*; essential points about, 152–53; glacial and interglacial changes in, 14, *16*; sedimentary deposits and, *147*, *148*. *See also* Drowning; Paleo-shorelines
Siberia hydrocarbon megaprovince, 42
Silicate minerals, 53, 140, **154**
Siliciclastic sediments: beach areas of, *136*, 137; carbonate platforms buried by, 53, *54*, *55*, *62*, 63, *64*, 65, *65*, 143–46, *144*; defined, **154**; essential points about, 152–53; transport process for, *144*, 146, *147*, *148*, *149*, 149–50; transport process for, summarized, 151–52. *See also* Quartz sand
Silver Bluff remnant, 202
Sinkholes: in exposed limestone, *120*; formation of, 20, 110, *112*, 113, 116, *117*, 118; on seafloor and on shelf, *123*, 127–28. *See also* Cenotes
Soil, 116, 140, *141*. *See also* Paleosol; Sediments
Solar insolation (incident solar radiation), **208**
Sound energy. *See* Seismic reflection profiling
South American Plate, *19*, 42, 96. *See also* Gondwana
South Atlantic Ocean, 38
South Carolina: chemical weathering in, 140; coastline of, *142*. *See also* Appalachian Mountains
South Florida: carbonate succession continuing in, 66; emergence of, 185, *186*, *187*; quartz-carbonate transition and, 187–88; quartz-rich and limestone units underlying, 146, 149
South Georgia Rift Basin. *See* Georgia Rift Valley
South Pole: Florida's basement rocks at, 25, *25*
Spain: gypsum outcrop in, *61*
Springs, 109, *110*, 118, 143
St. Mary's River, 152
Straits of Florida: Bahama and Florida platforms separated by, 18–19; essential points about, 104; as foreland basin, *100*, 101, *102*, 103; formation of, 78; oil drilling in, 215; reef development and, *194*; sediments carried to, 146
Stratigraphic succession, *13*
Stratigraphic units: cycles in formation of, 191, *192*, *193*, 193–94, *194*; defined, **68**, **155**, **182**; of phosphate, 159; of quartz sand, *137*, 138; variability of, 60, 65–66
Strip-mining, phosphate: ecological problems of, *162*, 165–66; importance of, *164*, 164–65; mechanics of, 159, *160*, *161*, 163; remnants of, 157, *158*, 159
Stromatolites, **68**
St. Vincent Island, *134*
Subaerial exposure: defined, **21**, **130**, **208**; of Florida Platform, 128, *190*, 191, *192*; of sand ridges, 197, *198*
Subduction: areas resistant to, 99; defined, **33**, **50**, **105**; Iapetus Ocean consumed through, 29; of island arcs, *28*, 96–97, 98, *98*
Submarine canyons: essential points about, 88; formation of, 81–82, *82*, *83*
Sulfates, 60, *61*, **68**, 173–74
Sulfuric acid, 165, 173–74

Sumatra-Andaman earthquake (2004), 97
Sun: centrality of, 23; precession, obliquity, and eccentricity orbital cycles of, 14, *15*; reflection of (albedos), **207**; solar insolation of, **208**
Sunniland Field, 108, 109
Supercontinent cycle: current direction of, 29; essential points about, 31–32, 47–48; oceans created in, *41*, 41–42; theories about breakups in, 35, *36–37*, 38; transition signs before breakup, 38–41, *39*, *40*, *41*. *See also* Plate tectonics; Wilson Cycle
Supercontinents: defined, **50**; reassembly in, 29–31, *30*. *See also* Gondwana; Laurasia; Laurentia; Pangea; Rodinia
Superia, 29
Surface aquifer, 116, *116*, **130**
Suspended sediment load, 140, *142*, 143, **154**
Sustainability, 4. *See also* Climate change; Terrestrial ecology
Suturing process and suture zone: defined, **33**; described, 29–31, *30*; possible weakness in, 40–41
Suwannee Basin Block: basement rocks of, 44; in breakup of Pangea, *36–37*; Florida Straits Block in relation to, *30*, 42; origins of, *30*, 40–41, *41*; structural boundary of, 45
Suwannee Channel. *See* Georgia Seaway Channel
Suwannee River, 118, 152

Tampa: fertilizer exports of, 163; phosphate mining near, 157, *158*, 159
Tampa Bay: bathymetry and topography terrain map of, *204–5*, 205; as karst-generated basin, 150, 202; limited river discharge in, 152; origins of, 125, *125*, 179; sediments of, 127, *127*
Tectonic movement: carbon dioxide levels and, 188–89; defined, **105**; obduction, **105**; subsidence, *62*, 62–63, **68**; uplift, 145, **155**. *See also* Dissolution tectonics; Rifting and rift system; Subduction; Suturing process and suture zone
Tectonic plates: boundaries of, 38, *39*, *41*, **50**; defined, **21**; fracturing of, 123, *123*; underlying Florida, 19, *19*. *See also* Exotic terrain; Margins; Plate tectonics; Supercontinent cycle; *and specific plates*
Temperature: of poles vs. equator, 74–75, *75*. *See also* Climate; Greenhouse Earth; Icehouse Earth
Tennessee Embayment, 25, *27*
Ten Thousand Islands, 132, *134*, *135*
Terrestrial ecology: defined, **182**; local and global impacts on, 5; strip-mining's impact on, *162*, 163, 165–66. *See also* Climate change; Food chain; Human population
Tethys Ocean: closing of, 145; defined, **155**; essential points about, 67; formation of, *41*, 42, 54, *55*, 56
Texas: evidence of extinction event in, 86; oil-bearing areas of, 107–8, *108*
Texas Promontory, 25, *27*
Thermal uplift, 35, **50**. *See also* Rifting and rift system
Thermocline layer, 75–76, **90**
Thrust sheets/thrust faults, **105**
Tidal currents: crust and water movement due to, 123; sand bodies carried by, 194, *196–97*, 197, *198–99*, 199
Time: geologic, 5–7; of K/Pg boundary extinction event, 83. *See also specific eras*
Tobacco growing, 93, *95*
Tohoku-oki earthquake (2011), 97
Tongue of the Ocean (seaway), 78
Topographic relief, **21**
Topographic steering, 173, **183**
Topography: of Alaska, 2–3, 16; of Florida Platform, 77; geology underlying, 20; satellite imagery of, 212; of Tampa Bay, *204–5*, 205; of West Florida Escarpment, 16, *17*, 18. *See also* Karst topography
Trade winds. *See* Ferrel cell; Hadley cell; Winds
Transcurrent margins, 46, **50**
Transform faults, 41–42, 46, **50**
Transmissivity, 124–25, **130**. *See also* Permeability
TransOcean, Inc. See *Deepwater Horizon* disaster
Triassic: rifting in, 39–40
Trophic levels, 169, **182**
Tsunamis, 85–86, 97
Turbidites, 85, *85*, **90**
Turbidity, 144, **155**

Unconformity: defined, **68–69**, **91**; marine, 175, *175–76*, 177, *178*. *See also* Middle Cretaceous Sequence Boundary
Undersaturation, 118, **130**
United States, 10–11, *11*
Universe: age of, 6
Uplift: of collision margins, 99, *100*, 101, *101*, *102*, 103; thermal, 35, **50**
Upwelling: defined, **91**, **182**; nutrients resurfacing in, 76, 169–70, *170*, 173; phosphatization events linked to, 174–75, 177
Uranium as by-product, 165

Venice: color of beaches around, 174; fossil hunting at, 166
Vent-type communities, 82–83, **91**
Vinales area mogotes (western Cuba), *95*
Virginia Promontory, 25, *27*
Virgin Islands: as part of Caribbean Plate, 99
Volcanic arc, **33**. *See also* Volcanic island arcs
Volcanic eruptions: degassing in, *73*, 73–74; in volcanic island arc, 97. *See also* Hot spots
Volcanic island arcs: defined, **105**; origins of, 29–30; subduction along, 96–97. *See also* Greater Antilles Volcanic Arc
Volcanic rocks, **105**

Waste-water injection, 114, *115*
Water: as alkaline or acidic, 115; density and contact of fresh- and sea-, 118, *119*; fluoridation of, 163; as rainfall, 115–16, *116*, 150; turbidity of, 144, **155**. *See also* Aquifers; Groundwater; Ocean(s); Rivers and streams; Sediments
Water table. *See* Surface aquifer
Weather forecasting, 212. *See also* Climate
Weathering: carbon dioxide levels and, 188–89; coastlines and, 140, *141*, *142*, 143, *144*; physical type of, 140, **154**. *See also* Chemical weathering; Erosion
Western boundary current: defined, **90–91**, **182–83**; factors in, 74–75, *75*; phosphatization events linked to, 174–75, 177. *See also* Gulf Stream; Loop Current
West Florida Escarpment: carbonate rock section of, 52; erosion and drowning of, 70, 76–77, 78–79, *79*, *80*, 81, *81*; essential points about, 87–88; exploration oil drilling on, 109; ramp of, 71, *72*; reefs of rudist bivalves along, 56, *57*; seascape and characteristics of, 70–72, *71*, *72*; structural boundaries of, 12, *12*, *13*;

submarine canyons of, 81–82, *82*, *83*; topographic relief of, 16, *17*, 18. *See also* Florida Platform
West Florida margin: eroded surfaces forming terraces, 175, *175–76*, 177, *178*; essential points about, 87–88; potential hydrocarbon accumulation on, 214–15; scalloped embayments along, 78, *80*; submarine canyons of, 81–82, *82*, *83*
Wetlands and marshes: eroded sediments in, 140, *142*, 143; surficial karst in, *116*, 119, *120*, *121*. *See also* Big Bend area; Everglades
Wilson, J. Tuzo, 28
Wilson Cycle, *28*, 28–29, **33**
Windley Key, 194, *195*
Windley Key Fossil Reef State Park, 194
Winds: African dust carried by, 93, *95*, *190*, 191, *192*; factors in circulation, 74–75, *75*; sand movement due to, 151. *See also* Currents; Erosion
Winter Park sinkhole, 110, *112*
Woolly mammoths, *167*

Yeats, William Butler, 4
Yucatan Block: in breakup of Pangea, *36–37*; carbonate megaplatform of, *55*, 56, 63; mogotes of, 93, *94–95*, 96, 119; origins of, *30*; southward movement of, 41
Yucatan-Florida-Bahama Platform: paleogeographic map, *85*; passive margin of, and collision with Greater Antilles, 99, *100*, 101, *101*, *102*, 103. *See also* Bahama Platform; Florida-Bahama Platform; Florida Platform; Yucatan Block
Yucatan-Loop-Florida Current, *178*
Yucatan Peninsula: extinction event on, 83–87, *84*, *85*, *86*, 88, 211; seaward extension of (Campeche Bank), *12*, 77, *98*
Yugoslavia: Karst Plateau of, 118

Zooplankton, 75, **91**, 169, **183**. *See also* Secondary productivity

ALBERT C. HINE is professor of geological oceanography in the College of Marine Science at the University of South Florida. He has participated as co-chief scientist on over seventy-five research cruises at numerous sites around the world, including two legs on the *DSRV JOIDES Resolution* scientific, ocean-drilling vessel. He is a winner of the prestigious national Francis P. Shepard Medal for outstanding research contributions to marine geology.

About the cover image: This image encapsulates many of the broader themes of the book. (1) The inky blackness of space contrasts with the curvature of Earth vividly illustrating the isolation and remoteness of our planet in the vastness of space and time. (2) The thin blue haze reveals the atmosphere without which we would have few of the geologic processes, such as the wind that drives ocean circulation and the hydrologic cycle, which both continuously shape Earth and sustain life. (3) The ocean dominants Florida by nearly surrounding the state and covering 50 percent of the Florida Platform. (4) The projection of peninsular Florida suggests that it originally was not part of North America but came from somewhere else. (5) Florida connects to the rest of North America indicating the important influence from the larger continental mass.

The image was captured by the Moderate Resolution Imaging Spectroradiometer (MODIS) onboard NASA's Aqua satellite and is courtesy of Norman Kuring, NASA Goddard Space Flight Center. For a high-resolution version of the image (you can make out the bridges across Tampa Bay), see http://oceancolor.gsfc.nasa.gov/cgi/image_archive.cgi?i=333.

The University Press of Florida is the scholarly publishing agency for the State University System of Florida, comprising Florida A&M University, Florida Atlantic University, Florida Gulf Coast University, Florida International University, Florida State University, New College of Florida, University of Central Florida, University of Florida, University of North Florida, University of South Florida, and University of West Florida.